LTE SELF-ORGANISING NETWORKS (SON)

LTE SELF-ORGANISING NETWORKS (SON)

NETWORK MANAGEMENT AUTOMATION FOR OPERATIONAL EFFICIENCY

Edited By

Seppo Hämäläinen, Henning Sanneck, Cinzia Sartori

Nokia Siemens Networks

A John Wiley & Sons, Ltd., Publication

This edition first published 2012
© 2012 John Wiley & Sons, Ltd

Registered office
John Wiley & Sons Ltd, The Atrium, Southern Gate, Chichester, West Sussex, PO19 8SQ, United Kingdom

For details of our global editorial offices, for customer services and for information about how to apply for permission to reuse the copyright material in this book please see our website at www.wiley.com.

The right of the author to be identified as the author of this work has been asserted in accordance with the Copyright, Designs and Patents Act 1988.

Library of Congress Cataloging-in-Publication Data

LTE self-organising networks (SON) : network management automation for operational efficiency / edited by Seppo Hämäläinen, Henning Sanneck, Cinzia Sartori.
    p. cm.
    Includes bibliographical references and index.
    ISBN 978-1-119-97067-5 (cloth)
  1. Self-organising networks. I. Hämäläinen, Seppo, 1969– II. Sanneck, Henning, 1968– III. Sartori, Cinzia, 1960–
    TK7872.D48L74 2012
    681'.2–dc23   2011032030

A catalogue record for this book is available from the British Library.

Print ISBN: 9781119970675

Set in 10/12pt Times by Thomson Digital, Noida, India

Printed and Bound in Great Britain by Antony Rowe Ltd, Chippenham, Wiltshire.

To Leevi, Lina-Maria and Terja

In memory of Dr.-Ing. Hugo Sanneck (1932-2011)

To Nikos and Marika

Contents

Foreword

When we designed the first generation of digital mobile networks, the concept of a self-organising network (SON) was not in focus. Now, with hindsight we can say that a lot of self-organising features already exist such as power control, handover between cells and efficient customer management based on central administration of SIM cards.

Why has the vision of self-organising networks gained importance in our industry? There are two obvious drivers: cost reduction and increasing complexity. Network operators urgently need much more automation in order to efficiently manage large networks consisting of tens of thousands of base stations with hundreds of settings each. Optimising several network layers providing a multitude of customer services with high service quality would be highly complex and labour intensive, and therefore would be very costly without the mechanisms and intelligence in our networks to 'organise themselves'.

While it was common practice in the early years with much smaller mobile networks to execute many tasks manually on site, now network operators are able to handle all operational activities remotely from one or few central locations. This trend was enabled to a large extent by the significant progress in the IT industry.

In order to guide the industry in developing automation functionality that is relevant for operators Deutsche Telekom together with their partner in the NGMN Alliance (Next Generation Mobile Networks) have taken the initiative to drive 'Self-Organising Networks' (SON).

Within the vision of SON the operators have defined the most relevant use cases for the automation of operational tasks: typically, those tasks that require significant manual effort, or tasks that are highly complex in nature, and are therefore prone to error. NGMN presented these use cases to the vendor industry, and 3GPP standardisation bodies were requested to develop and standardise respective self-organising solutions.

There are some prominent examples of SON solutions which are already implemented today. For instance the 'Plug-and-Play' deployment of a base station requiring only the physical installation on sites, where all complex and specific configuration settings as well as software management are executed automatically. Similarly the 'Automatic Neighbour Relationship Configuration' (ANR) reduces effort but improves also the perceived network quality.

With LTE we have a great opportunity to bring the value of SON into our networks. But the potential and ambition clearly includes legacy networks such as UMTS and GSM also.

This book is an excellent introduction to the world of SON for technicians in research as well as developers in the industry providing solid knowledge and motivation to push SON solutions forward in the telecommunications sector.

Dr. Klaus-Jürgen Krath
Deutsche Telekom AG

Preface

Mobile network operators will meet many challenges in the coming years. It is expected that the number of people connected, wireline and wireless, will reach five billion by 2015. At the same time, people use more wireless services and they expect similar user experience to what they can now get from fixed networks. Because of that we will see a hundred-fold increase in network traffic in the near future. At the same time markets are saturating and the revenue per bit is dropping.

To meet the increase in demand for a wide range of content services with high bit rate requirements, the Third Generation Partnership Project (3GPP) is standardising the next generation of cellular networks called Long Term Evolution (LTE). When LTE is introduced by the operators, it leads to parallel operation of LTE together with existing 2G and 3G networks that are not phased out for a long time to come. LTE represents a major advance, designed to meet needs for high-speed data and media transport as well as high-capacity voice support for carriers. This includes support for new types of network elements, such as relay and femto nodes, and different cell layers. Due to that fact, a significantly increased number of base stations is required to assure coverage and capacity, all of which have to be managed properly. Also many complex radio network parameters have to be maintained and optimised.

Mobile Network Operators' vital interest is to minimise operational effort and cost. The concept of Self Organising Networks (SON), introduced by the Next Generation Mobile Networks (NGMN) alliance on 2007, is a key enabler for simplifying operation and maintenance in next generation mobile networks. SON aims at:

- Reducing operating cost by reducing the degree of human intervention in network design, build and operate phases.
- Reducing capital expenditure by optimising the usage of available resources.
- Protecting revenue by reducing the amount of errors introduced by humans.

This is accomplished by simplifying operational tasks through automated mechanisms such as self-configuration, self-optimisation and self-healing. SON can be seen as an approach in which many functions which have earlier been done manually as a part of the ('offline') network planning and optimisation tool chain are now moved to be executed ('online') in the network elements and their OAM system.

While NGMN has set requirements for SON use cases, 3GPP has made technical specification and standardisation for them. However, not all SON functions require standardisation. In this book, both 3GPP-standardised SON use cases and functions, but also functionality not using standardised interfaces or signalling is discussed.

The book focuses on LTE as for this new technology SON features can be designed from the start and thus take full effect. Where applicable, however, similar concepts are described for 3G and 2G. As the main operational challenges are seen in the management of radio networks, the focus of the book is on radio access. The end to end view is touched by covering some of the core network and transport (backhaul) aspects. Core network aspects are treated in a separate chapter and related transport aspects are treated where relevant, however, self-organisation of the transport network as such is beyond the scope of the book.

The book is organised as follows. The network management challenges that demand automation and thus SON are discussed in Chapter 1. In addition, the motivation behind applying SON for LTE networks is discussed.

An overview of 3GPP as well as LTE requirements and specifications are given in Chapter 2. Also LTE radio access network scenarios and their evolution are covered.

In Chapter 3, a vision for SON addressing the foreseen challenges is discussed and NGMN and 3GPP SON use cases presented. Typically, when benefits of SON are discussed, the first benefit is seen in saving in operational expenses. However, this is not the only benefit SON offers; SON will also have impact on for example, capital expenses and network quality of service. Such SON business benefits for selected use cases are discussed in Chapter 3. In addition, Chapter 3 presents the foundations for SON, that is, technologies on which SON is based as well as previous research projects, architectural considerations for SON-enabled systems and the operational and technical challenges of SON.

The operational life-cycle of a mobile network consists of design, build and operation/ maintain phases. The two latter phases can be automated by SON. The build phase can be automated and thus simplified through auto-connectivity, -commissioning, and dynamic radio configuration (Chapter 4). In the operational phase, self-optimisation and self-healing functions automatically change the network configuration based on the network performance and incidents in the network. Self-optimisation and self-healing are discussed in Chapters 5 and 6, respectively.

Minimisation of Drive Tests (MDT) functionality is planned for 3GPP Release 10. MDT supports autonomous collection of UE measurements and positioning of the UE. This information, together with information available in the radio access network can be used to visualise in detail the network performance and health. Thus MDT, described in Chapter 7, is an enabler for both self-optimisation and -healing of the network. NGMN use cases exist also for core networks. Here, SON concepts are also applicable and closely linked to SON in radio access (Chapter 8).

When many different SON functions are active in a system, interactions between them may occur. Therefore mechanisms to operate SON at a system level are needed. This includes mechanisms for preventive coordination between different SON functions to avoid conflicts on one hand but assure efficient operation (parallelisation) on the other hand. Additionally, it is crucial that human operators can interact with the SON-enabled system and stay in control. SON operation is discussed in Chapter 9.

Chapter 2 already introduced network scenarios relevant for SON. A particular scenario is a 'Heterogeneous Networks' scenario in which a network is made of several different cell types, technologies or layers. Such a network will impose stronger requirements for SON with regards to scalability, interoperability but also improved functionality. Therefore a separate chapter, Chapter 10, is dedicated to SON for heterogeneous networks.

Finally, Chapter 11 gives an outlook to future SON related topics, such as cognitive radio networks, and novel technological enablers for future SON.

Concepts to automate network operations have recently gained significant interest to improve an operator's cost position. The book addresses particularly the *novel* SON components in the network elements and the OAM system. While a number of research publications (in addition to NGMN requirements and 3GPP standards material) have appeared, no comprehensive single source on the LTE self-configuration and -optimisation topic has been available. While including the latest status in 3GPP, the book aims at providing a comprehensive picture of a SON-enabled system.

For more information, please visit the companion website, www.wiley.com/go/Hamalainen.

List of Contributors

Tobias Bandh
Gyula Bódog
Yves Bouwen
Christoph Frenzel
Jürgen Goerge
Seppo Hämäläinen
Anssi Juppi
Risto Kauppinen
Raimund Kausl
Ilkka Keskitalo
Krzysztof Kordybach
Jaroslaw Lachowski
Daniela Laselva
Andreas Lobinger
Henrik Martikainen
Szabolcs Nováczki
Klaus Pedersen
Johanna Pekonen
Miikka Poikselkä
Simone Redana
Dirk Rose
Henning Sanneck
Cinzia Sartori
Christoph Schmelz
Markus Stauffer
Paul Stephens
Clemens Suerbaum
Péter Szilágyi
Haitao Tang
Malgorzata Tomala
Eddy Troch
Ingo Viering
Achim Wacker
Richard Waldhauser
Bernhard Wegmann
Jeroen Wigard
Volker Wille
Osman Yilmaz

Acknowledgements

The editors would like to acknowledge all the colleagues who enthusiastically contributed to the writing of the book (cf. 'list of contributors' above) not only as authors but also reviewers. SON is a very diverse technical area requiring many different competences in the radio and distributed systems fields. Hence, we are very grateful that it has been possible to bring together such a great team of close to 40 contributors from Nokia Siemens Networks, partner companies and universities.

We would like to thank the following colleagues for their help to set up the book project and valuable comments during the book's review process: Kari Aaltonen, Guillaume Decarreau, Richard Fehlmann, Nadine Herold, Günther Horn, Matthias Kaetzke, Patrick Marsch, Peter Merz, Wolf-Dietrich Moeller, Olaf Pollakowski, Raphael Romeikat, Mikael Rutanen, Dariusz Tomecko, Ville Tsusoff and Marcin Wiczanowski.

We appreciate the fast and smooth editing process and all help provided during the process of writing the book by Wiley-Blackwell and in particular: Mariam Cheok, Richard Davies, Lynette James, Abhishan Sharma, Sophia Travis and Mark Hammond.

We are grateful to our families and contributors' families for their understanding and patience during the long evenings spent when writing and editing the contents of the book.

Our employer made it possible to write this book by providing support and encouragement during the process of writing and editing. Therefore special thanks are for Nokia Siemens Networks.

Finally we are grateful for the interactions with the SON research and 3GPP SON standardisation community comprising mobile network operators, vendors and academia. The industry-wide effort to make SON happen has been clearly the basis for this book.

We welcome any proposals and suggestions for improvements of the contents of the book in forthcoming editions as well as pointing us to any possible mistakes. The feedback is welcome to the editors' e-mail addresses: seppo.hamalainen@nsn.com, henning.sanneck@nsn.com and cinzia.sartori@nsn.com.

List of Abbreviations

3G	3rd Generation (Cellular Systems)
3GPP	3rd Generation Partnership Project
AAA	Authentication, Authorisation and Accounting function
AAS	Active Antenna System
ABS	Almost Black Subframe
AC	Admission Control
ACI	Autonomic Computing Initiative
ACS	Auto-Connection Server
AF	Amplify and Forward
AGP	Automatic Generation of Initial Parameters for eNodeB Insertion
AI	Artificial Intelligence
ANDSF	Access Network Discovery and Selection Function
ANR	Automatic Neighbour(s) Relation/Relationship setup
AP	Application Protocol
APN	Access Point Name
ARCF	Automatic Radio Configuration Function
ARPU	Average Revenue Per User
ARRM	Advanced RRM
AS	Access Stratum
ATM	Asynchronous Transfer Mode
B3G	Beyond 3G
BB	Basic Biasing
BCCH	Broadcast Channel
BH	Busy Hour
BFD	Bidirectional Forward Detection
BLER	Block error rate
BN	Bayesian Network
BP	Blocking Probability
BPM	Business Process Management
BS	Base Station
BW	Bandwidth
CA	Carrier Aggregation
CA	Certificate Authority
CAC	Composite Available Capacity
CAPEX	Capital Expenditure
CBL	Case Based Learning
CBR	Case Based Reasoning
CBRA	Contention-Based RA
CC	Component Carrier
CCO	Coverage and Capacity Optimisation

CDF	Cumulative Distribution Function
CF	Cooling Factor
CIO	Cell Individual Offset
CM	Configuration Management
CMDB	Configuration Management Database
CMP	Certificate Management Protocol
CN	Cognitive Networks
COC	Cell Outage Compensation
COM	Cell Outage Management
CoMP	Coordinated Multipoint transmission and reception
C-plane	Control Plane
CP	Collision Probability
CPC	Cognitive Pilot Channel
CPE	Customer Premises Equipment
CPICH RSCP	Common PIlot CHannel Received Signal Code Power
CQI	Channel Quality Indicator
CR	Cognitive Radio
C-RAN	Cloud RAN
CRN	Cognitive Radio Networks
CRS	Common Reference Symbol
CSFB	Circuit-Switched FallBack
CSG	Closed Subscriber Group
CSI	Channel State Information/Indicator
CSL	Cognitive Specification Language
CSP	Communication Service Provider
CQI	Channel Quality Indicator
CRS	Common Reference Symbols
DDDS	Dynamic Delegation Discovery System
DF	Decode and Forward
DIR	Dominant Interference Ratio
DL	Downlink
DM	Domain Manager
DMP	Detection Miss Probability
DNS	Domain Name System
DRC	Dynamic Radio Configuration
Ec/No	chip energy per noise
ECGI	E-UTRAN Cell Global Identification
EDGE	Enhanced Datarates for GSM Evolution
eICIC	enhanced Inter-Cell Interference Coordination
EIRP	Equivalent Isotropic Radiated Power
EM	Element Manager
EMS	Element Management System
eNB or eNodeB	Evolved Node B
EPC	Evolved Packet Core
EPS	Evolved Packet System
ES	Energy Savings

ETSI	European Telecommunications Standards Institute
E-UTRA	Evolved Universal Terrestrial Radio Access
E-UTRAN	Evolved Universal Terrestrial Radio Access Network
FBS	Flexible Base Station
FDD	Frequency Division Duplex
FM	Fault Management
GA	Genetic Algorithm
GBR	Guaranteed Bit Rate
GNSS	Global Navigation Satellite System
GPS	Global Positioning System
GSMA	GSM Association
GU	Globally Unique
GUMMEI	Globally Unique MME Identifier
GUTI	Globally Unique Temporary Identity
GW	Gateway
HARQ	Hybrid Automatic Repeat reQuest
HeMS	Home eNodeB Management System (LTE)
HeNB	Home eNodeB (LTE)
HetNet	Heterogeneous Networks
HII	High Interference Indicator
HMS	Home NodeB Management System (3G)
HNB	Home NodeB (3G)
HNB GW	Home NodeB Gateway
HO	HandOver
HOF	HandOver Failure
HOO	HandOver Optimisation
HPO	Handover Parameter Optimisation
HSDPA	High Speed Downlink Packet Access
HSPA	High Speed Packet Access
HSPA+	High Speed Packet Access evolution
HSS	Home Subscriber Server
HSUPA	High Speed Uplink Packet Access
HW	Hardware
HW-ID	Hardware Identity
ICIC	Inter-Cell Interference Coordination
ICO	Interference Coordination Optimisation
ID	Identifier
IE	Information Element
IMPEX	Implementational Expenditures
IMS	IP Multimedia Subsystem
IP	Internet Protocol
IRP	Integration Reference Point
Itf-N	Interface-Northbound
IS	Information Service
ISCP	Interference Signal Code Power
JRRM	Joint Radio Resource Management

k-NN	k-Nearest Neighbour
KPI	Key Performance Indicator
KQI	Key Quality Indicator
LAC	Location Area Code
LBO	Load Balancing Optimisation
LIPA	Local IP Access
LTE	Long-Term Evolution
LTE-A	Long-Term Evolution-Advanced
LIPA	Local IP Access
MAC	Medium Access Control
MAPE	Monitoring – Analysis – Planning – Execution
MBMS	Multimedia Broadcast and Multicast Service
MCC	Mobile Country Code
MCS	Modulation and Coding Scheme
MDT	Minimisation of Drive Tests
MGW	Media Gateway
MLB	Mobility Load Balancing
MIMO	Multiple Input Multiple Output
MME	Mobility Management Entity
MNC	Mobile Network Code
MNO	Mobile Network Operator
MPLS	Multi-Protocol Label Switching
MSC	Mobile Switching Centre
MRO	Mobility Robustness Optimisation
MU-MIMO	Multi-User MIMO
NAS	Non-Access Stratum
NB	Node B
NCBRA	Non-Contention-Based RA
NCL	Neighbour Cell List
NE	Network Element
NGMN	Next Generation Mobile Networks
NLM	Network Listening Mode
NM	Network Management
NMS	Network Management System
NN	Neural Network
NNSF	NAS Node Selection Function
NR	Neighbour Relation
NRM	Network Resource Model
NRO	Neighbour Relationship Optimisation
OAM	Operation, Administration and Management
OFD	Operational Fault Detection
OFDM	Orthogonal Frequency Division Multiplexing
OFDMA	Orthogonal Frequency Division Multiple Access
OMC	Operation and Maintenance Centre
OPEX	Operation Expenditure
OSG	Open Subscriber Group

OSPF	Open Shortest Path First
OSS	Operations Support System
OWL	Web Ontology Language
PBCH	Physical Broadcast Channel
PBLA	Push-to-Best Layer Algorithm
P-CCPCH RSCP	Primary Common Control Physical Channel RSCP
PCEF	Policy Control Enforcement Function
PCI	Physical Cell ID
PCO	Protocol Configuration Options
PCRF	Policy and Charging Resource Function
P-CSCF	Proxy-Call Session Control Function
PDCP	Packet Data Control Protocol
PDSCH	Physical Downlink Shared Channel
P-GW	Packet Data Network Gateway
PHR	Power Headroom Report
PHY	Physical Layer
PKI	Public Key Infrastructure
PM	Performance Management
PMI	Precoding Matrix Indicator
PRACH	Physical Random Access Channel
PRB	Physical Resource Block
P-RNTI	Paging Radio Network Temporary Identity
PS	Packet Scheduling
PSF	Power Supply Factor
QCI	Quality of Service Class Identifier
QoS	Quality of Service
RA	Random Access
RAB	Radio Access Bearer
RAC	Routing Area Code
RACH	Random Access Channel
RAN	Radio Access Network
RAS	Remote Azimuth Steering
RAT	Radio Access Technology
RB	Resource Block
RBS	Radio Base Station
RE	Range Expansion/Extension
RET	Remote Electrical Tilt
RF	Radio Frequency
RFID	Radio Frequency Identification
RI	Rank Indication
RIM	RAN Information Management
RLF	Radio Link Failure
RLM	Radio Link Monitoring
RNC	Radio Network Controller
RRC	Radio Resource Control
RRH	Radio Resource Head

RRM	Radio Resource Management
RSRP	Reference Symbol Received Power
RSRQ	Reference Symbol Received Quality
RSSI	Received Signal Strength Indication
SAE	System Architecture Evolution
SAN	Software Adaptable Network
SBN	Smooth Bayesian Network
ScM	Self-Configuration Management
S-CSCF	Serving-Call Session Control Function
SDR	Software Defined Radio
SEG	Security Gateway
SGSN	Serving GPRS Support Node
S-GW	Serving Gateway
SI	Study Item
SIB	System Information Block
SINR	Signal to Interference plus Noise ratio
SLA	Service Level Agreement
SOAP	Simple Object Access Protocol
SOM	Self-Organising Maps
SON	Self-Organising Networks
SS	Solutions Set
SU-MIMO	Single User MIMO
SW	Software
SWG	Sub-Working Group
TA	Tracking Area or Timing Advance
TAC	Tracking Area Code
TAI	Tracking Area Identifier
TAU	Tracking Area Update
TCE	Trace Collection Entity
TCO	Total Cost of Ownership
TDD	Time Division Duplex
TDM	Time Domain
TPM	Trusted Platform Module
TrE	Trusted Environment
TRX	Transmission and Reception Unit
TS	Technical Specification (3GPP)
TSG	Technical Specification Group
TU	Typical Urban
TX	Transmission
TXP	Transmission Power
UE	User Equipment
UL	Uplink
UMA	Unlicensed Mobile Access
UMTS	Universal Mobile Terrestrial System
U-plane	User Plane
USIM	Universal Subscriber Identity Module

UTA	User Throughput-Based Algorithm
UTRA	Universal Terrestrial Radio Access
UTRAN	Universal Terrestrial Radio Access Network
VET	Variable Electrical Tilt
VLR	Visitor Location Register
VoIP	Voice over IP
VoLTE	Voice over LTE
WAN	Wireless Access Network
WCDMA	Wideband Code Division Multiple Access
WG	Working Group
WI	Work Item

1

Introduction

Cinzia Sartori, Henning Sanneck, Jürgen Goerge, Seppo Hämäläinen
and Achim Wacker

The number of mobile subscribers has impressively increased during the last decade; at the same time wireless data usage continues to accelerate at an unprecedented pace even when (for developed countries) subscriber numbers reach saturation.

With the adoption of the Global System for Mobile Communication (GSM), mobile phones have become indispensable devices for voice communication and, nowadays, mobile networks are available for 90% of the world population. However, GSM was mainly designed for carrying voice traffic and some data capability was only added subsequently. The 'mobile data explosion' is a quite recent phenomenon driven by the introduction of the 'Third Generation' (3G) mobile system with Wideband Code Division Multiple Access (WCDMA), High Speed Packet Access (HSPA) and its enhancements called High Speed Packet Access Plus (HSPA +). The introduction of HSPA has marked the beginning of the transformation from voice-dominated to packet data-dominated mobile networks. These 3G evolution technologies are crucial to allow upgrading the network at relatively low costs and hence those technologies will be still important for a long period of time to come. However, it is clear that only a new Radio Access Technology (RAT) comprising a new air interface together with a new network architecture can cope with the described data explosion in the longer term. Long-Term Evolution (LTE; Holma and Toskala, 2011) is this technology which at the time of writing had been rolled out and put into commercial use in several countries already. Chapter 2 introduces the key technical concepts and radio access network scenarios of LTE.

The exponential growth of mobile broadband traffic is certainly caused by both, the increasing demand for known and new data services, such as mobile Internet access, social networking, location-based services/personal navigation, and so on, and the data processing and storage capabilities of state-of-the-art terminals, such as smartphones and, most recently, tablets (Figure 1.1). Such 'always-on' devices used by humans as well as network usage by machines (Machine to Machine; M2M) also put strong requirements on the capabilities of the network control plane.

LTE Self-Organising Networks (SON): Network Management Automation for Operational Efficiency, First Edition.
Edited by Seppo Hämäläinen, Henning Sanneck and Cinzia Sartori.
© 2012 John Wiley & Sons, Ltd. Published 2012 by John Wiley & Sons, Ltd.

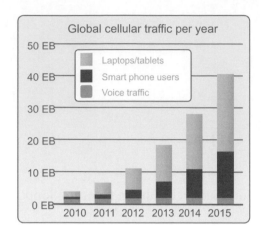

Figure 1.1 Data volume growth. *Source:* Nokia Siemens Networks.

As a next step, the use of tablets may increase the demand for wireless video applications to a large extent and put tremendous stress on the wireless network infrastructure. This is the case, because high resolution displays and powerful processors enable the transmission of high-definition video. This, in turn, produces a demand for high data rates required 'everywhere' to satisfy the expectation of the end customers. Such 'data hungry' applications ask for more capacity and higher quality of service, which can only be satisfied with the introduction of LTE and its evolution called 'LTE-Advanced' (LTE-A).

To cope with such a huge demand for data traffic transmission, the wireless network operators need to significantly upgrade their networks and use these resources most efficiently. Traditional methods like macro site 'densification', along with improved receivers and higher order sectorisation, will not be fully sufficient to provide the desired capacity for the predicted traffic growth. The deployment of small cells as an additional layer to the macro layer is definitely the most promising solution for building improved spectral efficiency (and thus capacity) per area. Thus, the migration from macro-only to *Multi-Layer* topology as part of a 'Heterogeneous Network' scenario are expected to further accelerate in the near future. Also, LTE will run for a long period of time in parallel with existing 2G and 3G networks (*Multi-RAT*).

The described requirements for wireless service providers to upgrade their networks, to deploy LTE and to integrate their existing RATs have the effect that the network infrastructure as a whole will be rather complex and heterogeneous. Thus, operators face significant operational challenges in terms of work effort and cost. Unfortunately, those costs will not be compensated by additional revenue due to the decreasing average revenue per user (caused by pricing schemes like e.g. flat rates, induced through fierce competition in the market). Hence, the cost position as a vital interest of operators, in particular the *operational expenses* (OPEX), has gained much more attention recently. Especially in the early deployment phase, the efforts to set up and optimise the network are significant and traditionally lead to substantial delays before an optimal and stable system setup can be reached. In order to minimise such delays and in general reduce the network operation expenses, the *Self-Organising Networks* (SON) concept is considered to be an integral part of LTE.

1.1 Self-Organising Networks (SON)

The concept of SON became frequently used after it was adopted by the *Next Generation Mobile Networks (NGMN)* alliance to address challenges foreseen due to management of several radio access technologies along with the LTE network introduction. Chapter 3 provides an introduction to the SON vision (Section 3.1), key SON concepts and benefits and their foundations.

One of the aims of operators is to keep their operational burden at the currently existing level, that is, manage the multi-RAT (including LTE), multi-layer infrastructure as described above with their existing operational staff and cost structure. Operators have to maximise their return on investment, they need to optimise the resource utilisation in order to minimise their huge, necessary investments hence, efficiency is essential in order to be able to manage the additional network without increased workforce.

Network operation today is based on a centralised *Operation, Administration and Maintenance (OAM)* architecture. Configuration and optimisation of network elements is performed centrally from an OAM system (also called the Operations and Maintenance Centre: OMC) with support of a set of planning and optimisation tools. Planning and optimisation tools are typically semi-automated and management tasks need to be tightly supervised by human operators. This manual effort is time-consuming, expensive, error-prone and requires a high degree of expertise (Laiho *et al.*, 2006).

Increased *automation of network operations* is seen as a proper means to cope with the described rising complexity of the network infrastructure in order to utilise deployed network resources in an optimised way. At the same time automation aims at:

- keeping the operational effort at an acceptable level;
- protecting the network operation by reducing the probability of errors during the overall process of rolling out a network and the permanently ongoing process of managing the network;
- speeding up operational processes.

Self-organisation is an advanced mechanism to enable such automations. It is crucial that automated features are properly integrated with the existing operator processes and embedded into the architecture of the overall OAM tool chain. Automation is achieved by adding (SON) features to network equipment which facilitates network operation processes and delivery of professional services related to the network. Hence SON is a contributor to the '*operability*' and '*serviceability*' characteristics of a network.

3GPP has created the actual SON standards upon NGMN long-term objectives for a 'SON-enabled mobile broadband network' by defining the necessary use cases, measurements, procedures and open interfaces to support better operability in a multi-vendor environment. SON standardisation is still an ongoing activity (Figure 1.2). SON standardisation has started with LTE in 3GPP Release 8 and continued in Release 9 and 10 (Release 10 was completed in June 2011). Release 11, which will contain additional SON features and enhancements to existing ones, is in definition phase at the time of writing this book.

SON Use Cases (NGMN, 2008), cf. Section 3.2, are categorised into functional areas along the key OAM areas of configuration, optimisation and troubleshooting (cf. Figure 1.3):

- **Self-Configuration** (Chapter 4);

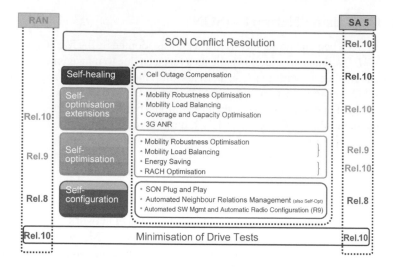

Figure 1.2   Roadmap for SON standardisation in 3GPP.

- **Self-Optimisation** (Chapter 5) including traffic steering between different type of radio resources; and
- **Self-Healing** (Chapter 6).

A common characteristic is that the degree of 'human-in-the-loop' for OAM use cases is reduced as much as possible reaching even fully '*closed loop*' automation for some of the use cases.

'*Minimisation of Drive Tests' (MDT)* functionality has been specified in 3GPP Release 10 for LTE and Universal Terrestrial Radio Access Network (UTRAN). MDT addresses the issue that often drive tests have to be executed to monitor and assess mobile network performance. Such drive tests are very expensive since the actual testing needs significant human operator involvement. Key characteristics of MDT are measurements collected on

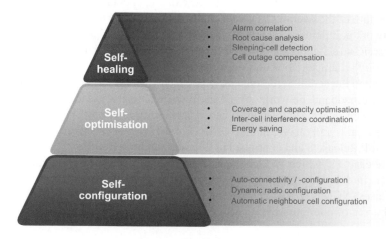

Figure 1.3   SON use case examples.

User Equipments (UEs), which may contain location information, thereby allowing to have a much more fine-grain view of a cell's performance. Because such a view is useful not only for a human operator but also for the automated SON functions, MDT is considered to be an important enabler for SON. MDT is discussed in detail in Chapter 7.

SON research and standardisation is mainly focused on the radio access domain, due to its intrinsic complexity (high number of widely distributed network elements) and thus significant cost share of the overall network infrastructure and its operations. Nevertheless, *SON for Core Networks* (Chapter 8) is also relevant from the perspective of properly configuring and optimising the network end-to-end. Note that backhaul aspects contributing to the end-to-end view are treated where relevant in Chapters 4–6 which discuss the SON functional areas.

Figure 1.3 shows some examples for SON use cases. There exists a significant number of different SON use cases which have partially conflicting goals, overlapping input or output parameters. Examples for such *SON function interactions* as well as technical solutions to control the interactions are discussed as the main topic for *SON Operation* in Chapter 9.

Like mentioned above, on one hand LTE needs to be integrated with existing RATs; on the other hand even the resource capabilities of LTE macro cells will not be sufficient in the long term but need to be complemented by smaller cells for capacity. In Heterogeneous

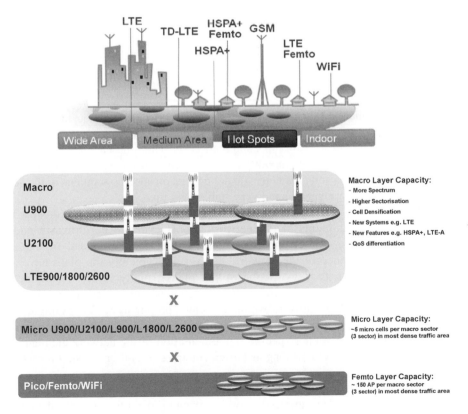

Figure 1.4 Heterogeneous Networks.

Networks (cf. Figure 1.4) operators will have to deal with handovers in inter-technology and macro/femto scenarios; interference management of macro/pico and macro/femto is definitely an outstanding issue. At the same time network capacity needs to be optimised via an efficient utilisation of all available resources (multi-RAT, multi-layer) while assuring desired end user experience with appropriate Quality of Service (QoS) and Quality of Experience (QoE). *SON for Heterogeneous Networks* and related challenges are described in Chapter 10.

While most of the concepts of the 'classical' SON use cases are now getting assessed and operators start their deployment, the SON concept keeps evolving by integrating new use cases and solutions based on existing and novel technologies. Chapter 11 describes that evolution which may lead to a true *'Cognitive Network'*.

1.2   The Transition from Conventional Network Operation to SON

While SON concepts are very appealing for network operators on the one hand, on the other hand they need to be carefully integrated into existing tool chains realising OAM processes. Hence, in the following, basics of 'conventional' network rollout and operation are introduced and the potential automation possibilities, which are the targets for SON, are discussed.

Business targets must be broken down to an optimal deployment of the network infrastructure and to an optimised setting of every individual parameter in each network element. Therefore, operators typically employ a *layered set of tools* as depicted in Figure 1.5. On the left hand side the classes of tools are depicted. The corresponding department in an operator's organisation, time scale of resulting plans, and class of algorithms and parameters are listed on the right hand side.

Traffic forecast, capacity planning and site planning translate above business targets to a proper deployment of network elements. They take bordering conditions into account like available budget, sites/transmission links and their costs. The timescale for these plans is in the year range, but can go down to monthly intervals. Scope of the plans is usually the entire network. Tools, algorithms and parameters do not depend on specific suppliers of the network elements, so these tools are generic and very stable over time.

Radio, transport and link planning and optimisation processes periodically try to optimise the different domains of the network. Tools evaluate performance data, run simulations and provide, for example, a set of optimised parameters, or a plan for optimised roll out sequence of the network as a result. Typically the plans have a time horizon of months down to days. Scope of the plans usually is the overall network, a large region of a network, or certain RATs. Planning and optimisation uses vendor independent, generic algorithms to simulate wave propagation for example. Those algorithms typically operate on standardised parameters. Since neither algorithms nor parameters depend on the vendor-specific implementation of the network elements, those algorithms and tools are stable as long as basic principles of call processing for a RAT do not change.

Many operators have a strict separation between planning departments and departments that operate the live network. The interface between planning departments with their vendor-agnostic tools for network or service management and the network operations department with

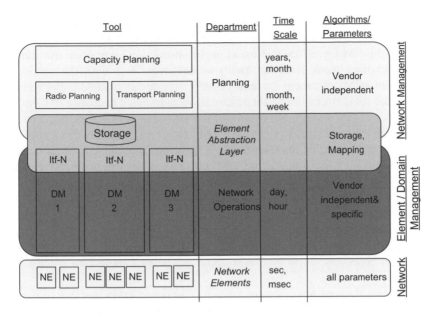

Figure 1.5 Model of a layered OAM tool chain.

mainly vendor-specific tools (Element Managers (EM), Domain Managers (DM)) is very often
established by a kind of element abstraction layer.

On the one hand the element abstraction layer acts as central repository, which is used to
collect, store and distribute all parameters in the network, that is, all standardised parameters
and many or even all vendor specific parameters of the overall multi-technology, multi-
vendor network. Usually, the element abstraction layer does not understand the semantics of
vendor-specific parameters, so the element abstraction layer is not able to check the
correctness of vendor-specific parameters, nor is this layer able to optimise them. However,
this repository is needed to transfer network data in a coordinated way between planning
department (or more generally: departments concerned with network management tasks) and
network operations department. According to the OAM reference architecture as defined by
3GPP SA5 this repository function would be 'counted' to the Network Management
(NM) layer.

On the other hand the element abstraction layer is used to map standardised parameters
between the vendor-specific representation used by network elements and the generic
information models used by NM functions. The mapping between vendor specific data and
generic data has to be modified most probably with any new release of network elements. Also
new features, for example, in planning tools, require mapping to additional parameters. Thus
maintenance effort for those mappings is high.

In the context of SON NM tools that collect alarms and performance data in
multi-vendor, multi-technology networks act as an element abstraction layer for Fault
Management (FM) and Performance Management (PM). They collect the data from the
vendor-specific DMs, and provide an abstracted view for the higher-level, generic tools.

According to the 3GPP OAM reference architecture the mapping is part of the DMs where the so-called 'northbound interfaces' (Itf-N) of the DM act as facade to hide this mapping from the NM layer. Tools of the element abstraction layer often use standardised interfaces (like those from 3GPP or TMF) to exchange data with the vendor-specific network or Domain Management Systems (DMS). Besides automated collection of alarm and performance data, already these interfaces allow automated exchange of Configuration Management (CM) data and remote control of the DMs and network elements by the tools of the element abstraction layer to a certain extent. However, in reality this mapping of CM data often is not performed by the DMs but by a dedicated tool that combines mapping and repository. Communication towards the DM and EM is then performed by proprietary interfaces.

In context of the introduction of SON it is worth emphasising that this element abstraction layer not only transforms information models. It often currently defines a strict boundary between departments and between different time scales of operation. Thus, introduction of SON is not only a technical challenge, but also might influence on the overall organisational and operational processes, as SON cycles could go beyond these two strictly separated departments.

Vendor-specific Element Management System (EMS) and DMS are able to handle most vendor-specific parameters, check their correctness and optimise them to a certain extent. The time scale of usage varies from days down to hours. Usually, the spatial scope of these tools is a vendor's radio network (or single radio technology) in a larger region. Definitely, those tools must be adapted with each new release of network elements.

Local craft/maintenance terminals, site managers, and so on are used on-site to commission and install network elements. They are able to manipulate the configuration of hardware and boards down to the lowest possible level. Those tools usually do not have a standardised interface to higher-level Network Management Systems (NMS). Although those tools are able to handle all data of a Network Element, they are usually not able to perform any kind of optimisation. Most of the time, these tools are specific to a certain type of Network Element, thus a field engineer has to cope with a multitude of such terminals. Local craft terminals typically connect to exactly one network element at a time.

Network elements are subject to management and in the past were not actively managing themselves at all. Even in the past, the network elements already implemented some kind of 'low-level' troubleshooting and optimisation. For example, the 3G UTRAN uses a hierarchy of layered optimisations: a radio network planning department provides quality values via a DM (e.g. for the block error rate, BLER) of an UTRAN cell to the Radio Network Controller (RNC), for example, once a month. Based on this input, the RNC uses proprietary algorithms to calculate optimised target values for the 'signal to interference plus noise ratio' (SINR) for each bearer connection under its control (outer loop power control, timeframe 10–100 ms) and sends this optimised SINR values to the WCDMA base stations (NodeBs). The NodeB in turn uses the given target SINR of outer loop power control as a bordering condition for its inner loop power control to adjust the actual transmit power for the individual connections every 666 ms. This example shows the degree of automation in Radio Resource Management (RRM) which should also be brought to the management of the network.

1.2.1 Automation of the Network Rollout

In order to maximise utilisation of invested capital, the number and location of base stations must be planned carefully. Coverage, capacity and quality must meet the business targets as well as regulatory obligations. Optimised deployment of the network is not only driven by the specifics of the air interface and utilisation of the available spectrum, but also depends on setting up a corresponding backhaul and aggregation network needed to route the traffic towards the core network. Also, long-term business processes like acquisition of base station sites (site lease) and site preparation with all the required construction and supplies such as electricity have to be taken into account. Further, the outcome of each step must be documented, for example, for proper payment of subcontractors and bookkeeping/inventory. Installation and commissioning of a base station as well as its registration for service are just small steps in this overall *business process*. A proper automation of this business process supports managing the overall project of rolling out a network.

Each individual step of the business process is a process (or 'workflow') of its own. Figure 1.6 shows the differentiation between the layer of the business process and the embedded workflows within the NM layer. This book focuses on the automation of the workflows in NM domain and element management and in the elements itself. Enterprise architecture and business process integration are not covered, except when northbound interfaces of domain management and NM are described.

Bringing a base station on air means planning the corresponding parameters, installing the base station physically and configuring the software logically (commissioning). Installation

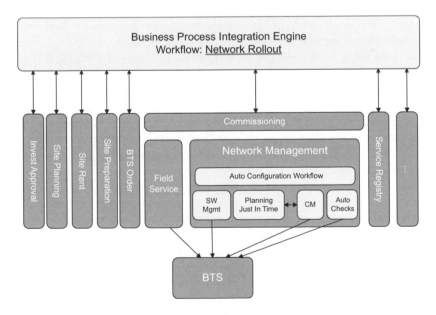

Figure 1.6 Workflows for installation and commissioning a base station (BTS).

and commissioning require field engineers with different skills, thus usually at least two site visits are necessary. Additionally, the commissioner needs to communicate with the planner in order to equip the base station with correct software and configuration data. Automation, which might include automated planning or planning on demand, may in most cases obsolete the site visit by the 'commissioner' resulting in bringing a base station on air faster, with fewer errors and with less workforce and thus significantly reduced costs. Chapter 4 discusses such 'self-configuration' concepts in detail.

1.2.2 Automation of Network Optimisation and Troubleshooting

Operators need to optimise revenue. This is the one, ultimate, business-level Key Performance Indicator (KPI). However, revenue not only depends on the network, but also on other instruments like marketing and sales, that is, attractive products at attractive prices, properly advertised with good customer relationship management. The best network does not help if the other instruments do not work. But also the opposite is true; even the best marketing and sales on their own will not generate satisfied customers.

Qualitatively, the overall performance of the network can be described by a 'Super KPI' like

$$P = x \cdot \text{``Coverage''} + y \cdot \text{``Capacity''} + z \cdot \text{``Quality''} \tag{1.1}$$

where the weights x, y, and z depend on the business targets of the operator as well as on specific area, maturity of the network layer, time of day, and so on. In the following, the different components of the overall performance indicator are introduced.

- *Coverage* is required to allow customers to use mobile services on all relevant places. It also is important, for example, not to lose roaming subscribers to competitors. So coverage might be important even in areas where no significant traffic occurs.
- *Capacity* is the ability of the network to carry traffic. It is important to note that only traffic translates into revenue, but not capacity as such. Providing capacity on areas without demand is waste of investment, while a shortage of capacity means losing traffic and thus revenue.
- Good *network quality* is important to catch new customers and to reduce churn, although, quality does not immediately translate into revenue. Quality is similar in effect as marketing. Good quality in the long run will positively influence revenue, since customers are attracted. In contrast, bad quality will negatively influence revenue, which might be compensated by advertisement or lower tariffs.

Because expenses for equipment and operation are limited, already this initial, even very coarse analysis shows that conflicting targets exist. For example, from a traffic-only point of view, it might be better to invest in more capacity in the city instead of closing coverage holes in rural areas: better earning money in the city from 1000 roaming business travellers from abroad than to cover 1000 own subscribers who have flat-rates anyway. On the other hand, this strategy definitely would be perceived as 'bad quality' from the 1000 own subscribers' perspective and drive them to competitors. Also legal obligations to cover especially rural areas (e.g. with the high bandwidth capabilities of LTE) might be in conflict with this strategy.

Any optimisation and troubleshooting activity (manual or automated), central in the OAM system or decentral in network elements, must optimise for the overall performance of the network according to the business strategy of the operator. Optimisation refers here to the activity to bring the network performance from a performance operating point P_1 to a point P_2 (with $P_2 > P_1$) triggered by the demand to increase revenue from an existing investment. Chapter 5 introduces a range of use cases and their solutions in 'self-optimisation'. Troubleshooting is in fact a similar activity, yet the trigger comes from addressing faults in the system. The system operates only at an inferior performance operating point P_1 (again with $P_2 > P_1$) as opposed to a 'normal' operating point P_2, which had been reached already before during fault-free network operation. Automated troubleshooting, also called 'self-healing', is introduced in Chapter 6.

If conflicting targets shall be optimised by different functions, those functions in the end must to be coordinated such way that overall performance is maximised. Not coordinating individual optimisation functions according to an overall strategy as, for example, given by a kind of 'Super KPI' might result in reaching local optima for individual functions, which might be far away from the global optimum. It also might result in 'ping-pong' and other unwanted effects of self-organisation. Chapter 9 will discuss several mechanisms of such coordination in detail.

1.2.3 SON Characteristics and Challenges

'Self-organisation is a process where the organisation (constraint, redundancy) of a system spontaneously increases, that is, without this increase being controlled by the environment or an encompassing or otherwise external system' (Heylighen, 2009). On the one hand, SON in LTE is based on the general definition like the one above used in philosophy, physics, biology, and so on. On the other hand, self-organising principles have already been applied to some extent in the IT (Autonomic Computing) and general networking domain (ad hoc networks). With regards to networking, the clear differentiator of SON in LTE is the *application of self-organisation to an infrastructure network*, which is desirable due to the inherent complexity of both, the network and its OAM system (Section 3.3).

Because SON functions are embedded into the OAM system and the network elements themselves, *SON architecture* is an important and often controversially debated topic. It is closely related to the general tradeoffs of *distributed versus centralised system architecture* in addition to the strong link to the 3GPP legacy OAM architecture (Section 3.4). The major driver for SON is the reduction of operational costs (rather than revenue increase, for example, by introduction of new services). Hence, it is crucial to understand the *business value* which can be generated by addressing a use case by SON (Section 3.5). Finally, the transition process towards SON described above brings technical *challenges* mainly caused by moving from an 'offline' planning and optimisation tool chain to embedding 'online' SON functions into the OAM system and the network elements. Furthermore, even if those challenges have been successfully addressed, SON functions need to be integrated with the corresponding, existing operator processes and the people implementing those processes (the human operator interacts with the system at a higher-level, setting policies and targets for SON functions, rather than directly changing config-uration parameters, cf. Section 3.6).

References

Heylighen, F. (2009) Self-Organisation, Principia Cybernetica Web http://pespmc1.vub.ac.be/selforg.html. [accessed 30 June 2011].

Holma, H. and Toskala, A. (eds) (2011) *LTE for UMTS: Evolution to LTE-Advanced, Revised Edition*, John Wiley & Sons, Chichester, Chapter on SON.

Laiho, J., Wacker, A. and Novosad, T. (2006) *Radio Network Planning and Optimisation for UMTS*, 2nd edn, John Wiley & Sons, Inc., New York.

NGMN (2008) Use Cases related to Self-Organising Network, Overall Description, NGMN Technical Working Group, *Self-Organising Networks*, (ed. F. Lehser), December 2008.

2

LTE Overview

Cinzia Sartori, Anssi Juppi, Henning Sanneck, Seppo Hämäläinen
and Miikka Poikselkä

LTE encompasses a set of aggressive requirements that aim at improving the end-user throughput, the cell capacity and reducing the user plane latency. These requirements, together with full mobility, will bring substantial benefits to user experience.

LTE is designed to support all kind of IP data traffic and voice is supported as Voice over IP (VoIP) for better integration with multimedia services. LTE aggressive requirements lead to the definition of a new Network Architecture, the Evolved Packet System (EPS), which comprises the Enhanced RAN (E-UTRAN or LTE) and the Evolved Packet Core (EPC). Both data and voice services are supported over the same packet switched network. E-UTRAN and EPC have been defined in 3GPP Release 8 and enhanced in further 3GPP Releases.

LTE paved the way to a new standardisation approach. In Release 8 LTE network and OAM have been standardised at the same time, yielding tremendous opportunities to design an overall optimised system with built in SON features.

The scope of this chapter is to give first a short introduction, without digging into technical details of the EPS network architecture, both E-UTRAN and EPC (deep inside is addressed in LTE technology specific books, for example, (Holma and Toskala, 2011) for LTE Radio Access Network). The chapter continues with a brief description of LTE-Advanced (LTE-A is specified in 3GPP Release 10), while the last part of the chapter is dedicated to LTE Radio Access Network Scenarios, their evolution and potential SON component.

2.1 Introduction to LTE and SAE

Long-Term Evolution (LTE) is a 3GPP project that provides extensions and modifications of the UMTS system allowing high data rate, low latency and packet optimised radio access networks. System Architecture Evolution (SAE) is an associated 3GPP project working on

LTE Self-Organising Networks (SON): Network Management Automation for Operational Efficiency, First Edition.
Edited by Seppo Hämäläinen, Henning Sanneck and Cinzia Sartori.
© 2012 John Wiley & Sons, Ltd. Published 2012 by John Wiley & Sons, Ltd.

3GPP core network evolution. The new air interface and network architecture aim at providing decreased cost per transmitted bit, achieved by:

- Advanced modulation techniques that allow optimised use of radio frequency.
- Flat architecture that minimises the number of network elements and optimises the usage of the transmission network.
- Capability to serve high quality, low latency real-time traffic, allowing both voice and data services to be provided over a single all-IP network.

The present chapter gives an overview of 3GPP as well as LTE Requirements and Specifications.

2.1.1 3GPP Structure, Timeline and LTE Specifications

The LTE standard, as well as WCDMA and the latest phase of GSM Evolution, have been developed by the 3rd Generation Partnership Project (3GPP). 3GPP is a collaborative standardisation model, uniting telecommunications standards bodies. It was formed by the European Telecommunications Standards Institute (ETSI), which defined the successful GSM standard, together with its counterparts on the other continents to be a global partnership which today includes more than 300 individual member companies worldwide.

3GPP standardisation covers specification work for GSM based 2G and WCDMA based 3G radio technologies in addition to 4G LTE technology. Further, 3GPP does standardisation for services and system aspects and core networks. To cope with such a huge dimension 3GPP has defined a very structured working procedure; each Technical Specification Group (TSG) is further sub-divided in Working Groups (WGs), each of those covering a specific aspect. For example RAN TSG is split into RAN Radio layer, Radio Layers 2 and 3, Radio performance and protocol aspects, and Mobile terminal conformance testing related working groups, as shown in Figure 2.1.

LTE requirements were defined in the first half of 2005 and they have been the basis for the LTE Study Item (SI). 3GPP Study Items are feasibility studies that are carried out for bigger topics before actual standardisation starts in Work Item (WI) phase. The focus of LTE SI was defining of the new LTE radio access technology in terms of both multiple access and system architecture which can satisfy such requirements. The LTE Study Item was formally closed in September 2006 after which LTE work item was started. The first LTE specifications were contained as a part of the 3GPP Release 8.

LTE was further enhanced in Release 9 and Release 10. The recent Release 10 provides a big step forward with LTE-Advanced features which significantly improves the user data rate, the coverage extension as well as reduces latency at the same time. LTE-Advanced will be briefly introduced in Section 2.1.8. The 3GPP roadmap is shown in Figure 2.2.

The outcome of the work item is a technical specification describing standardised technology. Technical specifications are grouped in categories, each category focusing their specific technology area. The specifications for the *E-UTRAN* are contained in the 36 series of Release 8, Release 9 and Release 10 and divided into the following subcategories:

- 36.100 series covering radio specifications and evolved Node B (eNB) conformance testing.
- 36.200 series covering layer 1 (physical layer) specifications.

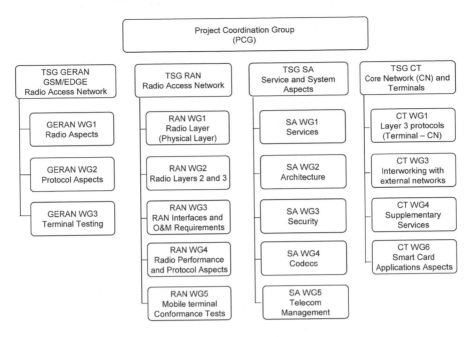

Figure 2.1 3GPP working structure. Adapted with permission from 3GPP.

- 36.300 series covering layer 2 and 3 air interface signalling specifications.
- 36.400 series covering network signalling specifications.
- 36.500 series covering user equipment conformance testing.
- 36.800 and 36.900 series, which are technical reports containing background information.

The specifications for *SAE* are scattered to many different specifications. The following documents cover the high level architecture of the SAE:

- 23.401 GPRS Enhancements for E-UTRAN access.
- 23.402 Architecture enhancements for non-3GPP accesses.

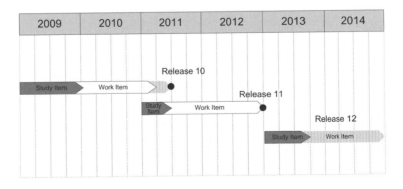

Figure 2.2 3GPP Roadmap.

Telecom Management LTE relevant specifications are:

- 32.100 series covering management principles, architecture and requirements, Fault Management Integration Reference Points (IRPs).
- 32.200 series covering charging management.
- 32.300 series: Common Management IRPs.
- 32.400 series: Performance and Trace Management IRPs.
- 32.500 series: Self-Organising Networks IRPs.
- 32.600 series: Configuration Management IRPs.

The latest versions of the LTE and SAE specifications can be found at the 3GPP site (http://www.3gpp.org/ftp/specs).

2.1.2 LTE Requirements

The LTE, the long-term evolution of the 3GPP radio-access technology, started with a Study Item in 3GPP with the scope of ensuring competitiveness for the next 10 years and beyond. LTE Requirements are described in (3GPP TR25.913, 2009). A key requirement set for LTE was that its performance should be superior if compared with 3G HSPA.

The LTE key performance requirements have been defined as comparison with HSPA R6:

- Spectral efficiency (bits/sec/Hz/site) in a loaded network two to four times more than HSPA R6 downlink and two to three times more than HSPA R6 enhanced uplink.
- Peak user throughput should be minimum 100 Mbps in downlink and 50 Mbps uplink within a 20 MHz spectrum allocation. The peak data rate should scale linearly with the size of the spectrum allocation.
- Frequency bandwidth flexibility from below 1.5 MHz up to 20 MHz allocations.
- Enables round trip time <10 ms.
- Packet switched optimised.
- Seamless mobility.

Table 2.1 summarises the main LTE requirements.

2.1.3 System Architecture Overview

As mentioned in the introduction 3GPP has defined a new system architecture for LTE. The EPS consists of the Evolved UTRAN (E-UTRAN), the Evolved Packet Core (EPC) and the connectivity to 3GPP and non-3GPP access systems. EPS solutions for 3GPP access are typically selected by operators who want to introduce EPS as smooth evolution to their existing 2G/3G infrastructure. EPS solutions for non-3GPP access are typically selected by operators who want to maximise the deployment of generic, non-3GPP protocols and to minimise the deployment of 3GPP specific protocols. The EPS system architecture is described in Figure 2.3. In this figure involved logical elements and interfaces between them are shown for the basic E-UTRAN configuration. In addition to this basic configuration, there are various architecture reference models specified in (3GPP TS23.402, 2011).

Table 2.1 LTE main requirements (3GPP TR25.913, 2009). Adapted with permission from 3GPP

Metric	Requirement	Conditions
System		
Round trip time	<10 ms	
Connection set-up latency	<100 ms	for idle to active
Operating bandwidth	1.4–20 MHz	
Coverage (cell sizes)	5–100 km	with slight degradation after 30 km
Mobility support	Up to 500 kmph but optimised for low speeds from 0 to 15 kmph	
Downlink		
Peak spectral efficiency	>5 bps/Hz	LTE in 20 MHz FDD
Peak transmission rate	>100 Mbps	2 × 2 spatial multiplexing
Cell edge spectral efficiency	>0.04–0.06 bps/Hz/user	2 × 2 spatial multiplexing
		Interference Rejection Combining receiver (IRC)
		10 users per cell (high load)
Uplink		
Peak spectral efficiency	>2.5 bps/Hz	LTE in 20 MHz FDD
Peak transmission rate	>50 Mbps	Single antenna transmission
Cell edge spectral efficiency	>0.02–0.03 bps/Hz/user	Single antenna transmission Interference Rejection Combining receiver (IRC)
		10 users per cell (high load)

In the EPS architecture only the Radio Access and the Core Networks are new, while the Service Connectivity Layer, that is the UE and Services, remains unchanged and is functionally the same as the other 3GPP systems.

The scope of EPS is to provide the IP Connectivity Layer and protocol layers in such a way that the elements are highly optimised for IP connectivity. EPS uses the concept of EPS bearer, with which an associated Quality of Service (QoS) parameters define the pipe where the IP traffic is routed from UE to the packet data network and vice-versa. One of the targets for EPS was simplified QoS scheme if compared to 3G.

The EPS architecture has three key aspects which address the performance requirements for LTE/EPC:

- Reduction of the number of network elements on the data path, compared to GPRS/UMTS;
- Streamlining of RAN functionality, by providing it in a single node.
- Separation of the control and user plane network elements (MME and S-GW).

The EPS system applies flat network architecture leading with two big architectural changes compared to EDGE and WCDMA. In the LTE access network all radio functionalities are collapsed into one network element only, the eNB and Controllers such as the RNC in

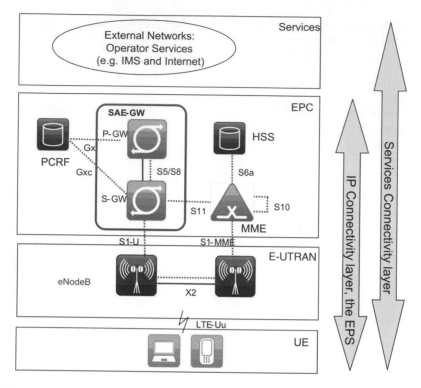

Figure 2.3 LTE System Architecture (Holma and Toskala, 2011). Reproduced with permission from John Wiley & Sons, Ltd.

WCDMA or the BSC in GERAN are not used. Also, the Core of the EPC does not contain the circuit switched domain.

2.1.4 Evolved UTRAN

LTE is based on Orthogonal Frequency Division Multiplexing (OFDM) for downlink radio transmission and data are carried simultaneously by narrow-band subcarriers. The signal is organised into subframes of 1 ms each as shown in Figure 2.4. This small duration, together with the flat network architecture, enables very short latency for both data and signalling.

For the uplink the requirement to limit as much as possible the power consumption of the UE transmitter leads to the use of Single Carrier FDMA.

LTE is very scalable with system bandwidth ranging from 1.4 MHz up to 20 MHz and it allows both paired (downlink and uplink use different frequency bands) and unpaired spectrum (downlink and uplink use same frequency band), using Frequency Division Multiplexing (FDD) and Time Division Multiplexing (TDD) both sharing the same downlink subframe structure.

Table 2.2 LTE performances for FDD at 20 MHz (3GPP TR25.913, 2009). Adapted with permission from 3GPP

Requirement	LTE	HSPA+
Peak transmission rate	DL: 150–300 Mbps	DL: 42–168 Mbps
At 20 MHz BW	UL: 75 Mbps	UL: 11–54 Mbps
Spectral efficiency (average) 4-rx mobile	1.7–2.7 bps/Hz/cell	1.21–1.9 bps/Hz/cell
Coverage (link budget)	162 dB	162 dB

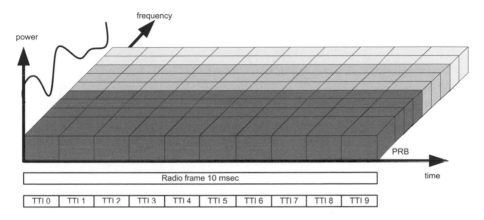

Figure 2.4 Time and frequency domain of the LTE system resources.

Multiple antennas are embedded in LTE. All UEs support at least two receive antennas, allowing downlink receive diversity. More advanced techniques, such as Transmit diversity, spatial multiplexing (Single-User and Multi-Users MIMO) and beam-forming are also supported.

LTE peak data rate increases by a factor of about 100 compared to HSPA+. Spectral efficiency also increases significantly while there are no substantial improvements in coverage. LTE and HSPA+ performances for a FDD system at 20 MHz are shown in Table 2.2.

2.1.5 E-UTRAN Functional Elements

The E-UTRAN consists of one node only, the eNB (eNodeB) that interfaces to the UE. The Radio Access Network is simply the meshed eNBs connected to neighbours eNBs through the X2 interface (Figure 2.5).

All radio functionalities are collapsed into the eNB, which means that the eNB is the termination point of all radio related protocols, the PHYsical (PHY) layer, the Medium Access Control (MAC) layer and the Packet Data Control Protocol (PDCP) layer that includes the user plane header compression and encryption.

The Control plane includes Radio Resource Control (RRC) functionality, with radio resource management, admission control and scheduling according to QoS policies.

Control and User planes in basic system architecture are shown in Figure 2.6.

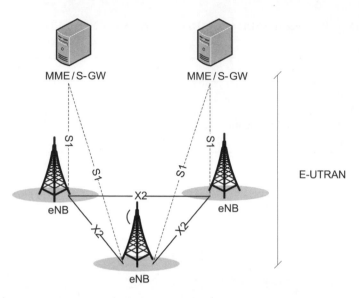

Figure 2.5 E-UTRAN Overall Architecture (3GPP TS36.300, 2011). Reproduced with permission from 3GPP.

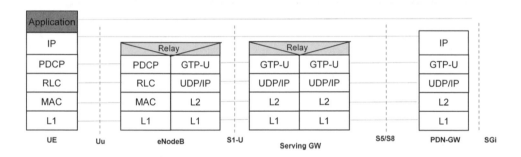

User Plane protocol stack

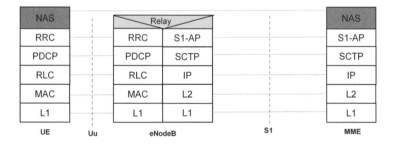

Control Plane protocol stack

Figure 2.6 User and Control Planes in basic system architecture (3GPP TS36.300, 2011). Reproduced with permission from 3GPP.

2.1.6 Evolved Packet Core

The core network, EPC, is optimised for IP connectivity and it does not involve support for circuit switched domain. EPC contains the following logical network elements for controlling EPS bearer and UE: Mobility Management Entity (MME), Serving Gateway (S-GW) and Packet Data Network Gateway (P-GW).

MME takes the role of SGSN in current GPRS networks. MME is the control plane element allowing the user plane traffic to bypass over a direct tunnel.

The role of GGSN to provide connectivity to operator service networks and the Internet is taken by two gateway elements in the EPC; S-GW is the user plane (U-plane) gateway to the E-UTRAN and P-GW is the U-plane gateway to the packet data network (for example, the Internet or the operator's IP Multimedia Subsystem, IMS). S-GW and P-GW can be colocated or separated network elements. When S-GW and P-GW are colocated the gateway is called SAE-GW.

One of the targets for EPC is to arrange optimised interworking with other 3GPP access networks and other wireless access networks.

2.1.6.1 Evolved Packet Core Functional Elements

The Evolved Packet Core is composed of several functional blocks (Figure 2.7).

For the establishment of EPS bearer and overall UE control the EPC includes the MME, S-GW, and P-GW.

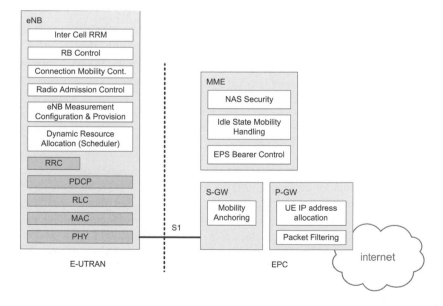

Figure 2.7 E-UTRAN and EPC functional split (3GPP TS36.300, 2011). Reproduced with permission from 3GPP.

For the user subscription data, both permanent and location-based, and user's policy and charging control the EPC includes the:

- Home Subscriber Service (HSS).
- Policy Control and Charging Rules Functions (PCRF).

The functional split between E-UTRAN and EPC is represented in Figure 2.7.

2.1.6.2   Mobility Management Entity

The MME is the central element in the EPC and it provides a logically direct Control plane connection (*Non-Access Stratum* or *NAS* signalling) with the UE. The MME manages and stores UE context, generates temporary identities and allocates them for UEs, authenticates the user, manages mobility and bearers, and is the termination point for NAS signalling. MME is responsible for:

- Mobility Management procedures:
  - Idle and active mode UE tracking in its service area and paging procedures.
  - MME controls the setting up and release of resources based on the UEs activity changes and participates to handover signalling.
- Authentication and security:
  - Interacting with the HSS MME verifies the UE authentication credentials and, to protect the UE privacy, it allocates each UE a temporary identity called the Globally Unique Temporary Identity (GUTI) so that the need to send the permanent UE identity, the IMSI, over the air interface is minimised.
  - Lawful interception of signalling for mobility between 2G-3G and LTE access networks through the S3 interface with the SGSN.
  - Providing control plane functions.
- Management of subscription profile and service connectivity. The MME retrieves the UE subscription profile from the home network, which determines the packet data network to which the UE should be connected at attachment.

2.1.6.3   Serving Gateway

S-GW is the U-plane gateway to the E-UTRAN. The primary function of the S-GW is a user plane tunnel meaning that all user plane packets go through it. The S-GW serves as an anchor point both for user moving across LTE cells and making inter-eNB handovers and anchor for mobility between LTE P-GW and other 3GPP technologies, meaning handovers to and from 2G/3G systems. The anchor point is the point which is common for all 3GPP accesses technologies, 2G, 3G and LTE. User data is routed to UE independently of underlying radio technology or changing radio technology due to handover.

S-GW is also responsible for packet forwarding, routing, and buffering of downlink data for UEs that are in LTE-IDLE state. In addition it terminates downlink data path for users in idle state and triggers paging when downlink data arrives for the UE. S-GW also replicates user traffic for lawful interception.

2.1.6.4   Packet Data Network Gateway

P-GW is the U-plane gateway to the packet data network (for example, the Internet or the operator's IP Multimedia Subsystem: IMS). The P-GW is responsible for policy enforcement, charging support, and user's IP address allocation. It also serves as a global mobility anchor for mobility between 3GPP and non-3GPP access, and LTE and pre-Release 8 3GPP access.

The P-GW is the edge router between EPS and external packet data networks. Typically the P-GW allocates the IP address to the UE and includes the Policy Control Enforcement Function (PCEF) which means it performs throttling, gating and filtering functions for user data including charging information reporting. A UE may have multiple and simultaneous connections to many external networks.

2.1.6.5   LTE/EPC Related Legacy Network Elements

The legacy network elements that interoperate with EPS (Figure 2.8) are the following:

- Serving GPRS Support Node (SGSN).
- Home Subscription Server (HSS).
- Policy and Charging Resource Function (PCRF).
- Authentication, Authorisation and Accounting function (AAA).

SGSN is responsible for the transfer of packet data between the Core Network and the legacy 2G/3G RAN. For EPS this node is only of interest from the perspective of inter-system mobility management.

HSS is the Core Network entity responsible for managing user profiles, performing the authentication and authorisation of users. The user profiles managed by HSS consist of

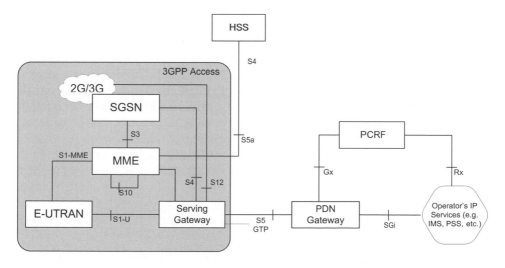

Figure 2.8   Non-roaming reference architecture for 3GPP (2G/3G/LTE) access within EPS using GTP based S5). Reproduced with permission from 3GPP.

subscription and security information as well as details on the physical allocation of the user. The HSS stores the user subscription data, indicating the services the user may use and the PDNs to which the user can connect to. In addition it records the location of the user of the last visited network MME.

The PCRF is responsible for Policy and Charging control and it makes decisions on how to handle services in terms of QoS and provides devices to the PCEF located in the P-GW.

AAA is responsible for relaying authentication and authorisation information to and from non-3GPP access network connected to EPC.

2.1.7 Voice over LTE (VoLTE)

Earlier in this chapter it has been explained that EPS is an all-IP technology. We can get to the conclusion that the voice service will have to be delivered in a new way. Voice in this IP world, would be implemented as Voice over IP (VoIP). 3GPP specified way to support VoIP is IP Multimedia Subsystem (IMS). IMS is an access-independent and standard-based IP connectivity and service control architecture that enables various types of multimedia services to end-users using common Internet-based protocols (3GPP TS23.228, 2011). The IMS based VoLTE solution puts the IMS in the centre of the voice core network, managing the connectivity between subscribers, taking care of charging and the implementation of policy control. The voice service (supplementary service, service continuity between LTE and CS, etc.) is then further managed by voice application servers on top of the IMS.

In addition to the 3GPP specifications defining the detailed architecture as well as protocol level requirements for these services, GSM Association (GSMA) has published more specific profile for *IMS profile for voice and SMS* as IR.92 document (GSMA PRD IR.92, 2010). The IR.92 contains agreed mandatory and optimum set of functionalities for the UE, the LTE Access Network, the Evolved Packet Core Network and the IP Multimedia Subsystem functionalities to manage voice and SMS in LTE based on 3GPP specifications. In a sense, this technical profile gives all industry stakeholders a level playing field on which to enhance their VoLTE service as they see fit, but most importantly a level playing field that enables the basic working, and interworking, of VoLTE across the entire industry landscape because over the time.

Although the road towards LTE, IMS and All-IP is clear, 3GPP has defined an alternative approach, CS-FallBack (CSFB) (3GPP TS23.272, 2011). In CSFB, whenever an LTE mobile generates or receives a voice call it is automatically transferred to the 2G or 3G networks. Once the call is finished the mobile reverts to LTE. This is a valid solution, but it relies on interrupting the LTE connection when the terminal is forced to move to the 2G or 3G network. This might be a big problem, depending on the application that is being used prior to the voice call. The CSFB solution is assumed to be interim VoLTE roaming solution. Furthermore, it fits to operators who plan to provide hotspot LTE coverage and/or wish to re-use their legacy networks for a while.

2.1.8 LTE-Advanced

Looking ahead, the exponential growth of data traffic will continue in upcoming years; enabling factors are the adoption of mobile broadband services, increase in usage intensity, great availability and choices of devices and machine-to-machine communications.

Table 2.3 LTE-A Requirements (3GPP TR36.913, 2011). Adapted with permission from 3GPP

LTE-A Requirement		Conditions
Peak transmission rate (20 MHz × 5) BW	DL: 1 Gbps UL: 500 Mbps	
Peak spectrum efficiency	DL: 30 bps/Hz UL: 15 bps/Hz	4 antennas BS and 2 antennas terminal
Average spectrum efficiency	DL: 2.6 bps/Hz UL: 2.0 bps/Hz	
Cell edge spectrum efficiency	DL: 0.09 bps/Hz UL: 0.07 bps/Hz	
Latency	User plane: 10 ms Control Plane: 50 ms	

In addition to these key drivers, analysis of data shows that data traffic is/will be distributed in an uneven way. These facts call for higher bandwidth and higher network efficiency which can be obtained by combining several tools, each of those tailored for specific network scenarios.

While Release 8 and Release 9 LTE have been optimised for conventional wide area deployment based on macro base stations and dual receiver and single transmit antenna single band terminals, LTE-Advanced targets more complex scenarios. In fact focus for LTE-Advanced is not to introduce a new air interface technology, but rather to extend features and capabilities of LTE to support new network deployments and ensuring optimal distribution of services.

LTE-Advanced is defined in Release 10 targeting IMT-Advanced requirements, as defined by ITU-R, while maintaining backwards compatibility with previous LTE versions; backwards compatibility means that an LTE Release-8 and Release 9 terminal will be able to operate in the LTE-Advanced network and vice versa. Key areas of enhancements are improved data-rate, coverage extension and latency reduction as well as inter-working with other technologies and global roaming (3GPP TR36.913, 2011), (Ghosh *et al.*, 2010), (Mogensen *et al.*, 2009). A summary is reported in Table 2.3.

LTE-Advanced includes several technology components as shown in Figure 2.9: Carrier Aggregation (CA) allows achieving peak data rate of 1 Gbps in downlink and 500 Mbps in uplink. Enhancements for uplink and downlink MIMO target at improving the spectral efficiency; Relay Nodes (in-band backhauling) and Coordinated Multipoint transmission and reception (CoMP; CoMP is not included in Release 10) cope with high-cell interference at cell-edge. Optimised interworking between cell layers is also introduced for Heterogeneous Networks. (see Section 2.1.8.5 and Chapter 10).

Each of the LTE-Advanced features is shortly presented in next paragraphs. As far as their SON component is concerned, a detailed description of Relay Self-Configuration and Automatic Neighbour Relations (ANR) is included in Chapter 4, and Chapter 10 is dedicated to Heterogeneous Networks.

At the time of writing SON applicability to the other LTE-Advanced feature is under analysis.

2.1.8.1 Carrier Aggregation

The very high peak data rate targets are fulfilled by means of bandwidth extensions.

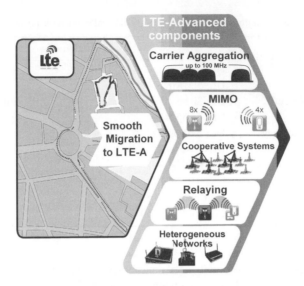

Figure 2.9 LTE-Advanced components.

Carrier Aggregation (CA) allows for combining up to five LTE Release 8 Component Carriers (CC) to achieve a transmission bandwidth of 100 MHz (5 × 20 MHz) and enhancing end-users peak data rates to 1 Gbps in downlink and 500 Mbps in uplink while maintaining backwards compatibility. Aggregation of Component Carriers is per user based, so that different users in a cell may be differently configured (Figure 2.10).

In order to enable operators to provide high throughput without wide continuous frequency band allocations, 3GPP defines, in addition to contiguous spectrum in a single band other two levels of aggregation, the 'non contiguous spectrum in single band' and the 'non contiguous spectrum in multiple band'. In addition, Carrier Aggregation allows asymmetrical bands into use with FDD since there can be uplink or downlink only frequency bands.

2.1.8.2 Improved MIMO Schemes

Multi-antenna technologies include spatial diversity and beam-forming. Antenna diversity is very effective in multi-path propagation handling, where the signal is reflected along multiple paths which may destructively interfere with one another, before being finally received.

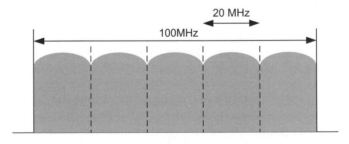

Figure 2.10 Carrier aggregation.

If the multiple transmissions is for a single user then the technology is called Single-User MIMO (SU-MIMO), for multiple users Multi-User MIMO (MU-MIMO).

MIMO performance is subject to a large number of parameters: the number of transmitting and receiving antennas, reference signals and algorithms for channel estimation, feedback of channel estimation data from the receiver to the transmitter and spatial encoding methods.

Release 8 and Release 9 support MIMO with up to four transmitting and receiving antennas in downlink, but with only one antenna in uplink. Release 10 extends downlink MIMO to support up to eight transmitting and receiving antennas, and uplink to support up to four transmit and eight receiver antennas.

2.1.8.3 Relay

The very high data targets of LTE-Advanced require a tight infrastructure (short distance between base stations) in order to be able to reach cell-edge users, which usually suffer from relatively low Signal-to-Interference plus Noise ratio (SINR). An attractive solution to this problem is provided by multi-hop technologies by means of Relay Nodes. The idea behind Relay Nodes is the improvement of the link budget by reducing the transmitter-to-receive distance for those users close to the cell edge.

In fact, deploying Relay Nodes near to the cell edge will help to increase the capacity or, alternatively, to extend the cell coverage area (Figure 2.11). The Donor eNB will 'donate' part of its air-interface capacity to providing backhaul connection for one or more Relay Nodes while still serving own users. Two types of Relay Nodes are envisaged. The simplest form is the conventional Amplify and Forward (AF) Relay, also called the repeater. AF simply amplify-and-forwards the signal received from Donor eNB. The drawback of such a solution is quite obvious, that is the AF Relays amplify both interference and noise in addition to the desired signal and are therefore not effective in presence of noise or interference.

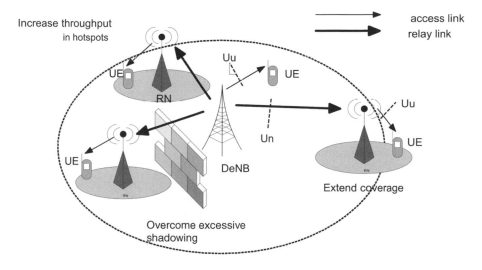

Figure 2.11 Donor eNB and relay nodes.

The most advanced Relay Nodes are the Decode and Forward (DF) ones, which detect and decode the desired signal and then re-encode and forward it. Therefore DF relays are applicable both in interference and noise limited environments and thus, they can be used to improve system capacity.

Relay Nodes can be seen as a special case in a Heterogeneous Network. For Relay Nodes the backhaul is being provided by the Donor eNB, the Relay node has some special configuration procedures and Automatic Neighbours relations (ANR) procedure is a bit different compared to other nodes. Relay self-configuration and ANR are described in dedicated paragraphs in Chapter 4.

2.1.8.4   Coordinated Multipoint Transmission and Reception (CoMP)

Coordinated Multipoint transmission and reception (CoMP) shows great potential to improve cell edge data rates and system capacity. It deals with low signal quality at cell edges and interference levels: due to frequency reuse one (i.e. systems where frequency channels are reused for achieving high level of area coverage), the cell-edge performance of LTE is limited by co-channel (inter-cell) interference.

Despite studies have shown high potential from CoMP for a single user, this was less evident in the case of large scale networks operation.

The Coordinated Multipoint processing requires close cooperation between a number of geographically separated eNBs which coordinate amongst them to provide joint scheduling and transmissions as well as performing joint processing of the received signals (see Figure 2.12). This leads to very high signalling load between the cells: as a consequence, the Intra-site CoMP deployment, where the communication is between the sectors of a single eNB, is likely to be the most feasible system solution.

Due to these reasons the technology was not seen mature enough to be included it in Release 10, and studies will continue in Release 11 with focus on practical concepts and real performance benefits, including transport technologies.

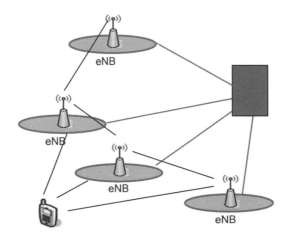

Figure 2.12   Coordinated multi-point.

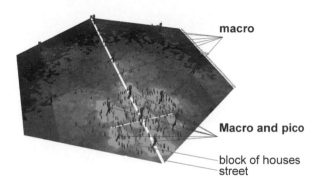

Figure 2.13   Heterogeneous Networks.

2.1.8.5   Heterogeneous Networks

Further spectral efficiency can be achieved with multi-layers network topologies; in this scenario macro cells provide the continuous and wide area coverage while small cells (served by low-power nodes) improve hot spot performance (Figure 2.13). Small cells can be served by a micro eNB, pico eNBs or Home eNodeBs (HeNB, also known as femto eNB).

Low-power nodes such as pico eNBs are usually deployed in a coordinated way that is under the operator's control by means of conventional network planning. While Enterprise HeNBs are installed by the operator, the Residential HeNBs are typically installed by the end-user without operator intervention. Therefore the latter is also called 'uncoordinated deployment'.

HeNB operating modes are *Open Subscriber Group* (*OSG*), *Closed Subscriber Group* (*CSG*) or *Hybrid* (meaning they operate with both OSG and CSG). CSG HeNB usage is constrained to its owner only or an otherwise limited set of users. Hybrid HeNB allows all UEs of the given operator to access the HeNB, but CSG members may receive preferential handling. Open and Closed Subscriber Group HeNBs are further explained in Section 10.2.

No interference problems exist if the operator is able to allocate different frequencies to different cell types (e.g. macro cell versus femto cell). If this cannot be done and the same frequency has to be used for the different cell types, the cardinal problem is the interference between users served by a macro eNB and the users served by a HeNB (see Figure 2.14).

Another problem with the HeNBs is that they may not have a fixed position (they can be moved inside the house) and/or be switched off at any time by the owner. In 3GPP Release 10, the Enhanced Inter-Cell Interference Coordination (eICIC) has been specified to mitigate interference. Heterogeneous Networks and the related interference mitigation mechanisms are described in Chapter 10.

Heterogeneous Networks have lately attracted very much attention from the operators. Pico-eNBs may be used both for coverage and capacity reasons (e.g. in shopping malls). In this constellation the cells are deployed in a coordinated way, that is, by the operators, as they operate in Open Subscriber Group mode, and can be placed both, indoor or outdoor.

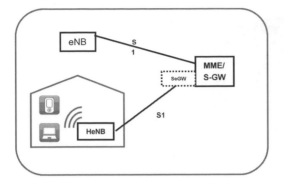

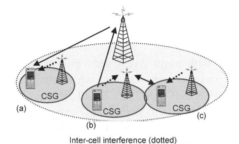 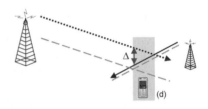

Inter-cell interference (dotted)   Rx power (solid, dotted->macro), 1/pathloss (dashed)

Figure 2.14 Residential HeNB and interference with macro (3GPP TR36.814). Reproduced with permission from 3GPP.

Actually multi-layer networks are already part of 3GPP Release 8 and 9 specifications. New in Release 10 are the mechanisms that very much improve the performances of such networks.

Chapter 10 analyses Heterogeneous Networks and their management in detail. Due to the complexity of this deployment and the lack of operator planning, a high degree of automation is needed to manage and properly use the network resources.

2.1.9 Network Management

LTE follows the same 3GPP management reference model as 3G as specified in 3GPP TS32.101 (2010) (Figure 2.15). It introduces several interfaces from operations systems to NEs:

- Itf-S (interface 1 in Figure 2.15): Interface between the Network Element (NE) and the Domain Manager (DM). This interface is vendor specific.
- Itf-N (interface 2 in Figure 2.15): Interface between the DM and the Network Manager (NM). This is a standardised open interface and thus facilitates multi-vendor management.
- The Itf-P2P (interface 4a in Figure 2.15), interface between the DM as well as the Element Manager (EM) embedded into the Network Element. Itf-P2P interfaces have been not used in real deployments.

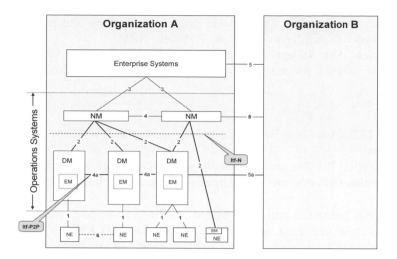

Figure 2.15 Management reference model (3GPP TS32.101, 2010). Reproduced with permission from 3GPP.

The same methodology for telecommunications management is employed for 3G and LTE as described in 3GPP TS32.103 (2011). The 3GPP's management is based on an interface concept known as Integration Reference Point (IRP). IRPs are defined with IRP Levels and IRP types as shown in Figure 2.16. IRP levels are:

- **Requirements.** The Requirements-level is intended to provide conceptual and use case definitions for specific management interface aspects and to define subsequent requirements for this IRP.

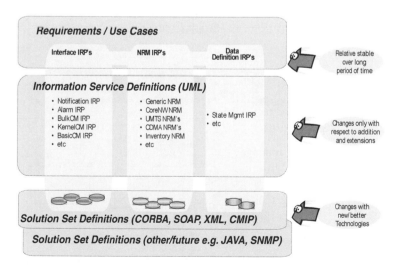

Figure 2.16 The IRP 3-Level Specifications Approach combined with the three IRP categories (3GPP TS32.103, 2011). Reproduced with permission from 3GPP.

- **Information Service (IS).** The IS-level provides technology independent definitions (protocol-neutral IS and Network Resource Models: NRM).
- **Solutions Sets (SS).** SS-level does mapping of IS definitions into technology specific solution sets. Solution sets are protocol-specific (CORBA, SOAP, XML).

IRP Types are:

- **Interface IRPs.** Interface IRP defines how information is shared (operations and notifications).
- **NRM IRP.** NRM IRP defines what can be managed (Network Resource Model).
- **Data Definition IRP.** Abstract data definitions to be used in NRM IRPs.

Each IRP type is partitioned into Requirements, IS-level and SS-level specifications.

SON functionality can reside in Network Management (NM), Element Manager (EM) or Network Element (NE) depending on the use cases and will use appropriate interfaces, as represented in Figure 2.17.

Decision on where SON functionality shall be located and which interface to use depends on requirements (e.g. multi-vendor capability, speed, scope) set for the use case in question. Architectural considerations for SON are discussed in Section 3.4.

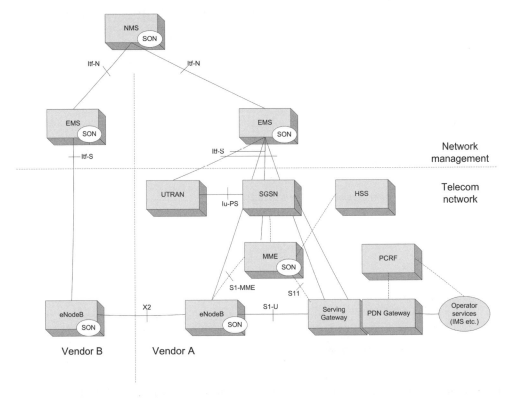

Figure 2.17   SON architecture, including core network, radio access network and network management.

2.2 LTE Radio Access Network Scenarios and Their Evolution

The LTE Radio Access Technology has been designed to be very flexible, allowing variable spectrum deployment in existing or future spectrum allocations; at the same time LTE is to be able to co-exist with existing 3GPP Radio Access Technologies without increasing UE and system complexity. As such, important requirements of a long-term evolution include reduced latency, higher user data rates, improved system capacity and coverage, and reduced cost for the operator.

LTE initial deployments have started in 2010, the focus of 3GPP is now gradually moving to further evolution of LTE, named as LTE-Advanced. As introduced in Section , one of the targets of such evolution is to meet, and even exceed, IMT-Advanced requirements as defined by ITU-R.

In the following the most relevant network scenarios are described. Scenario 2.2.1 and 2.2.2 are based on macro cell deployments, for both intra-LTE and inter-RAT, while scenario 2.2.3 introduces an underlay-overlay network where the overlay layer is composed of small cells.

SON use cases are specified by 3GPP mainly for LTE. Exceptions are (for the time being) Minimisation of Drive Tests (MDT), which is for both LTE and 3G, and 3G-ANR as a part of 3GPP Release 10. The LTE-A features are part of Release 10. As Release 10 has been just completed, solid SON concepts for features, such as Carrier Aggregation, CoMP and Improved MIMO techniques will come in later releases. On the other side, Heterogeneous Networks already include some SON components and Relay Nodes define basic ANR function in Release 10 (see ANR paragraph in Chapter 4).

Depending on operators' network and needs, LTE can be deployed in a number of ways; in the following the most relevant network scenarios and related SON components are described.

2.2.1 LTE Radio Coverage Scenario

In this scenario the radio coverage is provided by means of sparsely deployed macro cells. Macro eNBs have a typical output power in tens of watts and their radio coverage over varying distances depends on the frequency used and the physical terrain.

This case is of outmost importance due to the Digital Dividend in Europe (an example of a Multiple Layer network is shown in Figure 2.18). With the switchover of the television signal from analogue to digital, a significant amount of spectrum in the 800 MHz band can be freed. The European Commission is pushing the Digital Dividend very hard and the EU Member States are now in the process of switching over (or planning to do so) from analogue to digital TV signal. The scope of the Digital Dividend is to effectively create a single mobile broadband market across national borders.

In many markets, for example, in Germany, the license requires the operator to deploy LTE in rural area before he is allowed to build out LTE coverage in urban areas. The goal of this strategy is to provide LTE coverage to EU residents in areas where no broadband coverage is available, that is, regions with neither 3G HSPA coverage nor deployed DSL services.

Alternatively this scenario is the one of Greenfield operators with no previous mobile networks in the area, or an operator which does have another overlay network but there is no need to integrate it with LTE (e.g. standalone wireless broadband application, see Figure 2.19).

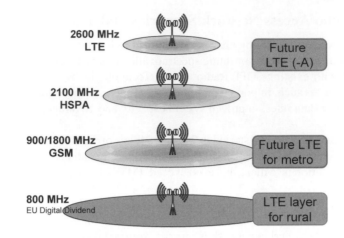

Figure 2.18 Digital Dividend and Multiple LTE layers.

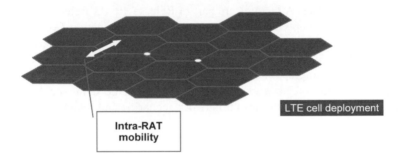

Figure 2.19 LTE Standalone deployment scenario.

2.2.2 *LTE for Capacity Enhancement in Existing GERAN/UTRAN*

This is the typical incumbent operator's scenario, and the main requirement is to effectively integrate LTE with existing UTRAN and/or GERAN networks.

Due to cost reasons the LTE coverage may be limited to high populated areas such as city centres, and very likely LTE will be co-sited with 2G and/or 3G base stations. The issue in such a constellation is the efficient use of LTE. This case is largely addressed in Chapter 5, both in terms of optimised handover thresholds (so called Mobility Robustness Optimisation which is introduced in Chapter 3) and Traffic Steering policies and mechanisms. Figure 2.20 shows some aspects of such implementations of LTE as an overlay network to existing UTRAN and/or GERAN infrastructure.

2.2.3 *Enhancing LTE Capacity, the Multi-Layer LTE*

As mentioned, the enormous data traffic growth requires more network capacity in locations where data demand is very high by means of deployment of small cells, in addition to macro cells. In addition, Relay Nodes may improve system capacity at cell edges.

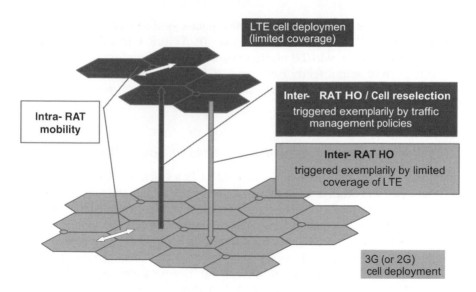

Figure 2.20 LTE as overlay to 2G-3G network.

In such cases the network will have multiple overlapping radio network layers, either using the same (co-channel deployment) or different frequencies (Figure 2.21).

In addition to multi-layer, the operator may also have a multi-RAT network, with installed GERAN and/or UTRAN network elements.

SON is vital for such networks, since the complexity of the network will significantly increase due to the large number of base stations and parameters that need to be configured and optimised. As previously mentioned, Heterogeneous Networks are addressed in Chapter 10.

2.2.4 Data Offloading, LIPA-SIPTO

3GPP has set requirement to provide local IP breakout for Home (e) NodeB subsystem and macro layer network (3GPP TR23.829, 2011).

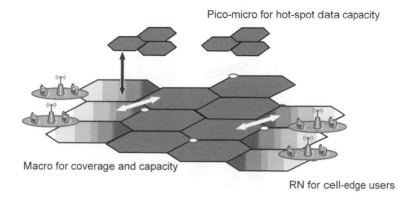

Figure 2.21 Multi-layer LTE.

LIPA is about local IP access to a residential/corporate local network for Home (e) NodeB Subsystem, while SIPTO considers Selective IP Traffic Offloading for Home (e) NodeB Subsystem and for 3G/LTE macro network. SIPTO therefore only considers how internet traffic can be sent directly from local 3GPP access to the internet. LIPA on the other hand defines access to private network resources in the user premises using Home (e) NodeB subsystem.

2.2.5 Multi-Radio Access Network Scenarios or non-3GPP

The emergence of so-called smartphones and Tablet PCs (such as iPhones, Android-based phones, iPads, etc.) and other devices supporting non-traditional mobile applications such as social media or gaming made the operator's mobile network drown in data traffic.

In previous scenarios the RAT selection between different cellular technologies (GSM/ GPRS/EDGE, 3G, HSPA and LTE) has been shown. In addition to 'pure mobile solutions', 3GPP defines three different fixed mobile convergence (FMC) architectures to alleviate the strain on the radio access network by offloading mobile internet traffic to non-3GPP networks, all utilising WLAN and IP access (3GPP TS23.234, 2011).

Unlicensed Mobile Access (UMA) technology is the oldest one, allowing full 2G and recently also 3G service offerings on top of WLAN (3GPP TS43.318, 2011). WLAN is more recent, providing packet based services on WLAN while retaining circuit voice in legacy 3GPP CS network.

The third option is Access Network Discovery and Selection Function (ANDSF) which is based on Release 9 and improves network selection and use (Figure 2.22).

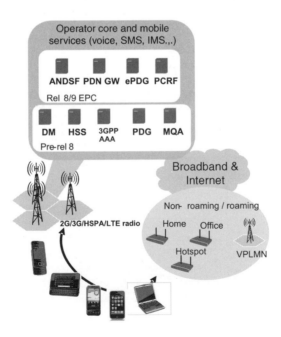

Figure 2.22 WiFi data offloading with ANDSF.

If access selection is left to the end-user, the traffic cannot be steered in a good way, that is, with load balancing and interference minimisation. To enable the operator to influence on 3GPP/ non-3GPP access selection, a set of mechanisms is introduced to transfer operator specific 3GPP/non-3GPP access selection policies to a device and the device is forced to consider those policies when selecting which 3GPP/non-3GPP access to use. 3GPP TS23.402 (2011) describes the access to EPC via non-3GPP access networks. The network discovery procedures are defined by the ANDSF function described in 3GPP TS24.302 (2011) and 3GPP TS24.312 (2011).

With ANDSF mechanism, the first UE access is through the 3GPP network for authentication and authorisation via AAA machinery. Once the UE is authenticated and authorised, the ANDSF policies set by the operator may prioritise WiFi hot-spot for that UE specific location. This is a quite powerful policy, which allows the operator to move data-hungry mobiles and applications to WiFi access points. Nevertheless WiFi suffers from several drawbacks, such as significant interference problems when the access point density grows. This is because WiFi access points cannot regulate their output power and access points in close proximity can drown each other (co-channel interference) leading to coverage black holes. In addition mobility is not supported and even with stationary users WiFi basically offers best effort service and a session can end if the access point is accessed by many users.

References

3GPP (2011) 3GPP: the Mobile Broadband Standard. Specifications. Available at http://www.3gpp.org/ftp/specs [accessed 02 September 2011].

3GPP TR25.913 (2009) Technical Report Technical Specification Group Radio Access Network, *Requirements for Evolved UTRA (E-UTRA) and Evolved UTRAN (E-UTRAN)*, ver.9.0.0., Release 9, 27 December 2009. Available from http://www.3gpp.org/ftp/Specs/latest/Rel-10/24_series/24913-900.zip [accessed 30 June 2011].

3GPP TR23.829 (2011) Technical Specification Group Services and System Aspects, *Local IP Access and Selected IP Traffic Offload (LIPA-SIPTO); System Description*, ver.10.0.0., Release 10, 29 March 2011, http://www.3gpp.org/ ftp/Specs/latest/Rel-10/23_series/23829-a00.zip [accessed 30 June 2011].

3GPP TS23.228 (2011) Technical Specification, Technical Specification Group Services and System Aspects, *IP Multimedia Subsystem (IMS); Stage 2*, ver.10.5.0., Release 10, 10 June 2011. Available from http://www.3gpp.org/ ftp/Specs/latest/Rel-10/23_series/23228-a50.zip [accessed 30 June 2011].

3GPP TS23.234 (2011) Technical Specification Group Services and System Aspects, *3GPP System to Wireless Local Area Network (WLAN) Interworking*, ver.10.0.0. Available from http://www.3gpp.org/ftp/Specs/latest/Rel-10/ 23_series/23234-a00.zip [accessed 30 June 2011].

3GPP TS23.272 (2011) Technical Report, Technical Specification Group Services and System Aspects, *Circuit Switched (CS) Fallback in Evolved Packet System (EPS); Stage 2*, ver.10.4.0., Release 10, 10 June 2011. Available from http://www.3gpp.org/ftp/Specs/latest/Rel-10/23_series/23829-a40.zip [accessed 30 June 2011].

3GPP TS23.402 (2011) Technical Specification, Technical Specification Group Services and System Aspects, *Architecture Enhancements for non-3GPP Accesses*, ver.10.4.0., Release 10, 10 June 2011. Available from http://www.3gpp.org/ftp/Specs/latest/Rel-10/23_series/23402-a40.zip [accessed 30 June 2011].

3GPP TS24.302 (2011) Technical Report, Technical Specification Technical Specification Group Core Network and Terminals, *Access to the 3GPP Evolved Packet Core (EPC) via non-3GPP Access Networks*, ver.10.4.0., Release 10, 14 June 2011. Available from http://www.3gpp.org/ftp/Specs/latest/Rel-10/24_series/24302-a40.zip [accessed 30 June 2011].

3GPP TS24.312 (2011) Technical Specification Group Core Network and Terminals, *Access Network Discovery and Selection Function (ANDSF) Management Object (MO)*, ver.10.3.0., Release 10, 14 June 2011. Available from http://www.3gpp.org/ftp/Specs/latest/Rel-10/24_series/24312-a30.zip [accessed 30 June 2011].

3GPP TS32.101 (2010) Technical Specification Group Services and System Aspects, *Telecommunication Management; Principles and High Level Requirements*, ver.10.0.0., Release 10. Available from http://www.3gpp.org/ftp/ Specs/latest/Rel-10/32_series/32101-a00.zip [accessed 30 June 2011].

3GPP TS32.103 (2011) Technical Specification Group Services and System Aspects, *Telecommunication Management; Integration Reference Point (IRP) Overview and Usage Guide*, ver.2.0.0., Release 10. Available from http://www.3gpp.org/ftp/Specs/latest/Rel-10/32_series/32103-a00.zip [accessed 30 June 2011].

3GPP TS36.300 (2011) *E-UTRA and E-UTRAN; Overall Description; Stage 2*, ver.10.3.0., Release 10, 22 June 2011. Available from http://www.3gpp.org/ftp/Specs/latest/Rel-10/36_series/36300-a40.zip [accessed 30 June 2011].

3GPP TR36.913 (2011) Technical Specification Group Radio Access Network, *Requirements for Further Advancements for Evolved Universal Terrestrial Radio Access (EUTRA)*, ver.10.0.0, Release 10. Available from http://www.3gpp.org/ftp/Specs/latest/Rel-10/36_series/36913-a00.zip [accessed 30 June 2011].

3GPP TS43.318 (2011) Technical Specification Group GSM/EDGE Radio Access Network, *Generic Access Network (GAN); Stage 2*, ver.10.1.0., Release 10. Available from http://www.3gpp.org/ftp/Specs/latest/Rel-10/43_series/43318-a10.zip [accessed 30 June 2011].

Ghosh, A., Ratasuk, R., Mondal, B. *et al.* (2010) LTE-advanced: next generation wireless broadband technology. *IEEE Wireless Communications*, **17**(3), 10–22.

GSMA (2010) *IMS Profile for Voice and SMS*, GSM Association, PRD IR.92, ver.1.0.

Holma, H. and Toskala, A. (eds) (2011) *LTE for UMTS: Evolution to LTE-Advanced*, John Wiley & Sons, Ltd, Chichester.

Mogensen, P.E., Koivisto, T., Pedersen, K.I. *et al.* (2009) LTE-advanced: the path towards Gigabit/s in wireless mobile communications. IEEE Wireless VITAE Conference, pp. 147–151.

3

Self-Organising Networks (SON)

Richard Waldhauser, Markus Staufer, Seppo Hämäläinen, Henning Sanneck,
Haitao Tang, Christoph Schmelz, Jürgen Goerge, Paul Stephens,
Krzysztof Kordybach and Clemens Suerbaum

3.1 Vision

Managing radio networks is a challenging task, especially in cellular mobile communication
systems due to their inherent complexity. This complexity arises from the number of network
elements that have to be deployed and managed, but also from interdependencies between their
configurations. The complexity will even increase when several radio networks should be
seamlessly integrated as will happen when LTE networks are introduced by operators now
operating their existing 2G/3G networks. In such Heterogeneous Network scenarios the variety
of deployed technologies and their specific operational paradigms will be difficult to handle.

Network management is usually based on a centralised operation, administration and
maintenance (OAM) architecture. Configuration and optimisation of network elements is
performed centrally from an OAM system (also called Operations and Maintenance Centre:
OMC) with support of a set of planning and optimisation tools. Today, planning and
optimisation tools are typically semi-automated and management tasks need to be tightly
supervised by human operators (cf. Figure 3.1). This manual effort by the human operator is
time-consuming, expensive, error prone and requires a high degree of expertise. Networks are
operated so that the human operator re-plans network configuration based on analysed
performance data. Involved processes and workflows in operating networks are; Performance
Management (PM) that provides performance measurements from the network elements, Fault
Management (FM) for alarms and Configuration Management (CM) for setting parameters in
the network elements. Particularly, correlating the knowledge established via the different
sensor PM/FM/CM paths and linking it to the subsequent configuration actions still incurs a
significant amount of manual work. For example, introducing new features through a software

LTE Self-Organising Networks (SON): Network Management Automation for Operational Efficiency, First Edition.
Edited by Seppo Hämäläinen, Henning Sanneck and Cinzia Sartori.
© 2012 John Wiley & Sons, Ltd. Published 2012 by John Wiley & Sons, Ltd.

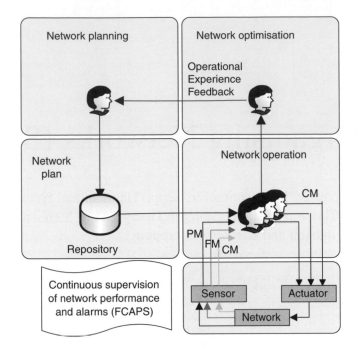

Figure 3.1 Manual network operation.

update or new network elements from different vendors requires changes in operational procedures. The operational staff use their operational experience to find optimised config-urations and they know when they have to diverge from standard procedures. In particular, fault management procedures are also based on operational experience.

Self-organising networks can be used to reduce operating cost by reducing manual workload and to protect revenue by minimising human errors. These challenges are addressed by bringing human interaction in operational procedures to a minimum by increasing automation in the Network Elements (NE), Domain Management (DM, also called 'Element Manage-ment': EM) and/or Network Management (NM, also called 'umbrella management') systems. 3GPP has addressed this with the standardisation of a framework for self-organising networks (3GPP TR32.500, 2011). The introduction of SON features aims at reducing the workload on the operation and maintenance staff through the automation in order to free them from time-consuming standard tasks so they are able to focus on crucial problems. The target is to gradually move towards a pure monitoring and steering of the SON-enabled system from manual planning and configuration. Finally, only high-level guidelines need to be inserted and updated in the system instead of supervising the actual changes via CM manually. Thus the PM/FM/CM sensor paths are much more integrated together and linked to the CM actions (cf. Figure 3.2). This leads to changed role for human that can be seen as a paradigm shift in how networks are operated. Along with SON, human effort is shifted to higher management level so that a human operator's role is to monitor SON processes and intervene only when needed. Therefore, the human's role is not to carry out frequent routine work anymore. Instead, the role is to design and decide policies that guide SON functioning.

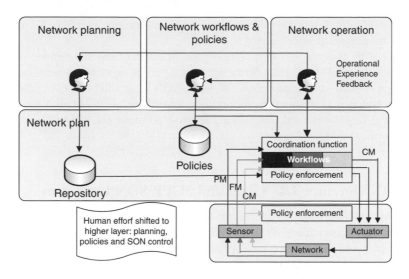

Figure 3.2 Network management with closed-loop automation.

The building blocks to reach a SON-enabled system are a workflow execution system and the actual management loop. The workflow execution system is a means to automate execution of SON functions and a policy system for automated decision making (Sanneck *et al.*, 2010). Policies control workflow execution and the adaptation of the control flow to the context. The usage of policies also allows to abstract from technical details and to change system behaviour at runtime. Several SON functions may be executed under control of the workflow execution system, each SON function performing their specific tasks independently. The workflow execution system coordinates independent SON functionalities in order to avoid conflicts between them. Both components together assure the control of the human operator as well as the stability of the SON-enabled system where many functions are executed concurrently.

The management loop is closed and automated and therefore it is expected to provide operational savings for network operators. Any re-configuration performed autonomously by SON must provide the operator a full understanding of the new configuration and allow him to revert to the original configuration set by the operator. Therefore the human operator can always overrule configurations made by SON.

SON functions in the different areas of *self-configuration, -optimisation and -healing* can reside at NE, DM or NM-level depending on their needs (cf. Section 3.4 on architecture). In practice, the integration of different SON functions leads to a 'hybrid' architecture due to different requirements for different functionalities. Some SON functionality may require a very fast reaction to changed network performance, thus such functionality should be implemented in the network element, for example, the eNB. On the other hand, some of the functionality could tolerate more delays but might require wider understanding of the network status and implications of changed configuration, thus such functionality should be implemented at the DM or NM level. Requirements for multi-vendor or multi-technology capability may also set requirements for the architecture, which should then be able to take benefit of standardised interfaces such as, inter base station interface X2 or the open northbound management interface Itf-N.

In practice SON systems will be a combination of standardised and vendor specific SON functionalities. The 3GPP standards define necessary, interfaces, measurements and signalling to better support multi-vendor capable SON, while SON logic will be always vendor specific. In addition, vendors may design their proprietary network management automation solutions beyond those defined by the 3rd Generation Partnership Project (3GPP) and Next Generation Mobile Network (NGMN) alliance.

3.2 NGMN Operator Use Cases and 3GPP SON Use Cases

In 2007 NGMN published a list of operator SON use cases (NGMN, 2007), followed in 2008 by a set of SON related use cases (NGMN, 2008). These use cases emerged from the knowledge about the problems an operator is typically faced with when deploying and operating a wireless mobile radio network. The NGMN use cases considerably influenced the SON related work in 3GPP standardisation, but also SON related research projects (cf. Section 3.3.3). Vice versa, NGMN has been influenced by the SON functions that have been introduced within the 3GPP standards and the related discussions. NGMN thereupon published operations efficiency recommendations (NGMN, 2010) to ensure that the corresponding operators' requirements are considered by 3GPP OAM standardisation.

This section first introduces the NGMN SON use cases from (NGMN, 2007) and (NGMN, 2008) at a high-level, then continues with the motivation of and recommendations for the individual use cases as described in (NGMN, 2010), and finally gives an overview of the use cases 3GPP has worked on, including a mapping to the corresponding NGMN use cases.

3.2.1 Operational Use Cases

The operator use cases of the NGMN are structured according to a workflow model. The work flow model is orientated at the different tasks that have to be managed during the planning, installation and configuration process for new equipment. These structure leads to the following categories for the operational use cases:

- Planning;
- Deployment;
- Optimisation;
- Maintenance.

Major elements of these categories are depicted in Figure 3.3. The interworking between these categories is not shown in detail in order to keep the figure simple. Nevertheless it should be noted that some of these categories overlap. For example, the deployment of a new network element may lead to the requirement of network re-planning or optimisation; or network monitoring within the Maintenance category may trigger activities of the Optimisation category.

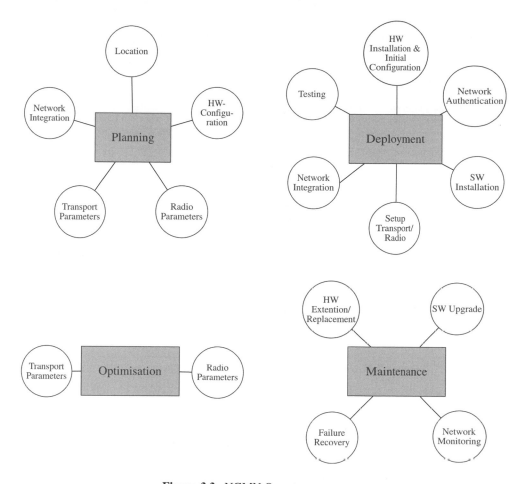

Figure 3.3 NGMN Operator use cases.

3.2.1.1 Planning

In classical management, network operators periodically rework the configuration of their complete radio network, to adapt to:

- Changed capacity, coverage and performance requirements, which may be based on long-term performance evaluation and failure analysis of the operational network, on the introduction of new services, or on business requirements from the operator.
- Changed technical requirements; for example, the exchange or modification of hardware and software of the network elements.

This leads to a network planning cycle as depicted in Figure 3.4. The network planning cycle typically starts with the long-term analysis and evaluation of performance and fault management data coming from the network (1). The requirements derived from this evaluation serve

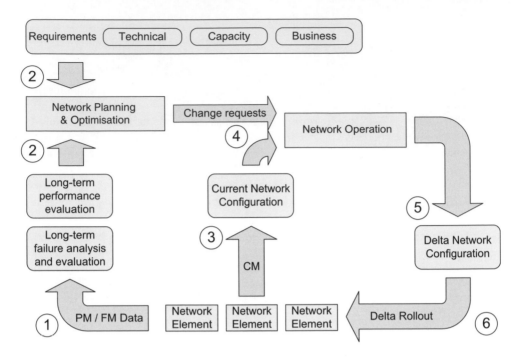

Figure 3.4 Network planning, optimisation and operation cycle.

as input to network planning together with higher-level technical, service or business requirements (2). From these input requirements, network planning creates a set of change requests. To determine a new configuration for the network, the current configuration of all network elements is uploaded to network planning (3). Based on this data and the change requests, the new configuration is created (4). Taking this new configuration and comparing it with the current configuration, network planning compiles a delta configuration that contains all modifications to be made (5). This delta configuration is finally rolled out to the network elements (6).

The creation (calculation) of a new configuration is performed by dedicated offline planning tools and may require several days to weeks, taken the timeframe from acquiring the current network configuration until the rollout of the delta configuration. It is very likely that within this timeframe, the 'current network configuration' changes due to modifications performed by regular network management (short-term configuration changes due to failures or performance issues), or due to removal/exchange of hardware and software, or the integration of new network elements. Therefore, the rollout of the delta configuration may cause inconsistencies within the network's configuration as the rolled out delta is not based on the actual 'current' configuration anymore. These inconsistencies may lead to failures, at least alarms, and degraded network performance.

Consequently, the operator use cases belonging to the planning category include preparing and provisioning of the right parameter settings for the new eNB. When brought into operation at the planned location the new equipment shall be enabled to smoothly cooperate with the already installed neighbour nodes and shall be accommodated to the local environmental radio conditions. The following operator use cases comprise the tasks of the planning category:

- **Location planning:** The location shall fulfil the coverage and capacity requirements. The required network coverage and capacity targets are determined based on customer feedback, the quality and requirements of existing services, and the requirements of planned services.
- **Planning of the HW Configuration:** Based on the requirements the most suitable HW configuration is determined for an eNB or antennas.
- **Planning of the radio parameters:** Typical examples are the power, tilt and handover settings. These shall fit to the equipment on spot or located in the neighbourhood. The neighbour relation list shall contain the significant candidate cells for handovers. Note that this shall also consider the special requirements introduced by scenarios including relay nodes and HeNBs, see Section 2.2.
- **Transport parameters:** This refers to the configuration of the transport network, which is required to connect the new eNB with the neighbour nodes. Typical examples for the task include the necessary QoS and address parameters of the transport layer.
- **Network Integration:** This comprises selecting reasonable neighbour nodes, Security Gateways, Serving GWs, MMEs and O&M nodes. The local databases of these nodes must be aligned with respect to the new eNB.

The long term goal for the last three NGMN use cases is to substitute these by SON functionalities that are provided by the use case categories deployment, optimisation and maintenance.

3.2.1.2 Deployment

This category comprises the use cases that are related to the deployment of a new eNB. This means installing the eNB hardware at the selected site and setting up the eNB's configuration in a status that is immediately preceding the commercial operation. The SON functionality supporting the operator use cases of this category should also involve the relevant updates in the neighbour eNBs and the affected nodes in the core network. The following use cases are assigned to the category deployment:

- **Hardware Installation and initial configuration:** Although this step predominantly is considered as a manual process, SON functionality is expected to reduce the effort by Plug&Play behaviour of all the components like antenna, line, boards and cages. Plug&Play of the antenna includes, for example, autonomous measurement of antenna loss and initial configuration of respective values in the nodes database. It shall be noted that the RAN nodes may require reconfiguration for the purpose of network optimisation. For example, in case an operator decides to use an eNB at another location. However, this is considered to be a quite rare situation. Another example is re-homing, but re-homing is not applicable for eNBs because in LTE the S-GW is dynamically assigned by the MME out of a pool. Therefore this is an issue for 2G and 3G networks rather than for LTE. From the workflow point of view these examples are reflected by the following two use cases:
  - **Re-location:** This use case spans the configuration-related workflow for the physical re-location of an eNB, that is, the eNB is de-installed and removed from the original location and reinstalled at another location.

– **Re-homing:** This use case contains the workflow for re-configuring a radio node such that either it is paired with a different radio node controller, or it shall connect to a different Element Manager (EM).
- **Network Authentication:** The eNB needs to discover the EM and perform mutual node authentication.
- **Software Installation, Setup of Transport and Radio Parameters and Network Integration:** The eNB gets equipped with the relevant software, retrieves a default data base, configuration data for transport and radio settings according to the planning phase. The configuration of the relevant neighbour eNBs, MMEs and Serving Gateways needs to be adjusted according to the newly installed eNB.
- **Testing:** The new eNB must accomplish a complete self-test before it can be admitted to commercial operation. Besides the hardware components this shall for example also include testing of feeder loss, the radio performance and the site acceptance.

The goal is to achieve a high degree of automation in this operator use case category.

3.2.1.3 Optimisation

The operator use cases of the optimisation category are necessary to support the need caused by the dynamic character of the mobile networks. New sites are deployed or capacity extensions are continuously installed. Also the environmental conditions may change through to new buildings or motorways. Other dynamic dimensions are network load and user distribution which may change significantly during the day. Network optimisation typically can be described as a repeating cycle of checking the network performance, detecting misaligned parameters and optimising the identified parameters.

- **Radio Parameter Optimisation:** Examples for radio parameter optimisation are neighbour cell list optimisation, interference control including HeNB deployments, the optimisation of the handover parameters, optimisation of the RACH and QoS parameter related optimisation.
- **Transport Parameter Optimisation** shall provide means for optimising S1/X2 associations and the data routing in a meshed network.

Some of these optimisation processes may require different behaviour depending on coverage driven, capacity driven or performance driven operator preferences. The goal for SON functionalities is to achieve significant simplifications with respect to these optimisation tasks. This implies to have less stringent requirements to define accurate default parameters for the operator use cases of the planning category.

3.2.1.4 Maintenance

This use case category comprises the operator tasks related to the daily operations of mobile networks. The use cases are assigned to the following sub-categories:

- **Hardware Extension/Replacement:** Typical tasks here are replacing of hardware with a minimum service outage, performance or capacity extensions by providing additional hardware and checking the hardware inventory and necessary hardware. The corresponding workflow strongly depends on the type of components to be added or exchanged: in case major components are exchanged that require an update of basic configuration data, a complete re-run of the 'installation' use case may be required.

- **Software Upgrade:** With minimum operator attention and service impact.
- **Network Monitoring:** This sub-category aims at the retrieval of measurements and analyses of the RAN performance in order to recognise insufficiencies of the network, seamless support of inter-vendor scenarios and sophisticated trace functionality.
- **Failure Recovery:** The recovery from network element outages shall not be a complex task that requires involving specially trained experts.

Concerning SON functionality the goal for the operator use cases of this category is to allow for a smarter operation and maintenance of the daily tasks. There's a high expectation on SON functionalities in reducing operational effort and costs.

3.2.2 NGMN SON Use Cases and Requirements

In the second half of 2010 NGMN issued a document (NGMN, 2010) describing the most important operator use cases, also providing recommendations, solutions and proposals to standardisation bodies in order to ensure that these requirements are considered by their specifications. An overview of these NGMN top operational efficiency recommendations for SON is shown in Figure 3.5.

NGMN's overall objective (NGMN, 2010) is to enable operating and maintaining multi-vendor networks with highest possible efficiency. The document also integrates former NGMN work expressing the TOP10 main recommendations in the operational area. With respect to SON the operational efficiency document formulates requirements aiming at 3GPP SON use cases and also considers more recent standardisation work as the Minimisation of Drive Tests or the Energy Saving study. The remaining part of this chapter provides an abstract of the operator SON use cases in the light of operational efficiency unless already stated in the previous chapter.

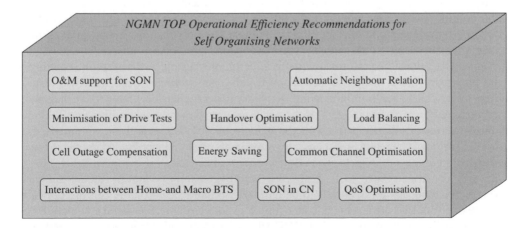

Figure 3.5 Overview of the NGMN TOP operational efficiency recommendations for SON.

3.2.2.1 O&M Support for SON

SON and related benefits are seen as an essential economical characteristic of LTE and are strongly asked for by all network operators. As a consequence it has to be ensured that network operator keep the control on all new SON functionality by implementation of appropriate policy control functions.

3.2.2.2 Minimisation of Drive Tests

Network operators strongly rely on manual drive tests to collect the field measurements that are needed to monitor and optimise the performance of their networks. Manual drive tests require a huge effort in terms of resources and time. Moreover, drive tests can be usually only be done in specific areas (e.g. roads), whereas users and traffic are also distributed on areas not accessible for drive tests (e.g. indoor). Therefore, it will be highly beneficial to automate the collection of field measurements and to minimise the need for operators to rely on manual drive tests.

3.2.2.3 Handover Optimisation

The handover optimisation considers the self-optimisation of the handover parameters like handover neighbour list, neighbour specific thresholds and hysteresis parameters. Therefore, this use case aims at reducing the occurrence of undesirable effects following handovers, such as call drops caused by too early handover, too late handover, or handover to wrong cells, but also reducing unnecessary handover and handover ping-pong effects between two cells.

3.2.2.4 Load Balancing

Load balancing promises the usage of given redundancy in the network to move load from the capacity restricted resource to these ones which have free capacity by sharing load information and appropriate reaction on this.

3.2.2.5 Energy Saving

For an integrated energy saving functionality network elements shall provide a stand-by mode with minimum power consumption and a possibility to switch on and off this stand-by mode. Switching between on and off modes shall be done remotely via the network management system or via the element management system using load thresholds to enter or leave the stand-by mode. This should be performed without affecting the customer, for example, entering to stand-by mode should not cause any dropped calls. An automatic capacity-driven energy saving mode can be realised in existing networks using higher-level network management systems based on performance data. However, a solution based on the evaluation of PM data requires shorter granularity periods for the delivery of PM data than those commonly used today. Therefore SON features that are integrated in network elements and element managers shall enable a dynamic temporary shutdown of unused capacity.

Several constraints have to be respected before going into stand-by mode: sufficient coverage for emergency calls needs to be ensured, no faults occur on the cell or co-located cell, site or co-located site is present, connection for wake-up command (if not automatic) must work, yo-yo effects should be avoided, 'VIP cells' may be excluded from going into stand-by, and so on. Similar considerations are necessary for leaving the stand-by mode.

3.2.2.6 Cell Outage Management

Cell or sector outages may have various reasons, including power supply failures, failed backbone connection, hardware failures, or software failures (the latter may lead to a 'sleeping cell'). Cell outages can be detected by statistical analysis, alarm or customer complaints. In some cases it may take several hours to days to detect the failure and to identify the underlying problems that caused the cell outage. This may also only refer to dedicated services of a cell (e.g. HSPA, GPRS). Cell Outage Management can be split into two main sub-functions: Cell Outage Detection (COD) and Cell Outage Compensation (COC). The goal of COD is to automatically detect failed cells or cells with poor performance. After detection, the COC function shall provide means to restore the failed service (in the case of software failures this may, for example, just require a reboot of the corresponding base station), or compensate the loss of service in the respective area while not degrading significant quality indicators of the surrounding neighbour cells. When the failure has been repaired an autonomous reconfiguration of the failed cell(s) shall take place.

3.2.2.7 Common Channel Optimisation

The main target is to optimise the parameters of downlink and uplink common channels based on UE and network measurements of the common channel performance. Based on these measurement SON functionality can optimise common channel performance.

3.2.2.8 Interactions Between Home- and Macro eNB

Scenarios including Home eNB (HeNB, femto nodes, cf. Section 2.1.8.5) may require special consideration with respect to interference management between HeNB and macro eNB, but also Mobility Robustness Optimisation and Load Balancing between HeNB and macro eNB. After applying corresponding SON functionality it is expected that a macro eNB will be able to control the level of interference that is introduced by the non-coordinated deployment of HeNBs without the need of operator involvement. The SON functions shall allow for automatic transmit power adjustment of the HeNBs to improve the interference level experienced by UEs served by a macro eNB. Another important target is the automatic distribution of the cell's physical resources to avoid interference between HeNBs and macro eNBs.

3.2.2.9 SON in Core Networks

The following main use cases are seen in core network area:

- Plug&Play support for eNB deployment by core network nodes.
- Plug&Play installation of core network nodes.
- Load balancing in core network and between eNB and core network.
- Operational use cases like improved Performance Monitoring, Configuration Management, Inventory and SW Management.

3.2.2.10 QoS Optimisation

It's a major target to provide to the customer high throughput and low delay. Mobile operators today are faced with very inhomogeneous QoS implementations. Several concepts could be used like user prioritisation (famous gold, silver, bronze user definition) or service prioritisa-tion via guaranteed bit rate bearer or packet/bearer prioritisation and even over-dimensioning of resources. SON functionality shall be used to optimise QoS related parameters. The efforts of this, often manual, tasks of experts shall be minimised to focus expert knowledge more on exceptional trouble cases.

3.2.3 SON Use Cases in 3GPP

Not all of the SON use cases that were defined by NGMN have a direct impact on the standardisation corresponding functions. Generally standardisation will be needed for func-tions that depend on information exchange between network elements provided by different vendors and if these information elements need to undergo a special, unambiguous and vendor independent treatment. This will guarantee for interoperability of the SON functionality even if the involved network elements are from different vendors. Otherwise vendor specific, proprietary solutions will serve the purpose. However, as some vendor-specific SON functionality needs to be controlled from the OAM system in a multi-vendor fashion, standardisation is needed for SON control, setting policies and targets and for monitoring.

The standardisation process concerning SON started in 3GPP already in 2008, quite soon after the NGMN use cases were issued and is continued since then. The work is split in three working groups according to their responsibilities (for additional information about 3GPP Structure, timeline and LTE specifications, see Section 2.1): RAN2, RAN3 and SA5.

RAN2 addressed only one use case: Minimisation of Drive Tests (MDT). In RAN3 the use cases to be solved were gathered in a technical report 3GPP TR36.902 [2011] They were:

- Coverage and Capacity Optimisation (CCO);
- Energy Savings (ES);
- Interference reduction;
- Automated configuration of PCI;
- Mobility Robustness Optimisation (MRO);
- Mobility Load Balancing (MLB);
- RACH (Random Access Channel) optimisation;
- Automatic Neighbour Relation (ANR) Function;
- Inter-Cell Interference Coordination (ICIC).

Most of the RAN2 and RAN3 use cases can be easily mapped onto the NGMN operator use cases: MRO addresses handover optimisation, RACH optimisation is an example of common channel optimisation, and CCO addresses the part of Cell Outage Compensation that does not assume quick corrective action. Automated configuration of PCI is one of the earliest use cases, introduced before the top NGMN use cases were defined, so it addresses a rather more general need for deployment automation. The same can be said about the two interference-related 3GPP use cases, though they are more recent than CCO use case. The work in RAN3 reached very different stages: while some use cases are considered already closed (e.g. RACH optimisation or ANR), others are still being enhanced (e.g. MLB, MRO and energy savings) and some are suspended or it has been decided to leave them aside for the time being

RAN2 and RAN3 activities

Self Conf.	R8	• Plug & Play, ANR: – Automatic PCI configuration, Detection of new neighbors and X2 setup

Self Optimisation Intra-LTE	R9	• Mobility Robustness Optimisation (MRO):
		   – Detection of radio link failures (RLFs), Reporting RLFsand possibly UE measurements
		• Load Balancing (MLB):
		   – Load and composite capacity reporting (including inter-RAT), HO negotiations
		• Energy Savings:
		   – LTE cell status reporting and wake-up request
		• RACH Optimisation
		   – Reporting RACH access statistics from UE, Exchanging PRACH configuration

Multi-RAT and MDT	R10	• Coverage and Capacity Optimisation (CCO):
		   – Current focus is on the detection of coverage problems
		• Enhancements to MRO and MLB:
		   – Inter-RAT and load/capacity information enhancements
		   – Support of unsuccessful re-establishment
		   – Possibly including consideration of home/macro interactions
		• Energy Savings study
		   – Inter-RAT energy savings
		   – Enhancements to the intra-LTE solutions,
		    e.g. in combination with coverage optimisation
		• Minimisation of Drive Tests (MDT)
		   – Main focus on coverage problems
		   – Both, immediate and non-real time reporting
		   – Based on RRC signaling and RRM events, with location information
		• ANR for 3G
		   – Study on methods to enable updating neighbourhood relation tables in 3G
		    (inter-RAT and intra-UTRAN)

Figure 3.6 SON related activities in RAN2 and RAN3 for different 3GPP releases.

(e.g. Interference reduction). CCO had a particularly difficult history: it was started, then suspended, restarted and finally considered to be out of scope of RAN3's competence. A summary of RAN2 and RAN3 SON related activities for different 3GPP releases is shown in Figure 3.6.

In SA5 the work similarly was guided by the operator use cases. First step was to define the configuration parameters for the ANR function in the 3GPP object model (called NRM: Network Resource Model) and to define how self-configuration (3GPP TS32.50x, 2011) and automatic software management work (3GPP TS32.53x, 2011) in a so-called Interface IRP (Integration Reference Point) which defines operations and notifications for specific functions. In Figure 3.7, a summary of SA5 related activities is shown for different 3GPP releases.

Next, Automatic Radio Configuration Function (ARCF) was added to Self-Configuration. ARCF allows considering the very latest radio network environment during Self-Configuration of network elements which are added to the radio network.

In 2009 SA5 also agreed on basic management principles for Self-Optimisation, namely target and policy controlled optimisation (3GPP TS32.52x, 2011). Targets were defined for HOO (HandOver Optimisation) and LBO (Load Balancing Optimisation), including new measurements to monitor target achievement.

Latest SA5 work in 2010 and 2011 defined more targets (for RACH optimisation), created NRM elements for control of Energy Saving and parts of cell outage compensation,

SA5 Activities

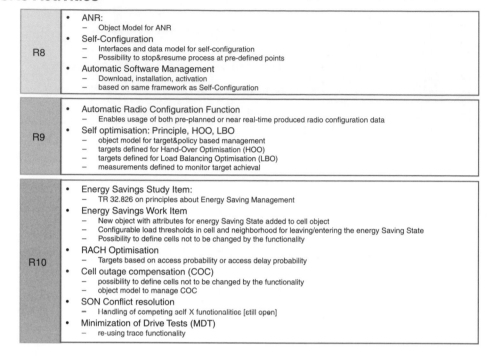

R8	• ANR:
	  − Object Model for ANR
	• Self-Configuration
	  − Interfaces and data model for self-configuration
	  − Possibility to stop&resume process at pre-defined points
	• Automatic Software Management
	  − Download, installation, activation
	  − based on same framework as Self-Configuration

R9	• Automatic Radio Configuration Function
	  − Enables usage of both pre-planned or near real-time produced radio configuration data
	• Self optimisation: Principle, HOO, LBO
	  − object model for target&policy based management
	  − targets defined for Hand-Over Optimisation (HOO)
	  − targets defined for Load Balancing Optimisation (LBO)
	  − measurements defined to monitor target achieval

R10	• Energy Savings Study Item:
	  − TR 32.826 on principles about Energy Saving Management
	• Energy Savings Work Item
	  − New object with attributes for energy Saving State added to cell object
	  − Configurable load thresholds in cell and neighborhood for leaving/entering the energy Saving State
	  − Possibility to define cells not to be changed by the functionality
	• RACH Optimisation
	  − Targets based on access probability or access delay probability
	• Cell outage compensation (COC)
	  − possibility to define cells not to be changed by the functionality
	  − object model to manage COC
	• SON Conflict resolution
	  − Handling of competing self X functionalities [still open]
	• Minimization of Drive Tests (MDT)
	  − re-using trace functionality

Figure 3.7   SON related activities in SA5 for different 3GPP releases.

extended trace functionality to support automatic drive tests and tackled the issue of conflicting targets.

3.3 Foundations for SON

'Self-organisation is a process where the organisation (constraint, redundancy) of a system spontaneously increases, that is, without this increase being controlled by the environment or an encompassing or otherwise external system' (Heylighen, 2009). Self-organising principles have been under research for decades already, for example, in physics (magnetism (Ising, 1925), crystal structures), or in biology (bird swarm intelligence). In (wireless) networking, self-organising principles have been applied typically in research for ad hoc/ mesh networks characterised by uncoordinated deployment of network nodes and the absence of a dedicated network operator organisation. Infrastructure networks like LTE, however, are coordinated deployments done by a dedicated network operator. Hence, it can be argued that an 'external system' according to the above definition can be employed to 'organise' the system. In fact, today's Operation Support Systems (OSS) have this capability, because all of the information required to control the system is available there, for example, due to the required site and coverage area planning, pre-operational neighbourships are computed which are in turn enabling to assign per-NE neighbour relations (cf. Section 4.2.3). So, in principle, there is no need to self-organise (with regards to the example ANR use case to establish a distributed, dynamic, automatic neighbour relationship setup function).

Nevertheless the SON vision (as outlined in Section 3.1) addressing the challenges (cf. Section 1.1) is motivated by several trends which make the application of Self-Organising principles attractive even for infrastructure networks:

- **The increasing complexity of the entire OSS** (the 'external system' according to the above definition): While the data required to organise the system is available in the OSS it may be not readily available at the right place (sub-system) at the right time.
- **The increasingly complex network infrastructure with numerous small cells** (cf. Section 2.1.8.5): Such infrastructures need to be managed in a more autonomous way to overcome OSS scalability limitations.
- **The demand to share network infrastructure:** Hence, there is not just one organisation involved in the network operation.

Roughly in the last decade already attempts have been made to address the identified challenges for mobile networks with similar principles, however, they have been typically:

- Built on top of unchanged NE and OAM systems, rather than embedding an automation solution into the network and the OAM system. While a good level of automation can be reached in the new system added to the existing one, the problem is in the interaction of the legacy system and the new one (i.e. if a lot of (manual) overheads are incurred by operating such a solution, the resulting overall improvement in automation may be very small). Hence, a key characteristic for a SON system is the shift of functionality from network planning to the OAM system and from the OAM system to the NE respectively.
- Addressing only an isolated area (e.g. just troubleshooting), rather than combining solutions for different Network Management areas into one system. In fact, often SON in NGMN/ 3GPP (cf. Section 3.2) is used as the summary term for self-configuration, -optimisation and –healing.
- Single vendor solutions.

Frequently, the characteristic of 'distribution' (versus centralisation) is associated with the Self-Organisation concept. However, it is emphasised here that from the perspective of the 'external system'/the human operator, internally the SON system can be completely centralised and still fulfil the above definition. In practice, this means that the centralised OAM system containing the SON functions is conceptually removed from the overall OSS (the 'external system') and is considered to be an integral part of the Self-Organising system.

In the following we introduce two key technical areas (control engineering and autonomic computing) which can be argued to form the baseline for SON. Of course, many more technical areas and technologies have some degree of relevance for SON (and will be touched in the later chapters of this book where appropriate).

Furthermore, existing research projects with a focus on SON are briefly introduced.

3.3.1 Control Engineering: Feedback Loops

Control theory is the mathematical analysis of the systems and mechanisms for achieving a desired state under changing internal and external conditions.

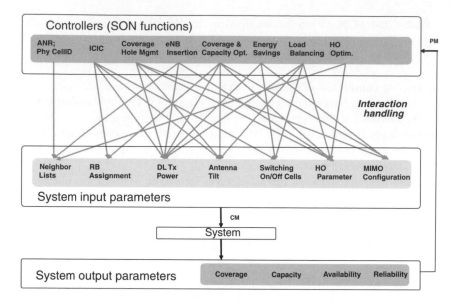

Figure 3.8 Control engineering view on a SON-enabled system (Bandh *et al.*, 2011). Reproduced with permission of © 2011 IEEE.

Figure 3.8 shows the mapping of SON in mobile networks to the classical feedback loop in control engineering. The 'system output parameters' in a mobile network are mainly coverage and capacity provided to users, including the availability and reliability of the service provisioning. Those parameters are (coarsely) established by a large set of network performance data which can be measured at the individual network elements.

The performance data is then consumed by a set of 'controllers' corresponding to the SON functions, which set the 'system input parameters'. As can be seen on the figure, there are rather few system input parameters which often have a relationship to more than a single SON function. Also, a single SON function may influence more than a single input parameter. Hence some handling of the interactions between the 'controllers' is required.

The 'system input parameters' are then imposed on the system resulting in some change of the 'system output parameters'. Here, again, a single parameter may influence more than one output parameter (e.g. power influencing both coverage and capacity). Also several input parameters may influence the same output parameter (e.g. power and tilt influencing coverage).

Control theory is not directly applicable due to the dependency between 'system output parameters' and the input to the controllers which cannot be described analytically at a sufficiently detailed level (hence simulation is typically used to derive the impact of configuration changes (CM) on the inputs to the controllers (PM data).

Also, traditional controller design cannot be applied as there are no simple 'measured output' (PM data) to 'system output' (e.g. coverage) relationships (in addition to the fact that many different controllers with some mutual influence are operating in parallel).

3.3.2 Autonomic Computing and Autonomic Management

Autonomic computing describes an approach to self-managed computing systems with a minimum of human intervention. The term derives from the body's autonomic nervous system, which controls key functions without conscious awareness or involvement. The autonomic computing vision describes a paradigm shift from a technology centric view towards a data and business centric view of computing in data centres. Autonomic does in this context not mean automated, as an automated system might, for example, simply specify that a network element is assigned to particular tasks for a certain time period. Autonomic systems go beyond this specification as they aim at a self-contained and self-reliant functioning, configuring and re-configuring themselves governed by high-level, business and quality derived rules and policies. They do not necessarily consist of a single, proprietary solution but may incorporate a bundle of different components and technologies.

Various initiatives have been started in recent years that developed a set of requirements and architectures for autonomic computing. Amongst these, the IBM Autonomic Computing Initiative (ACI), (see Kephart and Chess, 2003; IBM, n.d.) can be seen as a long-term strategy towards autonomy and a source for the basic terminology used in the area of self-management. Similar to IBM ACI that predominantly targets IT networks' systems management, the Sun N1 Architecture is a platform which is, for example, being implemented for data centres, computer grids, service provisioning systems and so on. Also Hewlett-Packard has developed an 'adaptive enterprise' strategy based on autonomic principles, named the HP Darwin Reference Architecture, which aims at connecting business processes with IT infrastructure. However, HP Darwin is to be seen more as a set of rules how to design an IT system that follows autonomic principles.

The vision of autonomic computing aims at the development of intelligent, open systems that are capable of running with minimal human intervention, can adapt to varying and unpredicted conditions in accordance with business policies and objectives, continuously tune themselves, can prevent and recover from failures and prepare their resources such that the requested workloads can be handled most efficiently.

Regarding the evolution path towards autonomic systems, different levels can be defined:

- **Level 1 (Basic):** Manual computing where all system elements are managed independently.
- **Level 2 (Managed):** Systems management technologies allow to collect and aggregate information from various systems elements into few consoles from where required configuration actions are initiated manually.
- **Level 3 (Predictive):** Pattern recognition technologies are used to recommend actions based on collected and aggregated information; the actions are executed on manual approval.
- **Level 4 (Adaptive):** The system automatically takes action based on collected and aggregated information, controlled through manually matching actual performance against SLAs.
- **Level 5 (Autonomic):** A fully integrated system is dynamically managed based on business rules and policies.

3.3.2.1 Autonomic Management

Autonomic management is often synonymously used for autonomic computing, especially in the area of communication networks. The focus of autonomic management is thereby on the following areas:

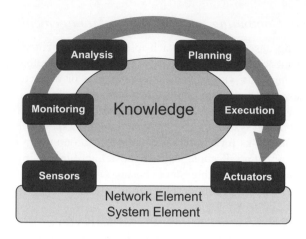

Figure 3.9 Autonomic cycle.

- Installation and deployment technologies and procedures.
- Common systems administration: management from a single viewpoint.
- Problem determination: uniform and standardised format of problem reports.
- Complex analysis: that is, using decision trees, Bayesian reasoning or rule engines to support the analysis and planning of actions.
- Policy based management: where common policies can be used to determine intended configurations or reactions to events.
- Autonomic control loop.
- Optimisation capabilities: where additional information (e.g. about current performance and utilisation) needs to be fed into the autonomic control loop.

The autonomic management cycle as depicted in Figure 3.9 implements the control loop, known as MAPE (Monitoring – Analysis – Planning – Execution) loop, through which correcting or optimising actions can be effected. The knowledge base depicted in the centre of the control loop supports the analysis of the events coming from the sensors and determining and planning the appropriate actions to be executed.

Autonomic management can be subdivided into four major domains (see Figure 3.10): self-configuration, self-optimisation, self-healing and self-protection.

- *Self-configuration* components increase the system responsiveness by adapting to environments as they change. The system can add and configure new features, additional network elements, and newly available software releases during runtime. The key to making this process autonomic is to require minimal human involvement.
- *Self-optimisation* components make the best use of available resources even though these resources and the requirements are constantly changing. A self-optimising system must continuously monitor and tune its components, network elements and sub-systems.
- *Self-healing* components improve business resiliency by eliminating disruptions that are discovered, analysed and acted upon. The system identifies and isolates a failed component. This component is taken offline, repaired or replaced, and then the functional component is

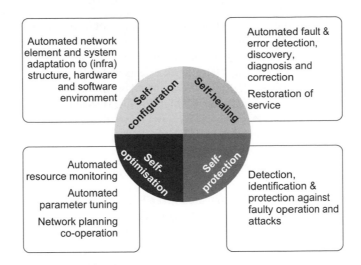

Figure 3.10   Self-management domains.

brought back online. Autonomic systems need to be designed with some level of redundancy so that this healing can occur transparently to users.

- *Self-protecting* components anticipate and protect against intrusions and corruptions of data. This includes the managing of authentication of users for accessing resources across an array of enterprise resources. Self-protection also includes monitoring who is accessing resources and reporting and responding to unauthorised intrusions.

3.3.3   SON Research Projects

Starting from 2004/2005 there have been various public research projects that at least partially worked within the area of self-management, autonomic management and SON. In the following, those projects having a clear relationship (at least in some work packages) to mobile networks SON, are summarised in short:

- CELTIC GANDALF (2005–2006);
- EU FP7 SOCRATES (2008–2011);
- EU FP7 E^3 (2008–2009);
- COST 2100 SWG 3.1 (since 2007).

3.3.3.1   CELTIC GANDALF

Within the European research and development programme Celtic, the Gandalf project (2005–2006) explored the potential of automating common management tasks in heterogeneous mobile networks (mostly GSM, UMTS and WLAN). Gandalf thereby focused on radio resource management issues, including Advanced and Joint RRM (ARRM, JRRM), auto-tuning and automated diagnosis in troubleshooting. Within the project, ARRM and JRRM algorithms have been developed for admission and load control as well as for mobility issues (e.g. vertical handover).

The approaches used in the Gandalf project for automation of management tasks were based on artificial intelligence and optimisation methods: Bayesian Networks for automated troubleshooting; and fuzzy logic, reinforcement learning, Q-learning and multi-agent optimisation algorithms for the auto-tuning process. The studies, tests and validations of the methods and techniques developed in the Gandalf project were carried out on two types of testing platforms: multi-system dynamic simulators including GSM, GPRS, UMTS and WLAN subsystems; and a multi-system test bed with WLAN and UMTS subsystems. Real network measurements (e.g. KPIs recovered from an Operation and Maintenance Centre) were used in the study on automated troubleshooting.

Major results were achieved for:

- ARRM and JRRM algorithms for statistical user load based admission control and load based vertical handover for UMTS-WLAN networks.
- Fuzzy-logic based auto-tuning algorithms with reinforcement learning methods that modify RRM parameters to adapt the network to traffic variations; a resource allocation auto-tuning algorithm has been implemented for UMTS networks and traffic balancing scenarios have been studied via mobility parameter auto-tuning in UMTS, GSM-UMTS, and UMTS-WLAN networks.
- Automated network diagnosis, where an existing Bayesian Network model for 2G has been adapted for automated fault cause diagnosis in UMTS networks, with an algorithm that includes automated model learning, segmentation and inference.

The results of the Gandalf project influenced, and were referenced by other research projects (e.g. the EU FP7 SOCRATES and E^3 projects).

3.3.3.2 EU FP7 SOCRATES Project

The SOCRATES (Self-optimisation and self-configuration in wireless networks, www.fp7-socrates.eu, January 2008 – March 2011) project developed solutions for dedicated LTE self-organisation use cases. The project followed a bottom-up approach with 3G manual configuration, fault and performance management solutions as starting point, and aimed at a short- to mid-term time horizon regarding the applicability of the project outcome into operational networks.

For a selection of eight SON use cases out of 24 initially described ones, solution concepts and algorithms have been developed and verified against previously defined requirements and assessment criteria through simulations. The use cases where partially derived from a survey made by the NGMN forum but also results and definitions out of the GANDALF and COST2100 projects influenced SOCRATES. The finally selected use cases were:

- Load Balancing Optimisation (LBO);
- Handover Parameter Optimisation (HPO);
- Home eNB (HeNB)Parameter Optimisation;
- Admission Control (AC)Parameter Optimisation;
- Packet Scheduling (PS)Parameter Optimisation;
- Interference Coordination Optimisation (ICO);
- Cell Outage Management (COM) (self-healing);
- Automatic Generation of Initial Parameters for eNB Insertion (AGP) (self-configuration).

Furthermore, a set of use case pairs that have a strong interrelationship regarding the modified configuration parameters have been analysed. Especially handover related configuration parameters have been used by several use cases, for example, handover offset. For some of these use case pairs it turned out that their simultaneous operation may cause unwanted effects on the allover network quality and performance as the output of the individual algorithms, that is, the modified configuration parameters, may be in conflict. Simulations for these use case pairs showed that an integration of the corresponding algorithms can considerably reduce the potential for conflicts and hence the impact on the system quality. In this context a functional architecture for the coordination and integration of various SON functions has also been developed that aims at avoiding conflicts by appropriately setting targets and policies for the individual SON functions and on the other hand can detect and resolve conflicts that occur due to the simultaneous operation of different SON functions.

The results and findings of the SOCRATES project influenced the work of NGMN but also 3GPP standardisation (e.g. on the LB use case) as SOCRATES continuously disseminated to these institutions. Furthermore, as the SOCRATES project had a high visibility in the mobile radio networks SON research community, the results and findings have been referenced in numerous publications.

3.3.3.3 EU FP7 E^3 Project

The general goal of the End-to-End Efficiency (E^3, https://ict-e3.cu/, January 2008 – December 2009) project was to transform wireless system infrastructures into an integrated and scalable managed Beyond-3rd-Generation (B3G) cognitive system framework. One of the E^3 objectives was thereby to increase the efficiency of wireless network operations, building distributed self-organisation principles including cognitive (self-learning) capabilities. This objective was mainly addressed by work package four on autonomous functionalities and algorithms in a cognitive radio/network context.

Regarding SON the E^3 project introduces the architectural concept of 'Cognitive Management and Control' as an enhancement of autonomic management, by including self-learning capabilities to improve the targets of self-optimisation engines in the management system. These self-learning capabilities shall furthermore improve the policies, the context acquisition and the profiles of the management system. In contrast to the SOCRATES project, E^3 followed a top-down approach, starting with the high-level concept and architecture, and subsequently applying the architecture to selected SON use cases. However, both projects cooperated regarding the selection of the use cases. E^3 had a mid- to long-term time horizon regarding the implementation of the results into operational systems and also analysed business aspects and impact of the implementation.

E^3 developed a set of autonomous and collaborative algorithms that cover SON issues. These algorithms were proven by a prototyping and demonstration environment.

- Autonomous algorithms for spectrum sharing and opportunistic spectrum access: alternative usage of spectrum in a cognitive radio context.
- Autonomous algorithms for attributing self-x principles and protocol reconfiguration for autonomous network elements.
- Collaborative algorithms for the optimisation of radio access networks, for example, outage compensation, inter-cell interference coordination, knowledge based reconfiguration of network segments.

- Collaborative algorithms for dynamic spectrum management exploiting cognition techniques, including spectrum sensing and information provisioning mechanisms.
- Collaborative algorithms for Joint Radio Resource Management (JRRM) enhanced with cognition techniques.
- Collaborative algorithms for the management of Flexible Base Stations (FBSs).
- Cognitive Pilot Channel (CPC) definition as enabler for autonomous and collaborative algorithms, for example, regarding message structures, bit rates required, mesh optimisation.

As with the SOCRATES project, the major impact of E^3 came through the participation of the project partners in and the corresponding dissemination of project results to 3GPP standardisation and the NGMN forum.

3.3.3.4   COST 2100 SWG 3.1

COST is a European, inter-governmental cooperation framework in the field of Scientific and Technical Research (http://www.cost2100.org). It is based on Actions, which are networks of coordinated national research projects in fields of interest to participants coming from different COST and non-COST countries. COST2100 is the Action on Pervasive Mobile & Ambient Wireless Communications and belongs to the ICT Domain. This Action basically addresses the various topics that are emerging in the area of mobile and wireless communications. The Sub-Working Group (SWG) 3.1 focuses on mobile wireless network optimisation aspects with special attention to the data used as an input to the optimisation process. The goal is to substitute artificially generated data from existing simulation- and prediction-based optimisation tools by measured data taken from the running system (e.g. OAM system, network elements and probe/drive testing tools). The SWG3.1 activities include:

- Network performance criteria development, measurement data definition, acquisition, filtering, classification, correlation, grouping and separation.
- Optimisation parameters definition and selection, parameter correlation with counters and quality indicators.
- Network modelling and model tuning with use of measured data, optimisation method development, algorithm implementation.
- Comparison between measurement and simulation based optimisation approaches including reference scenario definition, field tests for verification with practical trials.

The COST 2100 SWG3.1 has cooperated with other research projects in the area of SON and mobile network optimisation, for example, the GANDALF and SOCRATES projects. As COST2100 is actually a framework and not a funded research project, results are available to the participating institutions only.

3.4   Architecture

This section discusses the architecture of a SON system; in particular, general considerations are presented on where SON functions should be located in the OAM hierarchy. The considerations comprise both use case-specific and system-specific critera.

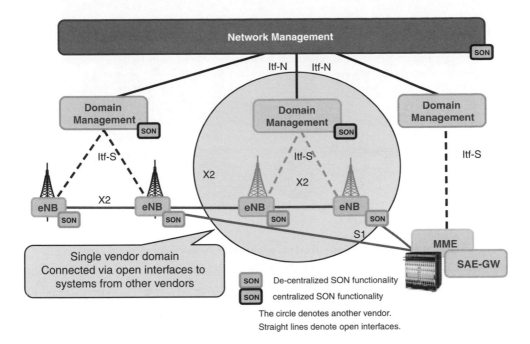

Figure 3.11 Location of SON functions in the 3GPP OAM architecture (Sanneck, Bouwen and Troch, 2010). Reproduced with permission of © 2010 IEEE.

Figure 3.11 shows the 3GPP OAM architecture as it has been introduced in Section 2.1.9. A network element (such as an eNB) is managed by its (vendor-specific) Domain Manager (DM). Different such vendor domains can then be managed in a uniform way via the northbound interface (Itf-N) at the Network Management (NM) level. In the figure, it is shown that SON functions can be located at the:

- **NM level:** *centralised SON* (using *standard* or proprietary interfaces).
- **DM level:** *centralised SON* (using a *proprietary* interface).
- **NE level:**
  - *distributed SON* (using *standard* or proprietary interfaces);
  - *local SON* (having no dependencies to other NE, hence requiring no interfaces).

A combination of a set of SON functions located at different levels can be called a *hybrid SON* system. A single SON function can also be hybrid in that different function components execute at different levels (an example being the ANR function, cf. Section 4.2.3).

It should be noted that SON functions themselves are entities which are 'managed' using the OAM architecture depicted in Figure 3.11. However, 'management' refers here to setting targets, configuring the SON function behaviour at a high level, and monitoring SON function results rather than directly changing the low-level configuration and monitor low-level performance indicators for the SON function (cf. the 3GPP SA5-related part in Section 3.2.3).

Data (KPIs, alarms) which drive the SON functions are obviously generated at the individual NEs. That means that for the distributed SON approach, information (data, requested

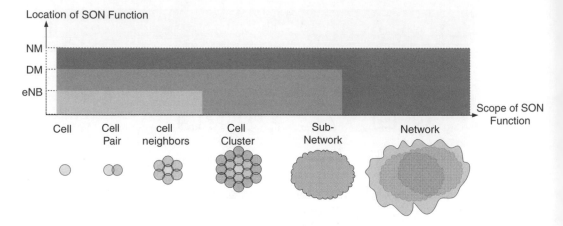

Figure 3.12 SON function spatial scope versus SON function execution location.

reconfigurations) needs to be exchanged between the NEs. In a centralised SON approach, first data needs to be transferred from the NE to the DM/NM-level. After a SON function has been executed the requested reconfiguration needs to be conveyed to the relevant NE in turn.

In the following sections, criteria for the choice to locate a SON function at any of these levels are discussed.

3.4.1 Use-Case Related Criteria

- **Spatial scope:** Figure 3.12 shows the relationships between the scope of a SON function (ranging from an individual cell to an entire network) and its execution location (NE/DM/NM-level). On one hand, it can be seen as natural to locate a function with a network-wide scope (covering multiple vendor domains) at the NM level, because the function is driven by data conceptually residing at that level (e.g. self-configuring a parameter which needs to be identical for all NEs in a network). On the other hand, it is also technically possible to execute such functions in a fully distributed way (for the example given above this would mean that an NE retrieves a network-wide parameter from the NM-level which is then propagated from NE to NE).

  For functions having a rather limited scope like a cell pair, also, on one hand it is natural to execute them on the NE, because the function is driven by data already available on two NEs. On the other hand, there is also no fundamental technical blocking point to execute it at the DM/NM level instead.

  Note that the key SON functions discussed in the following chapters of this book have typically a scope of up to a cell cluster level.
- **Timing requirements:** SON functions have very specific time intervals in which they acquire data, execute and request reconfigurations of NEs (this is elaborated in Section 9.1 on SON function interactions). These time intervals (ranging from minutes to days) are typically larger than those of RRM functions (cf. Section 5.6) and smaller than those of conventional network optimisation functions. The novelty with SON compared to choosing

conventional OAM function locations is that in fact these time intervals are somewhere 'in the middle' of these two extremes, hence the choice where SON functions should reside in the architecture is not straightforward.

It can be argued that SON functions requiring relatively frequent data acquisition and execution advocate more a distributed location of functions, whereas functions which rarely execute should be realised in a centralised way. Similarly to the spatial scope consideration above, this is just a rough guideline. If the loop of data acquisition, transfer, execution/ decision making and transfer of the reconfiguration requests is fast enough and sufficiently dimensioned in the concrete system instantiation, any function can be implemented in a centralised way and be scalable (this is the link to the system-level criteria presented below).

- **Standardised versus proprietary parameters:** In a typical cellular network there are tens of thousands of network elements, leading to millions of parameters to be maintained. Even though most of the parameters can be set to their default or vendor-defined values, or can be specified through network element and environment specific templates, the number of residual parameters is very high. So even if only few percent of all parameters need to be considered for daily configuration and optimisation work, still several hundred thousand parameter values are subject of daily management in a typical network. Parameters can be classified being either vendor-specific or standardised also contributing to the execution location choice for SON functions manipulating those parameters:
  - **Standardised parameters (3GPP, IETF, ATM Forum, etc.):** These parameters have been standardised in order to configure properties of standardised external interfaces (air interface, backhaul transmission line), cf. Section 4.1 (e.g. VLAN, IP addresses), Section 4.2 (e.g. frequency bands, cell IDs) and Chapter 5 (e.g. handover parameters in Section 5.1.3). They have common semantics across all vendors and thus can be configured and optimised independently from the vendor of the specific NE. This class of parameters accounts for about 10–20% of all parameters (note: a radio cell typically has about 500 parameters). Many of those parameters are touched during daily work in network operation.
  - **Vendor-specific parameters for call processing features:** For example, 3GPP intentionally did *not* standardise the logic within the elements to decide on handover (for handover the message sequence between the network elements is standardised, but in order to offer possibility for vendors to differentiate and to introduce enhanced features, the logic to decide on handover is not standardised; cf. Section 5.1). So most parameters that configure the decision logic for handover are vendor-specific. In most cases those parameters cannot be mapped across vendors, even if semantics might be similar. Thus they *cannot* be optimised in a vendor-independent way, but nevertheless must be tuned very carefully.
  - **Vendor specific parameters for low level hardware and internal properties of the NE:** Those parameters usually are set during commissioning and hardly ever are modified during the lifetime of a network element. Thus they are not further considered in the context of SON.

  In total, vendor specific parameters account for about 80–90% of all parameters.
- **Amount of data/processing required:** NE resources for processing are usually limited due to the relatively high cost of installing memory and processing power in NE when comparing to the DM/NM level. This advocates centralised realisation. On the other hand, the transfer of data mentioned above from the NE to the DM/NM level consumes OAM bandwidth and

causes delay. Again these use-case specific constraints need to be considered in the framework given by the concrete system instantiation.

- **Decision making:** Centralised decision making based on inputs from several NEs is usually straightforward, whereas distributed decision making needs to be carefully planned and controlled to avoid classical concurrency issues in distributed systems like oscillations, race conditions and deadlocks. Also, distributed decision making may work efficiently for spatial scopes like a cell cluster, but may not scale to, for example, domain-wide tasks.

3.4.2 System-Level Criteria

- **Scalability:** As mentioned above, SON functions usually have a scope up to a cell cluster-level, based on which a per-function scalability analysis for a distributed versus a centralised approach needs to be performed. Additionally (for both conventional OAM and SON functions), there is an upper bound on the number of NEs which can be treated within a centralised administrative OAM/SON domain. That means, for scenarios with a very high number of NEs (cf. Chapter 2.1.8.5) it may be required to introduce an additional (DM) level of hierarchy to improve scalability, whereas in a distributed SON approach, no additional such means are required.
- **Reliability and availability:** It can be argued that the centralised SON approach contains with the DM-/NM-level entities a single point of failure, whereas the distributed approach has some inherent redundancy being a distributed system. While the general argumentation is valid, of course, centralised SON functions are added to an OAM system which usually already contains redundancy mechanisms like server clustering to improve reliability and availability.
- **Multi-vendor capability:** Both the NE- and the NM-level provide for 3GPP-standardised multi-vendor integration whereas at the DM-level only a proprietary integration is possible. SON-related standardisation at the NE-level (3GPP RAN WGs) is orientated along the tight call processing standardisation, that is, the interoperability regarding the exchange of data is assured. However, this comes at the price of a rather long standardisation process. On the other hand, NM-level standardisation for SON could be (similarly to OAM standardisation in general) considered to be faster but to leave more room for vendor-specific interpretation. In general, the interoperability regarding (vendor-specific) algorithms inside SON functions is not directly covered in the standards, cf. 'standard versus proprietary parameters' above.
- **Management/controllability:** Particularly in the initial SON deployment phase (cf. Section 3.6 on operational challenges), operators want to exercise still a lot of control on the system and tightly monitor what SON functions are doing. New operational workflows (cf. Sections 3.2 and 3.6) have to be realised by the combination of SON functions with existing (human-level) workflows.

  Both requirements are facilitated by a centralised approach, because all the SON-related data and the automated SON decision making are co-located with the function via which the operator exercise control and monitoring. In a distributed approach, however, specific additional instrumentation of SON functions may be required (i.e. additional data may need to be transferred from the NE to the OAM system just to satisfy the operator's control and monitoring requirements, thereby assuring consistency of the distributed configuration changes with operator policy).

- **Extensibility:** Similar to the previous point, a single entry point (as in a centralised architecture) for doing upgrades for the existing SON functionality and adding new SON functions facilitates the operability of the SON-enabled system itself.
- **System legacy and lifecycle:** There are evolution paths to SON functions from both the RRM as well as the network optimisation tool domains. If an existing optimisation tool is evolved into a set of SON functions, the natural architecture choice will be centralised SON. If RRM functions (cf. Section 5.6) serve as the baseline for a SON function, the approach will typically be a distributed one.

There may also exist a distributed versus centralised distinction regarding the system lifecycle: in initial LTE deployments it may be sufficient to rollout individual distributed SON functions for basic network configuration and optimisation, requiring few or no support from the centralised OAM system. Later in the life cycle, when the full set of optimisation and troubleshooting functions are required, the OAM systems have been upgraded to provide centralised SON support thereby covering domain- and network-wide operational workflows.

In summary, concluding from the above points, it is crucial to select the SON function execution location and thus the SON architecture mainly on a *per use case basis*. The following chapters comment on the function execution location choice in the architecture (e.g. Section 4.2.4 for Dynamic Radio Configuration functions, Section 5.1.5.2 for Mobility Robustness Optimisation, Section 5.4 on Coverage and Capacity Optimisation) for the respective discussed SON function.

In addition, as outlined above, also the system-level, operational constraints need to be taken into account for a set of SON functions. This is elaborated further in Chapter 9 (SON operation).

3.5 Business Value

In this section an overview about the business impact of SON shall be given. The analysis will concentrate on the case of macro eNB deployments. Since the achievable benefit of SON depends very much on the specific situation of an operator and the deployment scenario, a universal quantification of the benefit of the different use case is impossible.

Thus this Section will start in Section 3.5.1 with outlining the general economical model of an eNB. In Section 3.5.2 the different types of benefit of SON will be described. Based on this information Sections 3.5.3 to 3.5.5 will provide a more detailed analysis of the expected benefit of selected SON features.

3.5.1 The Economics of eNB Sites

The costs of a mobile network as seen by a mobile operator go far beyond the pure purchase price of the individual network elements. Thus, it is commonly accepted to use the concept of the Total Cost of Ownership (TCO) in the context of telecom infrastructure equipment. The TCO aggregates all costs, which occur over the entire lifetime of technical solution (in this case an eNB) in one single figure. The TCO can be calculated as the sum of three components, which are Capital Expenditures (Capex), Implementational Expenditures (Impex) and Operational Expenditures (Opex).

An additional dimension for structuring the costs is the origin of the cost, for example, whether they are related to the equipment itself or to the site, which is used to host the equipment.

In the following paragraphs the composition of Capex, Impex and Opex will be described in more detail. Thereafter, also the revenue potential of a base station site will be explained, which finally provides a complete view on the eNB economics.

3.5.1.1   Capex

Capex related to the equipment is simply the purchase price, which is paid by the network operator to the equipment manufacturer. In case of an LTE base station site the equipment includes the eNB, the antenna and auxiliary material such as cables and mounting material.

Site related Capex includes the costs for the acquisition and the build out of the site, including civil works, planning, and project management. In case of a base station site, these costs might be quite substantial if, for instance, a dedicated tower has to be built or a rooftop has to be equipped.

3.5.1.2   Impex

The Impex covers all the costs, which are necessary to install the equipment. However, treatment of installation in cost models is not free of ambiguity.

In less formal considerations Impex is sometimes also referred to as one-time Opex owing to the character of the underlying activities, which predominantly consists of manual work of service personnel.

On the other hand, according to financial accounting rules, expenditures for transportation and installation of so-called property, plants and equipment (i.e. tangible assets, which are likely to generate positive revenue streams over several accounting streams) should be treated as Capital Expenditures (IASB, 2003). Note that this rule is independent from whether these tasks are executed by the operator itself or bought from a third party, for example, the equipment vendor. Therefore, Impex could also be considered as being part of the Capex.

The following discussion will not further elaborate on those issues related to accounting principles and possible formal arrangements of balance sheets, which are beyond the scope of this book, but will rather describe the different items belonging to the Impex.

Impex includes the transportation of the equipment to the site, the physical mounting of the device and establishing connection with power supply, antennas and backhaul link. These costs are above all related to the site and the physical appearance of the base station.

Furthermore the Impex also includes purely equipment related costs, that is, the effort for the initial configuration of the equipment, commonly referred to as commissioning. Also costs for the initial planning of the radio network parameters (e.g., hand-over parameters) and the corresponding verification (i.e. by means of drive tests) can also considered as part of the Impex. However, in case those activities are carried out in a repeated manner during the normal operation of the network, they should be regarded rather as Opex.

3.5.1.3   Opex

Site related operational costs cover above all site rental costs and if applicable also other house keeping costs like air conditioning. Equipment related Opex consists of electricity costs, transmission cost and costs for operations and maintenance (O&M). The costs for electricity are calculated based on the power consumption of the equipment and the electricity costs. Transmission costs occur in case leased lines are used for backhauling the base station traffic. Note, that in case of a self-built mobile backhaul (by e.g., microwave radio) the costs for such a build-out would be part of the operator's Capex, whereas the subsequent costs for operating the backhaul network are part of the Opex. However, within the scope of this study, which focuses on the benefit of LTE SON, this distinction is not directly relevant.

O&M costs cover all the activities necessary to keep the network up and running. One way of further refining O&M is to split it into field maintenance on the one hand and network operations, which out carried out within so-called network operation centres, on the other hand. Whereas field maintenance covers all activities, like replacement of faulty equipment or mechanical re-adjustments, which require a visit of the affected sites, the latter typically consists of network optimisation, fault management, and performance management. As a rule of thumb a headcount of one person is necessary per 100 base stations for field maintenance and network operations, respectively. Due to the rising costs of fuel travel to and from sites, and also drive testing are becoming an increasingly important aspect of the overall expenditure.

3.5.1.4 TCO Break Down

While the TCO for a base station site heavily depends on a number of circumstances, nevertheless some general figures shall be provided here. The numbers refer to the case of a base station site in central Europe.

Whereas Capex and Impex are one-time items, the Opex is a recurring cost and thus has to be aggregated over the lifetime of the equipment. A good starting point, which leads to realistic results, is to base the TCO analysis on a duration of five years, which is a typical economical life time of telecom equipment. Investments in infrastructure, like base station towers or fibre rollout, might exhibit a much longer lifetime: up to several decades. Thus those costs should be added only pro-rated.

Furthermore, to obtain a formally consistent TCO value, cash flows, which do not occur at the beginning of the lifetime, need to be discounted with an appropriate interest rates. In this way the so-called time value of money can be considered. Due to the qualitative nature of the analysis presented here the effect of discounting has been neglected. Rather the TCO is estimated to consist of Capex plus Impex plus five times the annual Opex.

Using this rule of thumb the TCO is found to be in the order of a low six digit Euro value. Equipment related Capex only accounts for a minority of the TCO and is in the order of 20%. Even though electricity consumption deserves a very high attention with respect to environmental protection the contribution of electricity to the TCO is less significant and lies in the order of few percent. Total O&M contribution to the TCO is below 10%, with field maintenance and network operations being in the same order of magnitude. Likewise, equipment related Impex accounts for a few percentage points of the TCO. That is, altogether the equipment related cost components mentioned so far sum up to around 30% of the entire TCO. The remaining 70% are related to site costs and to the costs for transmission.

3.5.1.5 Revenue Potential

To understand the economic role of a base station site also the revenue potential has to be considered. Again these figures vary from case to case and thus only a few rough calculations can be provided here. It is typically a fair assumption that on average one macro eNB is used to serve 1000 subscribers. Assuming a monthly ARPU of €20 the revenue generated by a base station site sums up to more than €1 million over the lifetime. Luckily the revenue potential is much higher than the total cost, which allows operator to cover other expenses like customer service or marketing and also to achieve some profit.

On the other hand it also gets obvious that SON is not only about reducing costs, but also has to unleash the full revenue potential of an eNB by, for example, optimising the capacity or increasing the availability. This will be outlined in the next section.

3.5.2   General Mode of Operation of SON

In this section an overview about the different mode of actions of SON for LTE is given. An analysis of specific use cases is provided in subsequent subsections.

It is obvious to most that SON should provide a productivity benefit to the operator by increasing the productivity of staff or reducing the number of staff required to do a specific job. It is perhaps less obvious that there are other SON benefits that deliver important economical value.

In further discussion, expected benefits will be structured according to the following three main categories:

- Saving of Operational and Implementational Expenses.
- Reducing the equipment needed to achieve certain network functionality.
- Increasing the value of a given network.

They will be discussed in the following one by one.

3.5.2.1   Saving of Operational and Implementational Expenses

This category can be further subdivided into productivity benefits, which stems from an increased efficiency in O&M, and reduced energy costs resulting from improving energy efficiency.

Productivity Benefit
The target of SON is to automate certain process steps such that the costs for effort, for example, human labour or tooling costs used so far can be completely avoided or at least reduced. The resulting benefit can be computed by comparing the costs of carrying out equivalent tasks on a legacy network with the likely costs for an LTE network supported by SON. For most SON tasks there are accepted approaches for manually achieving the same goals. Unfortunately, operators have different internal staffing models, different processes and different geographical considerations that make it difficult to consistently compare task costs across different operators even on legacy networks. Nevertheless a few general remarks shall be made here. The realised savings depend on three factors: the amount of work or equipment related to a single execution of a certain process step which is reduced through SON automation, the cost per unit of working time or equipment employed and finally a multiplier, which indicates how often the process is executed. This consideration leads directly towards two important conclusions. Firstly, process automation is more important in areas where labour costs are high. Secondly, even seemingly small improvements might sum up to large numbers if the

addressed process step is executed frequently. Vice versa, automation of a complex process might not pay off if the process is rarely executed. In this context the frequency of occurrence is given not just by the rate of repetition (e.g. number of occurrences per year) but also by the number of entities (e.g. number of eNBs or number of network management systems), which require separate execution of a process step.

Although computation of savings in human labour cost is straightforward in theory, practical savings might deviate from these simple predictions. For instance, any change in the level of expertise required needs also to be factored in. Cost savings due to a reduction of the amount of work might be partly offset by the fact that the remaining work is more sophisticated and thus needs to be paid higher.

Energy Efficiency
Due to the increasing prices for energy and the general awareness of environmental and climate protection, there is clearly a measurable benefit in reducing the power drawn by the network infrastructure.

Any reduction has to be achieved without sacrificing the necessary RF transmit power required to maintain an adequate level of service and coverage. Reducing the energy consumption of a mobile network and especially of an eNB is subject to constant research and development. To give just a few examples, higher spectral efficiency, energy-optimised RF amplifiers, optimised heat dissipation, or Remote Radio Heads can be mentioned here.

Despite this progress, there is still room for additional improvements based on SON functionality, like power optimisation or switching off temporarily unused eNBs (or parts of it) during periods of low demand. Although exact savings are dependent on a lot of factors, like development of electricity tariffs, technical realisation of SON solutions, or traffic patterns and are therefore very difficult to predict, rough estimates indicate that those savings might sum up to €1000 over the lifetime of an eNB.

3.5.2.2 Reducing the Amount of Required Equipment

It has been pointed out that one goal of SON is to automate certain activities in order to save personnel costs and related costs of material. A further goal of SON is to provide a network optimisation in order to decrease the number of equipment needed to achieve a certain level of network performance. In case of a mobile network this is, above all, a matter of the number of required eNBs.

One can further distinguish between coverage and capacity driven scenarios. The first scenario, achieving coverage for a given area with minimal number of eNBs, will be of primary interest during the initial rollout phases. Taking into account the predicted traffic forecast, the latter scenario, providing the required network capacity with a minimal number of eNBs, will dominate in the long-term.

When thinking about the benefit that stems from reducing the number of eNB, it has to be kept in mind that the Capex accounts for only a small fraction ($\sim$10–20%) of the overall eNB TCO. Thus, from an operator's point of view the main savings resulting from a reduction of the number of eNBs are related to these additional costs, which contribute to the TCO, rather than to the eNB Capex itself.

It should be noted that there might also exist cases where it is not possible to reduce the number of base stations, for example, because all eNBs are necessary due to coverage reasons.

However, even in this case optimal capacity utilisation might allow postponing of necessary network updates if the network traffic is going to increase later on.

Monetary benefit depends above all on the technical benefit, that is, on the degree of performance increase. Therefore, the benefit cannot be calculated by a business analysis only, but needs to be combined with the corresponding network planning and network dimensioning activities.

3.5.2.3   Increasing the Value of a Given Network

Finally, SON can also help to increase the value of a given network, or in other words SON can help to increase an operator's revenue. One aspect in doing so is to increase the coverage or the capacity of a given network. Since a performance increase is also possible by adding additional eNBs, the point is to avoid an increase of eNBs by means of SON based optimisation. Thus, the argumentation is a complete analogy to that mentioned above and shall not be repeated here.

Another possibility to increase the value of a network is to increase its availability, or vice versa, to decrease the times of unavailability. In this context, unavailability covers the initial time until first commercial operation as well as the down times and malfunction periods during operation.

Reducing Initial Time until First Operation
The first aspect, reducing the time initial time until first operation, enables an operator to generate revenue earlier and furthermore results in a leading position over competition with respect to time to market. Concerning a reduction of the initial time until first commercial operation, SON related improvements during the rollout process like auto-configuration or ANR should be mentioned. Besides saving of operational expenses those concepts also speed up the installation and commissioning process and, most important, allow commercial operation directly after physical installation independent from external network synchronisation events.

Another aspect not mentioned so far is the possibility to reduce the time for trial periods. Prior to a complete network rollout a new release is usually installed in a limited area first and subject to intensive testing. One goal of a trial is to optimise configuration parameters, which can be later on used as default values. The trial phases could be shortened significantly, if parameter optimisation is carried out in an automated fashion (e.g. based on an automated evaluation of performance management data instead of carrying out drive tests).

The initial downtime is the most uneconomical phase of a typical rollout scenario. Capex, Impex and to some extent also Opex have materialised already, but no revenue can be obtained yet. Thus it is of high importance for an operator and involved suppliers to minimise this period of cash burning. In this way SON also helps to reduce time until break even.

Reducing Down Times and Periods of Malfunctions
Network's availability can be improved by means of 'self-healing'. During down times and periods of malfunction little or no revenue can be generated. Besides this direct loss of revenue, down times and periods of malfunction impede the quality perceived by the subscribers. This increases the churn rate and in this way further impedes the competitive situation of an operator.

The nature of 'self-healing' is that problems that would otherwise take time and resources to detect and resolve, are instead addressed very quickly. This benefit can be measured by recording the service outages per network element against time. Examples of such SON cases are '*Cell Outage Detection and Compensation*' and '*Compensation for Outage of Higher Order Elements*', see also Section 3.2.2.6 and Chapter 6. In many of these cases, SON does not fix the original problem, but having quickly identified the existence of the problem, it is able to reconfigure the network to mitigate the effect of the outage and therefore improve the network availability until the original problem is solved.

Several possibilities to automate incident management, like alarm correlation, automated fault correction or sleeping cell detection exist. The automation of incident management does not only reduce the necessary effort and labour, but also shortens down times and periods of malfunction.

After this general overview the benefit of SON will be discussed in more detail in the next sections. The discussion will be structured along the three categories; Installation and Planning, Network Optimisation and Fault Management.

3.5.3 Installation and Planning

Concerning the rollout/installation phase, the main benefits of SON are the facilitation of both, the planning process and the actual eNB installation.

3.5.3.1 Network Planning

Due to the complexity and specific requirements coming, for example, from the physical appearance of eNB sites, a complete automation of the planning process is not possible. Important factors influencing the planning process and thus the potential benefit of SON are an operator's legacy network and the maturity of an operator and their network.

NGMN includes radio planning amongst the set of use cases that should ideally be addressed by SON (NGMN, 2008). NGMN does, however, acknowledge that the nature of the task is such that it is unlikely to be addressed by SON anytime soon. The reality of radio planning is that greenfield site planning cannot easily be automated, as the planner needs to consider many factors outside the scope of most SON capabilities. These include land ownership, planning designation, building topologies and proximity of existing sites.

Network operators with existing infrastructure across multiple technologies and frequency bands will by necessity have a database of sites with some indication of the coverage capability for these sites. For such operators, it may prove cost effective to create a SON based planning solution. The introduction of SON also challenges incumbent operators as they generally have the most entrenched organisational structures and processes. SON efficiency is maximised by adapting processes to accommodate the new approach to tasks.

The main point addressable by SON is an automated creation of configuration data based on the results of the initial coverage planning.

Auto-Configuration

Besides, aspects related to the physical design of an eNB, like form factor or cabling, the SON auto-configuration feature is a very important ingredient towards a Plug&Play style installation. Installation activity and cost is essentially reduced to a mechanical mounting and connection of the eNB. Separate site visits of a commissioner or extensive support from a remote commissioner are not necessary. The saving of IMPEX possible in this way depends on the exact processes of an operator and the salaries, but could typically be in the order of €100 per eNB.

Since the eNB software and all configuration data are downloaded to the eNB during the auto-configuration process, they do not have to be installed in advance (e.g. already during production). The circumstance that all eNBs leave the factory in the same default state improves the entire logistic of eNB distribution significantly, and allows for instance a short-term re-prioritisation of rollout schedules.

Automatic Neighbour Relations (ANR)
If planning and establishment of neighbourhood is carried out automatically, the corresponding costs for manual execution, which are in the order of €15 per eNB can be saved. Furthermore, with the help of ANR eNB is ready for commercial use directly after initial power-up and auto-configuration. Specific external synchronisation events, which might delay the time until commercial operation, are no longer required. The calculation of such a time-to-market benefit strongly depends on the exact business environment and cannot be generalised. Depending on the situation this benefit might significantly exceed the possible savings related to IMPEX. Assuming a time to market benefit of one month the additional revenues for an operator might be even above €1000 per eNB.

3.5.4 Network Optimisation

When discussing Network Optimisation two different aspects have to be distinguished. Firstly, Automated Network Optimisation targets to automate existing network optimisation by increasing the productivity of the involved staff.

Secondly, self-optimisation allows improving the efficiency of the existing infrastructure beyond a level possible with current manual optimisation procedures. In this way the efficient cost per bit can be reduced further on. An example for such a use case improving the network efficiency is mobile load balancing.

3.5.4.1 Automated Network Optimisation

Currently Network Optimisation is a more or less manual or tool-assisted change of network parameters, which influence the performance of the network. Parameters to be optimised include handover parameters, neighbour cell lists, or cell-individual parameters. The costs for manual network optimisation are significant and easily sum up to several hundred Euros per year and base station. Assuming further that introduction of SON increases the operational efficiency by a factor of two a value of €200 per eNB and year provides a realistic order of magnitude for achievable savings. The potential savings from automated network optimisation are even higher, if it is possible to avoid so-called drive tests, which are carried out to physically measure the network quality in a certain area by using vehicles with special test equipment. These tests require personnel and vehicles including complex test equipment and are thus a major cost factor. Depending on the frequency of drive tests before introduction of SON additional savings between €100 and €200 per eNB and year might be possible.

3.5.4.2 Mobility Load Balancing (MLB)

MLB enables overloaded cells to redirect a percentage of their load to neighbouring cells. In this way traffic load generated by subscribers is distributed on available cells such that the possible network throughput is maximised. This inherent trunking efficiency increases the average network utilisation and allows postponing deployment of additional network capacity.

Utilisation of SON to implement MLB does not target to improve staff productivity, but enables dynamic load balancing on short time scales, which could not alternatively be achieved by any reasonable addition to the headcount.

Thus, obviously the gain of SON MLB is highest in case the subscribers in the cells, which participate in the load balancing, exhibit different usage patterns with respect to time, that is, if the busy hours take place at different points in time.

The benefits of algorithms like MLB, which increase the performance of a network, depend first of all on the realised performance gain. While a general statement on this performance gain is not in the scope of this section, it should be noted that the possible impact on TCO and possible revenue is significant and might easily account for several thousands of Euros over the eNB lifetime.

3.5.5 Fault Management

The Fault Management covers activities like monitoring of the network, handling of alarms and resolving of faults. From an organisational point of view the activities can be split into a remote part, which typically is carried out in Network Operation Centres, and Field Maintenance. The latter part consists of repair and maintenance work, which has to be carried out on-site. Since site visits require a car and timely drives to and from the site, they are expensive and should be minimised.

Independent from SON the primary goal is to design products in such a way that fault situations never occur or can be resolved through built-in redundancy. Since this is very often not possible in an economical way, SON functionality can contribute to optimise the handling and resolving of fault situations.

3.5.5.1 Alarm Correlation

By means of alarm correlation it is not only possible to reduce the number of alarms, but also to perform automated analyses, which lead directly to the root-cause of a problem. In this way alarm correlation reduces the effort needed to manually evaluate alarms. Assuming an average of 0.2 ambiguous alarms per eNB and year and a human effort of 20 minutes to manually analyse such an alarm, the potential saving of human labour sums up to a few Euros per eNB and year. Even more important, the risk of a wrong or ambiguous diagnosis can be minimised, which in turn reduces the number of unsuccessful, and thus unnecessary, site visits. Depending on the number of saved site visits, the resulting operational savings might be in the order of €10 per eNB and per year.

The next step, beyond an automated analysis of fault situations, leads towards self-healing concepts, that is, automatic reaction on a fault diagnosis. A simple form of automatic reaction on faults consists of automatically triggering a reset of components or units, which have been identified as causes of a fault. These actions have to be embedded into a framework, which avoids infinite loops and still allows the operator to keep control (e.g. via policies). Again the resulting benefit is in the order of €10 per eNB and year.

3.5.5.2 Sleeping Cell Detection

A slightly different approach has to be used, if a fault situation occurs, but no alarm is produced at all. A typical example is the problem of sleeping or poorly performing cells. It is possible to identify cells, for instance, by means of regular drive tests or via the complaints from affected users. These solutions are either expensive or customer unfriendly. The intended solution should be proactive but cheap. From a SON perspective both goals can be achieved by regularly collecting suited Performance Data or Key Performance Indicators. By scanning these data intelligently for irregularities sleeping or poorly performing cells can be identified.

3.5.5.3 Cell Outage Compensation

Another form of reaction on faults is to temporarily compensate the consequences of a fault for instance by reconfiguring neighbouring cells. Reasons for a cell outage might be loss or energy a fault in the transmission network or a fault of the eNB itself.

The main benefit of Cell Outage Compensation is the purpose of revenue protection. Obviously a cell that does not work cannot generate any revenue. Furthermore, cell outages decrease the satisfaction of the affected subscribers and might even lead to an increased churn amongst those subscribers. Cell outage compensation mitigates those effects. For economical reasons this mitigation will be possible only to a certain extent (because costs for complete redundancy would be too high). Reconfiguring of neighbouring cells to compensate a cell outage naturally leads to a degradation of the performance in the original coverage area of these cells. Thus the logic behind cell outage compensation is that the benefit (in terms of value generated to the affected subscribers) generated in the area affected by the cell outage is higher than the harm caused to the cells, which are used for compensation.

As mentioned, postponing the permanent resolution of a cell outage is only possible to a certain extent and does not completely eliminate the need for repairing the fault. Thus the main benefit of cell outage compensation related to operational expenditures is to postpone the repair until it can be executed in the most efficient way. In urban areas with road traffic congestions this could be during night, when cell sites can be reached very easily. Alternatively, if the staffing model does not include night shift maintenance repair could also be postponed to the next morning or from the weekend to the next working day. Thus the main saving potential with respect to operational expenses consists of lowering the requirements for stand-by maintenance and avoidance of surcharges for night or weekend shifts.

Optimising the overall benefit of outage compensation is a trade-off between the operational savings related to shifting the repair, partial revenue protection (with respect to complete outage) through outage compensation and loss of revenue (due to imperfect compensation) until final repair.

3.5.6 Conclusions

As shown in this section, SON provides benefits to an operator in several ways.

In the long-term, features which increase the performance of a network and in this way decrease the number of required network elements to achieve a certain performance, are promising the highest value. These features do not only save the Capex for the network element itself, but all related implementation and operational costs as well.

This, however, should not be interpreted in such a way that optimisation of the operational costs should be disregarded. Finally, independent of the network size an operator will always try to reduce operational costs as far as possible, since they are directly influencing its profitability.

Even if in a specific scenario reducing human labour through automation is of lower importance, SON remains beneficial due to the improved time to market and better network availability that it provides.

3.6 SON Operational and Technical Challenges

Section 3.5 has highlighted the business benefits through SON. However, as one of the main goals of SON is to reduce the cost in network deployment and operation, it is clear that it is closely related to the way the associated processes are being performed in an operator organisation. Hence, the introduction of a SON function is more challenging than, for example, the introduction of a new radio feature via a software or hardware upgrade. Thus, the implementation of SON bears both operational and technical challenges which are inter-related. In particular the transition to a SON-enabled system is difficult as first effort has to be invested to implement the new SON functions and put them to use in a real network operation concurrently with effort to change existing operational processes. These are the typical short-term pain points in automating systems which nevertheless assure to receive the clear benefits in the longer term.

3.6.1 Transition of Operational Processes to SON

Figure 3.13 shows the high-level operational lifecycle of a mobile network (cf. also Section 3.2.1). Today configuration, optimisation and troubleshooting are based on manual and semi-automated workflows along the network build and operate/maintain phases respectively.

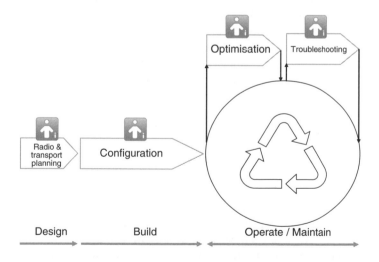

Figure 3.13 High-level operational lifecycle.

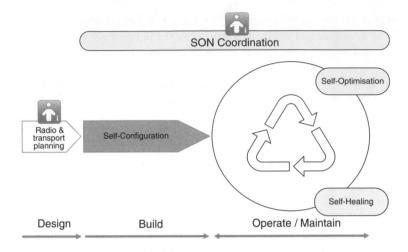

Figure 3.14 High-level operational lifecycle with SON.

In this lifecycle, the execution of the workflows is decoupled from the actual network operation. For optimisation, performance and configuration data are aligned to an offline tool where algorithms are executed on the data. Finally a new network plan is created and rolled out.

Figure 3.14 shows the corresponding operational lifecycle for a SON-enabled system at a high level. In this case configuration, optimisation and troubleshooting workflows are automated and realised by orchestration of SON functions. Self-configuration and -healing are triggered by incidents (e.g. deployment of a new NE or outage of a cell), hence there is still surely the notion of a corresponding workflow consisting of potentially several SON functions. For self-optimisation, individual SON functions are embedded into the ongoing network operation rather than being executed offline. While the operator should be still able to orchestrate and govern the behaviour of SON functions, fixed workflows as employed before are not mandatory any more. The coordination of individual functions' interactions to assure system consistency and stability is done via a dedicated coordination component (cf. Chapter 9), which also serves as the entry point for human operators to control the system as a whole.

As an example, Figure 3.15 shows the changes in the configuration process introduced by self-configuration. The setup of connectivity and the NE commissioning (cf. Section 4.1) are automated steps which are orchestrated at design-time and supervised at run-time via the SON coordination component. Thus, on-site activities are replaced by automated functions (supervised at a remote OAM system). Some functions are shifted from radio planning in the design phase to the actual network build phase (cf. Section 4.2).

This change from manual/semi-manual operations to SON-enabled automated operations will change the operational lifecycle. Challenges in the transition process can be summarised as follows (cf. also Kasinger *et al.*; 2006):

- **Knowledge acquisition:** Today, most of the (critical) 'domain knowledge' for OAM is held by the human operators. However, in order to design SON solutions, this domain knowledge is required. Hence, ways have to be found to acquire and capture all of this knowledge.
- **Operator acceptance ('human factor'):** One clearly non-technical, soft challenge is the acceptance of SON technologies/solutions by human network operators. This includes

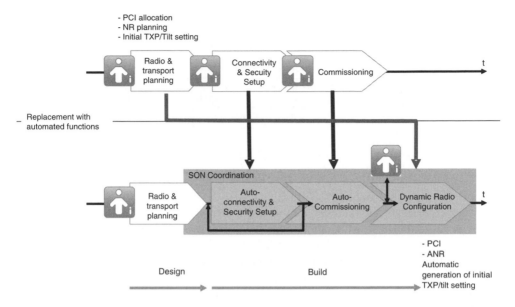

Figure 3.15 Changes in design and build parts of lifecycle with self-configuration.

autonomous actions by SON functions alongside human interaction. Human operators must gain trust in the SON-enabled system in a step-by-step process. That means it must be assured that all autonomous decisions and actions can easily be tracked by a human operator. The human operator must stay in control all the time and thus be able to override any decisions and actions taken by the SON system.

SON enables to shift human operator work from low-level repetitive tasks during operation to more high-level, mostly preparatory tasks, similar to automation concepts in other technical domains. Hence on one hand, the type of work is more creative and more challenging (but then requiring an increased skill set); on the other hand, employees may still perceive automation as a threat to their particular job. The new type of work has a learning curve on its own because new roles will be defined (for 'design-time' preparation of the automated system and for its 'run-time' supervision). During the transition phase, still some degree of lower-level management will be required, which adds then to the ongoing tasks of operators.

Note that the acceptance by human operators is directly related to the knowledge acquisition process mentioned earlier.

- **Changes of human-level processes:** The integration of the automated workflow parts into existing operator processes causes some initial complexity at the process level and thus needs to be assisted by change management. In particular, the change (reduction) of time intervals in which network planning, operation, optimisation and troubleshooting can be performed with SON requires the adaptation of corresponding human-level process cycles.

On one hand, automating low-level and tedious manual processes leads to less human-induced errors. On the other hand, when managing the network at a higher-level, for example, the change of a single policy may affect many operational workflows at the same time and hence may cause more damage in case a mistake has been made. Therefore, like stated above, a higher skill level is required for these crucial decision making tasks.

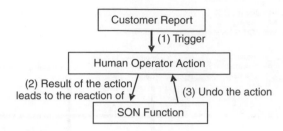

Figure 3.16 Action competition between operator and a SON function.

The embedding (cf. Figures 3.13 versus 3.14) of workflows formerly executed offline into the network operation implies that organisational boundaries (between a network planning, network operation, and network optimisation departments) are blurred or may even disappear. Additionally, as mentioned above, new roles for human operators are defined in a new organisational structure both requiring the proper change management.

Typically operators have fine-tuned their processes over the years and therefore they have been made very efficient. This may become as a hurdle for adopting some of the SON functionality, as in some cases it maybe difficult to show clear benefits from SON over those fine-tuned processes.

- **Conflict avoidance between human operator and SON actions:** There may be conflicts between workflows induced by a human operator (e.g. to perform changes triggered by a customer complaint) and automated workflows (realised by SON functions) such that a SON function reacts erroneously on an action performed by the human operator. For example, (Figure 3.16), in case a human operator tries to re-configure radio parameters of a cell to close a customer reported coverage hole and a SON function (e.g. Coverage and Capacity Optimisation: CCO) interprets the changes as sub-optimal and revokes them such that the coverage hole remains. Such problems can be treated in the framework of SON function interactions (cf. Section 9.1), where the human operator action can be seen as a specific (highest priority) SON function of its own.

3.6.2   Technical (Engineering) Challenges

In the following, technical challenges in the SON area are summarised. The relevant chapters in which those challenges are addressed are highlighted. In this book a structure according to the SON functional areas (configuration, optimisation, troubleshooting/healing) has been adopted, that is, SON functions within those respective areas are introduced. The functions can be seen as building blocks which can be employed in a rather uniform way across different networks. These functions can then be orchestrated (in different (operator-specific) ways) to realise the SON use cases as defined by NGMN (cf. Section 3.2). The use cases are then re-integrated into the operational processes as described above. Challenges are:

- Moving from offline planning/pre-configuration and optimisation to online SON (cf. Chapter 1 and Section 3.1):
  - **Event-orientation** also for CM, PM rather than only for FM. This characteristic has similarities to Business Process Management (BPM), however, the different requirements of Mobile Network Management need to be addressed (cf. Chapter 9).

- **Algorithms** need to be able to execute in an *incremental* way (event-triggered) (cf. Dynamic Radio Configuration, Section 4.2, 'one-shot'/trial-and-error optimisation, cf. Chapter 5).
- **Retro-fitting:** The choice is between adding SON solutions onto legacy systems (which may lead to inefficiencies in operation) versus re-designing system parts (which may lead to high implementation cost).
- **Degree of detail:** While offline network planning will still be necessary (e.g. for site planning) the required degree of detail of the planned data can be reduced, but a sensible solution has to be found therefore.
- Moving from centralised to distributed execution: *distributed algorithms* (cf. the general discussion in Section 3.4 and specific instantiations for self-optimisation, Chapter 5).
- *Acquisition and processing* of data:
  - Needs to be embedded into the regular system operation (rather than being a dedicated actions by humans), for example, drive testing, cf. Chapter 7 (this in particular includes the data acquisition from mobile terminals).
  - Needs protocol support and thus may need additional standardisation (e.g. X2 to support self-optimisation, vendor-specific DHCP options in self-configuration).
- *Knowledge management*:
  - **Acquisition of knowledge:** This is particularly difficult, for example, in the diagnosis part of self-healing (cf. Chapter 6): 'real world' knowledge on faults and their causes need to be acquired from human experts and embedded into the SON system.
  - **Representation and maintenance of knowledge:** Proper support in information models/ontologies is required (with implications to standardisation).
  - **Evolution of knowledge:** Learning methods allow the on-line acquisition of knowledge. Even if the actual operation is fully automated in a SON system, supervising the learning part may still require significant human actions (i.e. the human effort is shifted from the actual network operation to this supervision), cf. Chapter 11.
  - **Decision making:** Technologies like policies, probabilistic graph models, decision theory are required.
- *Security* (cf. auto-connectivity, Section 4.1): Security is relevant for SON with regards to the initial setup of the NE and its configuration, because the setup of the NEs security environment should not impair the level of automation reached. Setting up secure connectivity between a NE and its OAM system (as well as other NEs) is crucial because increasingly physical security (dedicated physical backhaul links, (macro) base stations installed in dedicated, locked cabinet) is replaced by virtual security (virtual backhaul links, pico/micro base stations installed in a public environment). Furthermore, automation means inherently reducing human interaction which from a security perspective means to reduce human monitoring of security-related aspects of the system. Hence, it is important to analyse if this reduction is acceptable and if a SON function even incurs new threat scenarios which in turn need to be addressed by security mechanisms.
- *Management* of the SON-enabled system (along the whole NE life cycle) (cf. Chapter 9):
  - **Instrumenting the system to ensure human operator control and maintainability** (cf. operational challenges): Logging, tracking, reporting. This also includes benchmarking the performance of a SON-enabled system versus a conventional one.
  - **Operability of the SON-enabled system:** As outlined above, the human operator is not required to work on low-level use cases any more (i.e the main part of work is shifted from

run-time operation to a design-time phase). However, the operability of these new higher-level tasks should also be assured (by providing the proper tools which allow reduction in the number of different operating policies and workflows to be maintained).

- *Reliability:*
  - Consistent behaviour in all network configurations and environments.
  - Controlled behaviour (e.g. escalation to human operator) in case of errors.
  - Feature interaction, (also considering the evolution of features over time) is addressed in Chapter 9. This relates to the interaction between SON features but also between legacy OAM features and SON.

References

3GPP TS32.500 (2011) Technical Specification Group System Architecture, *Telecommunication Management; Self-Organising Networks (SON); Concepts and Requirements*, ver.11.0.0., Release 11, 18 June 2011. Available from http://www.3gpp.org/ftp/Specs/archive/32_series/32.500/32500-b00.zip [accessed 30 June 2011].

3GPP TS32.50x (2011) series, Specification Group System Architecture, *Telecommunication management, Self-Configuration of Network Elements*, 3rd Generation Partnership Project (3GPP).

3GPP TS32.52x (2011) series, Specification Group System Architecture, *Telecommunication Management, Self-Organizing Networks (SON) Policy Network Resource Model (NRM) Integration Reference Point (IRP)*, 3rd Generation Partnership Project (3GPP).

3GPP TS32.53x (2011) series, Specification Group System Architecture, *Telecommunication management, Software Management (SWM)*, 3rd Generation Partnership Project (3GPP).

3GPP TR36.902 (2011) Technical Specification, Technical Specification Group Radio Access Network, *Evolved Universal Terrestrial Radio Access Network (E-UTRAN); Self-Configuring and Self-Optimizing Network (SON) Use Cases and Solutions*, ver.9.3.1., Release 9, 7 April 2011. Available from http://www.3gpp.org/ftp/Specs/archive/36_series/36.902/36902-931.zip [accessed 30 June 2011].

Bandh, T., Romeikat, R., Sanneck, H., and Tang, H. (2011) Policy-based coordination and management of Self-Organising-Network (SON) Functions. IFIP/IEEE Symposium on Integrated Management, Dublin, Ireland, May.

Heylighen, F. (2009) Self-Organisation, Principia Cybernetica Web. Available from http://pespmc1.vub.ac.be/selforg.html [accessed 15 June 2011].

Ising, E. (1925) Beitrag zur Theorie des Ferromagnetismus. *Zeitschrift für Physik*, Bd. **31**(S). 253–258.

Kasinger, H., Bauer, B., Sanneck, H. and Schmelz, C. (2006) A management automation framework for mobile networks. In Proceedings of the 17th World Wireless Research Forum, Heidelberg, Germany.

Kephart, J.O. and Chess, D.M. (2003) The Vision of Autonomic Computing, *IEEE Computer Magazine*, pp. 41–50.

IBM (n.d.) Autonomic Computing, http://www.research.ibm.com/autonomic/ [accessed 1 August 2011].

IASB (2003) International Accounting Standards (IAS) 16 – Property, Plant and Equipment.

NGMN (2007) NGMN Informative List of SON Use Cases, NGMN Technical Working Group, Self Organising Networks, (ed. F. Lehser), April.

NGMN (2008) Use Cases related to Self Organising Network, Overall Description, NGMN Technical Working Group, Self Organising Networks, (ed. F. Lehser), December.

NGMN (2010) Top OPE Recommendations, NGMN P-OPE PROJECT, (ed. F. Lehser), September.

Sanneck, H., Bouwen, Y. and Troch, E., (2010) Dynamic radio configuration of self-organizing base stations. 7th International Symposium on Wireless Communication Systems, York, September.

Sanneck, H., Schmelz, C., Bandh, T. *et al.* (2010) Policy-driven workflows for mobile network management automation. In 6th International Wireless Communications and Mobile Computing Conference, Caen, France.

4

Self-Configuration ('Plug-and-Play')

Henning Sanneck, Cinzia Sartori, Péter Szilágyi, Tobias Bandh, Christoph Schmelz, Yves Bouwen, Eddy Troch, Jürgen Goerge, Simone Redana and Raimund Kausl

In wireless access networks, the rollout of new network elements (NEs) or changes to the NE HW and SW cause considerable (re-)planning and configuration effort across the different deployment scenarios (cf. Section 2.1.8). The total number of NEs (in particular base stations) to be installed and configured is thus a significant expense factor. With the extension of existing networks and the introduction of radio access technologies like LTE the cell size will decrease (for capacity-driven deployment scenarios, cf. Sections 2.2.2/2.2.3), thereby raising the total number of cells and increasing the overall deployment cost. As outlined in Section 3.3, SON for LTE is still targeted to 'infrastructure networks', that is, while parts of the configuration may be generated on the fly or 'acquired' for example, from neighbouring base stations, still the control of an operator via a central Operation, Administration and Maintenance (OAM) system plays a significant role.

For LTE-Advanced (cf. Section 2.1.8) the importance of self-configuration is increased even further as the density of NEs in an area is again increased. Furthermore the NEs which are deployed in addition to an already existing macro layer (like pico base stations and relays) are inherently low cost devices. Hence, the absolute cost of installation and configuration per NE must be even lower than for the macro case. Also, for such deployments, the physical security of the NEs (like being well protected in a locked cabinet), as well as the feasibility of manual security-related processes are reduced. Taking this together with the presence of an IP-based, multi-purpose backhaul, it is even more important to integrate the initialisation of network security into the self-configuration process.

LTE Self-Organising Networks (SON): Network Management Automation for Operational Efficiency, First Edition.
Edited by Seppo Hämäläinen, Henning Sanneck and Cinzia Sartori.
© 2012 John Wiley & Sons, Ltd. Published 2012 by John Wiley & Sons, Ltd.

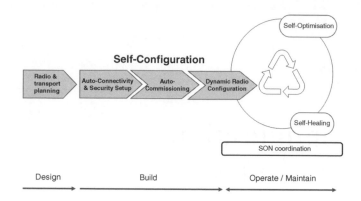

Figure 4.1 Self-configuration phases within the SON lifecycle (Sanneck *et al.*, 2010b). Reproduced with permission of © 2010 IEEE.

Self-configuration is the process of bringing a new network element or network element parts into service with minimal human operator intervention. The process encompasses three phases as illustrated in Figure 4.1. After the base station has been installed and switched on, the *auto-connectivity setup* (Sanneck *et al.*, 2007) function establishes a secure connection between the eNodeB and the network's OAM system. *Auto-commissioning* includes the automated provisioning and testing of the software and NE configuration data according to the installed hardware and required NE functionality for a specific site. Finally, *dynamic radio configuration* (Sanneck *et al.*, 2010b) complements the configuration data downloaded in the commissioning phase by generating and acquiring radio configuration parameters on the fly, that is, the NE adapts its configuration to the current state of the network deployment, thereby enabling an 'incremental' way of deployment eliminating the need for re-planning at every deployment step.

4.1 Auto-Connectivity and -Commissioning

Auto-connectivity and -commissioning aims at automatically connecting the NE to its Domain Management (DM) system, thereby avoiding manual intervention in the process as much as possible. The process also has to incorporate the security setup of the NE (and again avoiding as much as possible manual tasks due to security requirements). This is achieved by shifting manufacturer and operator activities to a preparation (rather than the actual rollout) phase and eliminating the interaction between them as much as possible (cf. Figure 4.2): the NE is delivered only with a minimal 'off-the-shelf' software and configuration installation. Only at the point in time when the NE is placed on site, the NE hardware-to-site mapping is executed. Together with authentication of the NE by the network it is possible to enable a very flexible and secure roll-out process. As can be seen from Figure 4.2 the flexibility gain comes from the reduced 'lead time' of the rollout process, that is, the network planning and corresponding NE configurations can be adapted to the very last minute until the actual physical rollout of NEs to sites starts.

The automated establishment of OAM connectivity aims at reducing the need for costly on-site personnel to enter initial configuration parameters. This is achieved by having the

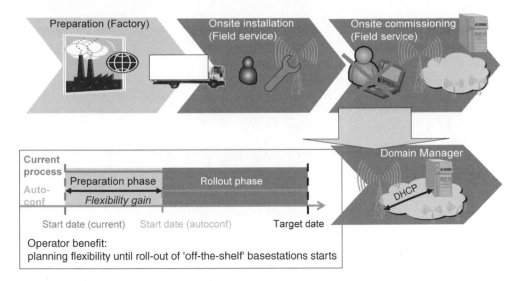

Figure 4.2 Auto-connectivity and –commissiong: high-level process (Sanneck *et al.*, 2007). Reproduced with permission of © 2007 IEEE.

eNodeB to acquire the initial parameters from a DHCP server and an Auto-Connection Server (ACS) in the following consecutive steps (Sanneck *et al.*, 2009):

- **Setup of basic connectivity:** The NE acquires an initial IP configuration for communication with the ACS, the operator's Certification Authority (CA) server and optionally a Security Gateway (SEG) by means of the Dynamic Host Configuration Protocol (IETF RFC 2131). The initial IP configuration may be replaced by a permanent OAM IP configuration during a subsequent auto-configuration step.
- **Initial secure connection setup:** Security is based on public keys for authentication, integrity and confidentiality protection. The NE will use a certificate for its communications with other NEs (directly and/or via the SEG). This certificate is generated by the operator's CA and needs to be downloaded to the NE before the secure link to the ACS can be established. Furthermore the NE needs an operator trust anchor (root certificate) for verifying the certificates received from other NEs while in normal operation. When the NE is delivered from the vendor's factory initially the operator certificate and trust anchors are not available. Thus, the NE uses its factory deployed vendor certificate to authenticate itself against the operator's CA for the operator certificate enrolment. As result of the successful automated credential enrolment the NE owns its operator certificate and the whole trust chain (operator root CA certificate and optional sub-ordinate CA certificates) is established. The NE is now able to establish a secure connection by IPSEC and/or TLS to the ACS (or the SEG) to continue with auto-connection.
- **Site identification** is required to define which off-line prepared configuration data shall be used for the NE. Typically dedicated configuration data for every site is required.
- **Download of final configuration and transport parameters.**
- **Secure connection setup with Domain Manager (DM):** The temporary secure connection with the ACS is torn down and a new connection is established. This connection is protected

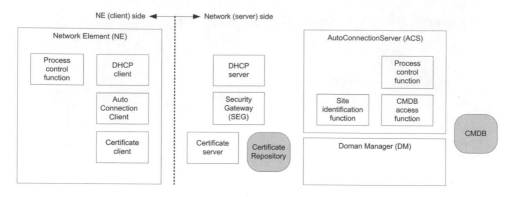

Figure 4.3 Auto-connectivity and security setup: entities.

by TLS and/or IPsec according to the operator's preference. Note that the implementation of IPsec to protect OAM connections is mandatory in 3GPP (except for cases where protection has already been established, e.g. by physically secure links).

The auto-connection and security setup is performed by self-configuration entities residing both at the NE (client) and the network (server) side, see Figure 4.3. The client side entities retrieve initial configuration data from the server side entities.

The self-configuration entities on the NE side fulfil the following functions:

- **Process control:** Control of the auto-connectivity and -commissioning sequence of steps;
- **DHCP client:** Retrieval of the initial IP configuration;
- **Certificate client:** Retrieval/storage of the NE certificates;
- **Auto-connection client:** Establishment of secure connectivity to the auto-connection server, registration request.

On the server side the following entities are present:

- **DHCP server:** Supplies the initial IP configuration and IP addresses of server-side functions.
- **ACS:**
  - **Site identification function:** Maps the presence of a specific NE (with a unique hardware identity) to a specific site (site-ID).
  - **Configuration Management Database (CMDB):** Contains the network configuration information (sites, topology) derived from the network planning.
  - **CMDB Access Function:** Supplies the initial configuration data and registers the hardware identity of the installed NE.
- **CA Server:** Authenticates the NE and provides certificates.
- **Certificate repository:** Holds the operator trust anchor and the NE certificates created by the operator Certification Authority for all NEs in service.
- The **Security Gateway (SEG)** separates the secure operator domain (containing the operator's OAM system and Evolved Packet Core: EPC) from the transport network domain to which the NE are physically attached.

4.1.1 Preparation

In order to support the auto-connection process some pre-configurations are needed by the manufacturer and the mobile network operator.

4.1.1.1 Preparation Activities by the Manufacturer

- The NE is assigned a hardware identity (HW-ID) which is a serial number of a HW module used by the manufacturer to identify the NE-HW for purchasing and service purposes. In addition the HW-ID may be used as a basis for creating vendor NE identities used in the NE vendor certificates. The HW-ID should be assigned for the lifetime of the NE-HW and should not be altered (yet, the HW-ID management is of course vendor-specific).
- NEs shall be shipped from the factory at least with an initial software and configuration data supporting the initial boot, the auto-connectivity and security setup and basic auto-commissioning. All other software packages, dependent on the hardware configuration and the mobile network where the base station is installed, may be downloaded on-site as part of a subsequent auto- or manual commissioning step. Note that this may even extend to the specific radio technology, that is, a generic, multi-RAT-capable basestation is shipped and then becomes a 3G or LTE basestation via the on-site software download and configuration.
- Vendor key/certificate generation and installation: During a certain manufacturing step the NE generates its own public/private key pair and stores it locally within a trusted environment (TrE), such that the private key does never need to be revealed outside the NE. The NE then requests a certificate for its public key from the factory CA and stores the certificate locally (vendor NE certificate including the vendor NE identity). When the NE is sold to an operator, the factory CA certificate used for signing the NE vendor certificates is handed over (in a secure way) to the operator and deployed into the operators CA pool of trust anchors. The generation of keys and the secure storage of the keys in the NE happens within the TrE which may be realised by a SW-based key storage (security by obfuscation) or HW-based solutions like a Trusted Platform Module (TPM), standardised by the Trusted Computing Group.

Pre-configuration of the NE in the factory with any operator- or site-specific parameters or software is avoided. This simplifies the manufacturing processes and allows an NE to be shipped to any customer and site.

4.1.1.2 Preparation Activities by the Operator

The operator preparation activities need to be split in tasks that are needed per project and per NE. A project could involve, for example, the rollout of a set of new base stations to improve radio coverage or network capacity. Per project preparations need to be executed only once for the auto-connection of a large number of NEs. Only few additional preparation activities need to be introduced for the auto-connection feature. These preparations are replacing the configuration activities that would otherwise be performed by installers and (remote) commissioners. The operational costs for the additional preparations are negligible in comparison to the operational cost savings of the site installers and commissioners.

Operator Preparations per Project

- Transport network preparation deals with the configuration of network equipment (e.g. microwave transmission) in the 'last mile' of the transport access network, as well as the configuration and traffic engineering of virtual LANs and Virtual Private Networks in the aggregation and core transport networks. Where the operator deploys third party transport networks this preparation activity is outsourced to the corresponding service provider. It should be noted that this (while being a prerequisite for the basestation self-configuration) is a use case of its own which itself can be heavily automated dependent on operator requirements).
- DHCP server configuration entails the provisioning of the initial IP configuration for the NEs, Auto-Connection Server (ACS) address, operator CA server and SEG addresses to be used by the NEs during an auto-connection/self-configuration. The ACS, CA server and SEG addresses are vendor-specific DHCP options which are only interpreted by NEs with the auto-connectivity feature implemented. Most commercial DHCP servers allow for the configuration and distribution of vendor-specific DHCP options.
- Auto-Connection Server configuration:
  - Addresses are configured to enable access to and from the ACS.
  - Installation of the operator root certificate (only required in case the ACS terminates a secure (TLS) connection from the NE).
  - Access to the Configuration Database is needed for retrieving auto-configuration data and updating the network topology.
- SEG: installation of the operator root certificate.

Operator Preparations per NE

- A transport and radio network planning needs to be performed prior to the installation, to allow for the insertion of new NEs in a network. This is not changed when auto-connection will be deployed (the amount of radio planning can be reduced in fact by the Dynamic Radio Configuration concept presented in the next section, however).
- Access Network Element planning is an activity of the operator's planning department supported by specialised planning tools. For the planning process, operators identify NEs by means of the 'Site-ID'. A Site-ID identifies the unambiguous planned (geographic) location of an NE, and is used as identifier for the NE configuration data. It may contain a postal address in conjunction with further location designations, for example, 'Site: Claudiusstr.1, 10557 Berlin Rack:x' and may include GPS coordinates.
- Transfer of planning data to the DM: the auto-connection process requires the output of the site planning (part of the network planning) for the site identification task. With the Site-ID it is then possible to find the final IP transport configuration and further configuration for the NE. Hence, as a preparation the planning output must be transferred from the planning tools to the DM. In many cases this is handled by a multi-vendor configuration management preparation tool. The operator may need to start a task on the DM to retrieve the planning files, or the DM may be ordered via a standardised interface to download the files. In particular the Site-ID is important for base stations in a mobile network as it refers to the geographic location and thus is inherently part of the radio planning done by the mobile network operator.
- Auto-Connection Server preparation: After the Domain Manager has stored the radio network planning files a subset of the data shall be extracted and transferred to the ACS.

In case the planning files are received by means of the 3GPP Bulk Configuration Management IRP (Integration Reference Point), this function can be automatically invoked. This way the data is readily available at the ACS for processing auto-connection requests.

- CA server preparation: every time a set of NE is purchased and delivered, the operator may download a list with the vendor NE identities (as contained in the NE vendor certificates), for example, from the vendor's web interface for customers. If an NE is asking for an operator certificate in the actual auto-connection procedure, the operator CA may compare the vendor NE identity received from the NE during certificate enrolment with the vendor NE identity of ordered and delivered NE and only assign a NE certificate if the identities match.

On-Site Activities by the Installer

- The main tasks left for the installer are the hardware installation of the NE and the physical connection of the NE to the antenna, the transport network and the power feed.
- Then the installer powers-on the NE and monitors from LED indications on the NE, whether the auto-connectivity setup proceeds successfully. When this is not the case, the installer shall call for assistance from a 'remote commissioner' for analysing the problem (very likely a preparation task has not been properly executed). Otherwise the installer can leave the site. Further configuration of the NE is handled by the remote commissioner or automatically by auto-commissioning.
- There is no longer a need for a commissioner (who has to be skilled and trusted by the operator) to travel on-site, connecting a laptop and entering the initial configuration parameters. This activity can be partially removed and partially shifted to Dynamic Radio Configuration, cf. Section 4.2 and the mentioned, cost-efficient 'remote commissioner' supervising the whole process.

4.1.2 Connectivity Setup, Site-Identification and Auto-Commissioning

The message flow in Figure 4.4 illustrates the steps of the authentication and security setup, more specifically for the case with automated site identification. Note that there are different alternatives on how to perform each step (e.g. employing standard DHCP versus using a proprietary protocol fulfilling the same requirements). The alternative which is included in the description is believed to show the best characteristics in the given scenario (e.g. DHCP being a proven, standard-based protocol with reliable implementations being readily available). The contained certificate enrolment procedure is standardised in (3GPP TS 33.310, 2010) and explained in detail in (Horn *et al.*, 2010).

Basic connection setup: When powered-up by the installer the NE executes a self-test and then starts the auto-connection process. The NE sets up Layer 2 (transport) connectivity and sends a DHCP request over the link. The transport connectivity needs to correspond to the transport network pre-configuration as mentioned above. For LTE typically this will be a default VLAN established across the transport network (yet the transport connectivity may also employ own auto-connectivity features). A DHCP server assigns an IP address from a reserved pool of addresses and replies to the NE with its initial IP configuration, ACS and CA server addresses and (optionally) the Security Gateway (SEG) address. If IPsec in tunnel mode is used, the initial IP address provided by DHCP is used as both 'inner address' for the operator network domain and an 'outer address' for the tunnel itself ('collapsed' IP address).

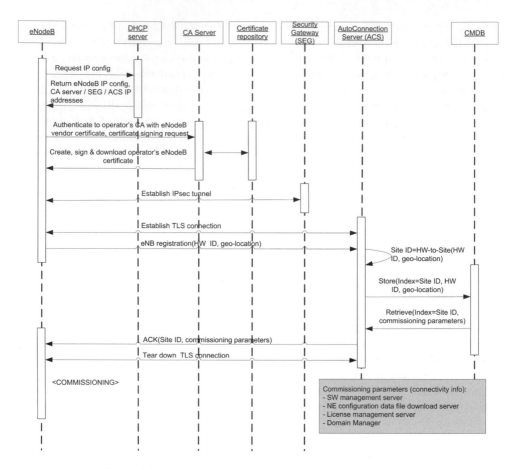

Figure 4.4 Auto-connectivity and security setup: message flow.

Certificate enrolment: The NE creates a new key pair and communicates with the CA server
using the Certificate Management Protocol (CMP) to get the public key portion signed by a
new operator certificate for this NE and to download further trust anchors such as the
operator's public signing CA certificate. If the list of serial numbers of ordered and delivered
vendor NE identities was downloaded before from the vendor the CA may verify optionally
if the NE belongs to the pool of ordered NE because the vendor NE identity is as well sent
within the CMP protocol exchange to the CA. Once the NE knows the final operator node
certificate to be used, it can establish secure connections to all of its communication partners
via IPsec or TLS (dependent on the properties of the certificate(s)) with the correct operator
node identity.

Secure connection establishment/authentication: Depending on the operators security
architecture the NE may establish an initial IPsec tunnel to the Security Gateway (at the
operator trusted network border) and/or a TLS connection between the NE and the ACS. In
case both IPsec and TLS are used the TLS connection is tunnelled within the IPsec tunnel.
The Security Gateway and/or the ACS verify the operator-signed NE certificate using the
operator trust anchors. During the handshaking the NE also verifies the security gateway

and/or ACS supplied TLS certificate by means of the operator trust anchors. After mutual authentication the peers start encrypting/decrypting all messages over that connection.

Site identification: The NE sends an announcement message to the ACS including the HW-ID and if available, the measured GPS coordinates. The actual site identification is performed at the ACS by matching the information supplied by the NE with the pre-configured information at the ACS. Then the ACS communicates with the Configuration Management Database (CMDB) for storing the HW-ID in the on-line configuration database related to the Site-ID, for updating the network topology database with the new NE and for retrieving some parameters for the further (auto-)commissioning.

The site identification task of the ACS involves the determination of the exact site where an auto-connection request is being issued. Once the site has been identified the hardware identity of the installed NE is linked with the Site-ID in the configuration database. This is further referred to as 'HW-to-Site mapping'.

Different mechanisms can be used for the site identification, ranging from solutions with some human assistance up to a fully automated one (Bandh and Sanneck, 2011), see Table 4.1:

- Before the self-configuration process is started the unique Site-ID may be entered manually by the installer and is then transferred with an auto-connection request to the ACS.
- Before the self-configuration the installer informs a remote commissioner about the HW-ID and Site-ID of the new NE. The remote commissioner then enters the HW-ID/Site-ID mapping in the configuration database of the Domain Manager. During the auto-connectivity setup the ACS queries the configuration database with the HW-ID to find the corresponding Site-ID.
- Semi-automated variant of the previous solution: the installed HW-ID is collected on-site from a sticker by means of a PDA with integrated bar code reader. The installer then also

Table 4.1 Site identification and HW-to-Site mapping options

Info Insertion	Manual	Semi-Automated	Automated
On-site	Planned Site-ID entered manually by installer	Measured Site Location using mobile GPS receiver (PDA) plugged into NE via Eth	Measured Site Location using fixed antenna mounted GPS receiver
Remote	HW-ID and planned Site-ID received in human-readable form via paper or phone call from installer, HW-ID entered manually into database	HW-ID read with barcode scanner into PDA, Site-ID entered into PDA, received from installer via SMS/ Web interface, HW-ID entered automatically into database	n.a.

enters the Site-ID and initiates an SMS, e-mail or web session for updating the configuration database. In contrast to the previous solution there isn't any remote commissioner support needed.

- The installer collects the geographic coordinates of the site by means of a handheld GPS receiver and transfers them to the NE via a proper interface (e.g. Ethernet). The ACS maps the GPS parameters to the Site-ID by means of a geographic matching algorithm.
- The site coordinates are automatically captured by means of a fixed antenna-mounted GPS receiver. This solution is cost effective when a permanent GPS receiver is needed at the base station for radio synchronisation purposes.

The actual matching/mapping of the IDs is always done at the server side obviously (matching of a measured location against the planned ones yielding a Site-ID, mapping from HW- or Site-ID to the correct planning data).

- **Provisioning of auto-commissioning information to the NE:** The ACS supplies the Site-ID and the auto-commissioning parameters within a reply (ACK) of the announcement message. In case the NE has communicated with the ACS over a separate secure (TLS) connection, the NE then tears down that connection and establishes a new secure connection to the Domain Manager/auto-commissioning servers. The NE is now ready for the actual commissioning.
- **Auto-Commissioning:** Figure 4.5 shows the considered steps within the self-configuration phases introduced in Figure 4.1.

After the initial boot self tests have been completed the connectivity to the OAM system is established immediately. Only then, the process can continue with auto-commissioning.
Auto-commissioning consists of:

- **An automatic inventory update:** The internal (installed HW boards) and external (antenna) components of a base station are identified automatically. The acquired data is sent to the OAM system where the CMDB for the NE is updated (and the data can be validated against the planning for the specific site).
- **Automatic software download:** The NE triggers the validation of its current SW version against the SW version requirements for the particular NE type and site imposed by the OAM system. Note that these requirements may even include the specific (multi-)RAT-type of the NE. As it is feasible due to the described process to install NEs only with default SW in the factory, usually the (latest) actual SW required for live operation is installed at this point (relatively late in the process).
- **Automatic Database Preparation and Download:** The basic configuration for the site/the NE is downloaded by the NE. The data has been prepared during the radio and transport planning process.
- The updated SW and configuration data is activated (e.g. by rebooting the NE).
- (Complementary to the 'Automatic Database Preparation and Download', radio parameters prepared on the fly at the time of installation are configured at this point; this is considered to be the separate phase of 'Dynamic Radio Configuration': DRC, cf. Section 4.2).
- After the NE is fully installed and configured, the corresponding license management procedures are performed.

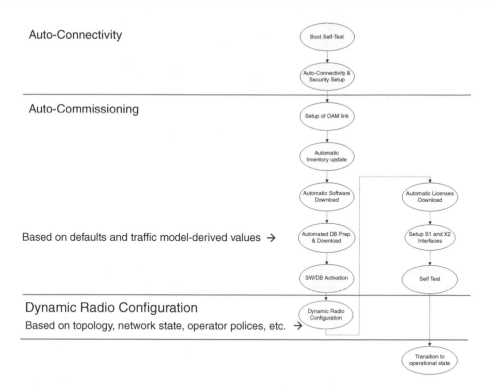

Figure 4.5 Self-configuration steps within the phases (see Figure 4.1).

- Finally, the call processing interfaces are established and the NE performs a final self-test.
- Dependent on the configured policies, the NE can now transit to operational state or a dedicated step with human intervention (acceptance into the network) is performed.

In contrast to auto-connectivity, the actual functions like SW management, configuration database preparation are typically legacy functions where the actual difference is only that they are triggered automatically (by the NE or the OAM system) as part of the auto-commissioning sequence.

It should be noted that the described sequence of auto-commissioning steps is not fixed, just as little standardised, as it is highly vendor-specific. However, there are some logical dependencies which advocate a particular sequence:

- SW installation may require knowledge on the installed HW components (i.e. a preceding auto-inventory step).
- NE database configuration may have dependencies to a specific SW version, or require a specific SW version to be running (i.e. a preceding SW download step).
- License management as well as the setup of interfaces requires the NE to be fully configured (i.e. preceding database download and DRC).

The described sequence of phases (Figure 4.1) and steps within those phases (Figure 4.5) needs some (machine-level) supervision, for example, to react in case of errors within one step.

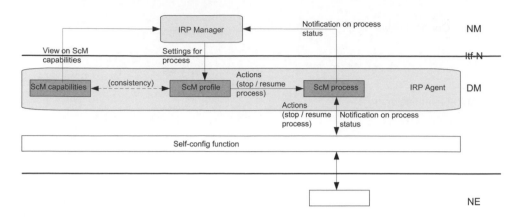

Figure 4.6   Self-configuration Management (ScM) IRP.

The corresponding control logic can be implemented at either the NE itself or the OAM system. Conceptually this control logic is part of the general 'SON operation' functionality (cf. Chapter 9).

While the exact sequence and details of the individual steps are not standardised (except for the authentication and certificate exchange procedure), 3GPP has standardised a Self-Configuration and Software Management Integration Reference Point (IRP) to enable high-level, multi-vendor-capable supervision of the self-configuration (3GPP TS32.501, 2010). This supervision relies on the capability to select 'stop points' before execution of a step in the self-configuration process. At a stop point the self-configuration process is suspended until the supervising entity requests to resume the process.

Figure 4.6 show the Self-configuration Management (ScM) objects exposed by an IRP agent which can be accessed by an IRP manager over Itf-N (3GPP TS32.501, 2010).

The objects contain the following functionality:

- *ScManagementCapability:* This object contains the sequence of the self-configuration steps and the possibility to select a stop point before executing the step. Also, the setting of different administrative states (locked/unlocked) at the end of the self-configuration process is offered.
- The *ScManagementProfile* contains the 'policy' of IRP manager wrt. which of the possible stop points offered in ScManagementCapability are actually used to suspend the self-configuration process of the specified NE type. Also the desired final administrative state can be selected.
- In the *ScProcess* the actual status of the self-configuration process for an NE is contained. Thus, the IRPManager can be notified about the progress of the self-configuration process. ScProcess is created automatically when the start of a self-configuration process is indicated to the IRP Agent.

In the preparation phase an IRP manager acquires the information on the self-configuration steps (ScManagementCapability) and then sets ScManagementProfile based on the NM-level requirements (e.g. full automation versus coordination of the steps being executed at different DMs). During 'run-time', that is, when NEs are actually being self-configured, the progress of the self-configuration can be monitored. At the pre-defined stop points the IRP Manager can take actions and then resume or abort further execution.

The Software Management IRP (3GPP TS32.532, 2010) is basically identical to the ScM IRP (i.e. the SW Management is considered to be an own 'process' within the self-configuration process). Similarly to the ScM IRP, a set of stop points are offered, selected and enforced. In addition, 'to-be-installed SW version' information is managed in a similar way (i.e. offered, selected and enforced).

4.1.3 LTE-A Relay Auto-Connectivity

Relay nodes (RN) are mobile network base stations, which connect to a radio network via an in-band wireless backhaul link instead of using a dedicated wired or wireless backhaul link, such as Ethernet or microwave link (cf. Section 2.1.8.3). Connecting through a wireless backhaul (i.e. the relay's regular air interface) instead of a dedicated link raises difficulties when it comes to accessing the mobile network operator's OAM system, since prior to the appearance of relay nodes in LTE-Advanced, wireless access has been reserved solely for user equipments and not used by network elements at all. However, after deploying a relay node, establishing an initial OAM access is essential in any kind of configuration, particularly when considering an automated configuration process according to the self-configuration principle.

Due to the demand for deployment costs to be minimised (Lang *et al.*, 2009), procedures related to deploying network elements should be automated to avoid costly human intervention. It is anticipated that a large total number of relay nodes will be deployed relative to the number of deployed regular LTE eNBs; hence, automation is also a requirement with regard to scale. Furthermore, it is expected that physical relocations may take place much more often for RNs than for regular eNBs in order to deal with changing operator requirements. Therefore, automation of initial configuration is perhaps even more important for RNs than for regular eNBs.

The auto-connectivity and -commissioning method outlined in Sections 4.1.1 and 4.1.2 can be used by all kinds of eNBs, but it is not suitable for RNs without modification. The reason is that the RN, being a wireless node, has no Layer-2 network to just being plugged in. Moreover, the Un interface (cf. Section 2.1.8.3) can only be used between the RN and its associated Donor eNB (DeNB) when the RN is in operational state, that is, when the RN has finished negotiating the resource partitioning with the DeNB concerning its relay link and access link and the RN is ready to accept incoming UE connections. When the RN is initially powered up, it is not yet recognised and authenticated as a network element, so the RN has to use the only access method over the radio interface that is standardised, namely the Radio Resource Control (RRC) establishment followed by the UE Attach procedure (3GPP TS23.401, 2010) (its purpose and what is exactly established will be described later). Using a standardised access method is important to avoid any standardisation impacts that would be an obstacle to relay self-configuration. For the UE Attach procedure to work, however, the RN needs to be able to act as a UE and have a UE identity.

Another concern is which eNB can or should be used by the RN to make its initial access through the UE Attach procedure to the network. One obvious choice would be the DeNB, since it is the eNB through which the RN will connect to the network after it has been configured. This choice, however, may not be fortunate or even feasible, as detailed in the list of choices in the following paragraph.

4.1.3.1 DeNB Selection Strategies

To understand the obstacles, consider the possible DeNB selection strategies based on where the decision for selecting the DeNB is made:

1. The DeNB is selected offline by the operator during network planning and the DeNB association is provisioned into the OAM system as part of the planning data. The selected DeNB association is downloaded when the RN is deployed and the configuration (self-configuration) takes place.
2. The DeNB is selected dynamically by the OAM system during the deployment process based on data acquired from the RN (e.g. the strengths of received radio signals or the signal to interference and noise ratio from different eNB) and additional knowledge about the network that is present at the OAM level but not at the network element level (e.g. planned locations of eNBs). The selected DeNB association is downloaded to the RN after it is computed in the OAM. The advantage of this approach is that the decision can be based on up-to-date parameters and a view of the current actual network status (i.e. not just the planned state which may differ particularly during the initial deployment process).
3. During the deployment process, the RN itself selects the DeNB. The advantage of this strategy is that it is scalable, since the decision is not escalated up to the OAM system but kept locally. The disadvantage (compared to the previous case) is that the decision can only be based on the knowledge present or obtainable at the RN, which usually does not have as broad view of the overall network as a central OAM system has. Note that currently there is no mechanism standardised that a RN could learn if an eNB in its vicinity is able to act as donor eNB.

It is clear that in the first two cases of the DeNB selection strategies, the RN does not have prior knowledge about its DeNB association before it has already connected to the OAM system. On the other hand, since the RN is a radio node, it has to connect to an eNB in order to access the core network and the OAM system even for its initial configuration. Therefore, the highest flexibility of choices is preserved if an intermediate eNB is used that can provide the RN with a connection to the OAM system for the time of its initial connectivity. This eNB will be referred to as the 'configurator eNB' or 'CeNB' for short (see Figure 4.7). After the self-configuration is

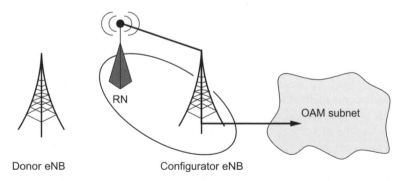

Figure 4.7 Initial connection though the configurator eNB (phase 1) (Szilágyi and Sanneck, 2011). Reproduced with permission of © 2011 IEEE.

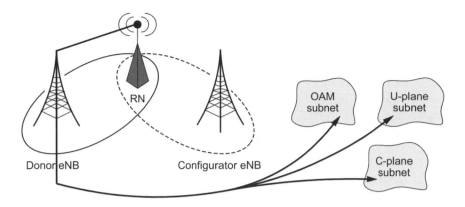

Figure 4.8 Final connection through the donor eNB (phase 2) (Szilágyi and Sanneck, 2011). Reproduced with permission of © 2011 IEEE.

completed through the configurator eNB (Phase 1), the RN can switch over to the DeNB (Phase 2) and enter operational state (see Figure 4.8).

4.1.3.2 Configurator eNB

The configurator eNB can be any of the accessible eNBs that provide coverage at the location of the RN. There is no special requirement on an eNB that is chosen by the RN as configurator, so the choice can be based on any suitable measure, for example, the eNB with the best signal to interference and noise ratio (SINR). Note that it is not true for DeNBs: a regular LTE eNB needs at least software upgrade in order to become donor-capable, hence, not all available eNBs are automatically potential donor eNBs too. It is unlikely and practically impossible that a network operator upgrades all eNBs overnight to support relay nodes in one step. Also, operators may choose only a subset of eNBs to act as donors anyway to reduce costs.

Later, when the RN has downloaded the initial configuration data from the OAM system and it is aware of its DeNB association, the OAM connection though the CeNB is torn down and re-established via the DeNB. Even if the CeNB and the DeNB are identical, the re-establishment cannot be avoided due to security requirements (different identities for the same node in the two phases of the configuration).

The configurator eNB concept has been accepted as standard by the 3GPP under the name of 'two-phase configuration', with Phase 1 corresponding to the initial attach of the RN to the CeNB and Phase 2 consisting of the establishment of the final connectivity through the DeNB. For DeNB selection, the second approach outlined in Section 4.1.3.1 is under consideration in 3GPP, with the addition that the RN receives a list of potential DeNB targets from which it can choose one based on other criteria (e.g. the one with the best signal strength is an obvious choice).

4.1.3.3 RN Auto-Connectivity Process

Now that the core configurator eNB concept has been presented, the details of the RN auto-connectivity process that is wrapped around this concept are presented.

The only way for the RN to connect without introducing new access methods is to first act as a UE via the Uu interface. Hence, relay nodes have two functions: a UE function and a RN function. This duality is also reflected in the two subscriptions (security identities: Universal Subscriber Identity Module USIM) needed for the RN, that is, USIM-INI for the initial connectivity as UE in Phase 1 and USIM-RN for the connection as a RN in Phase 2. The two USIMs can co-exist on the same universal integrated circuit card (UICC) but only one of them may be activated at a given time.

The auto-connectivity of RNs builds on some prerequisites needed to technically enable RN self-configuration in the first place. The following preparations are required to enable the actual auto-connectivity:

1. A specific Access Point Name (APN), which will be referred to as 'APN-OAM', defining the OAM subnet as a Packet Data Network (PDN) is configured in the Home Subscriber Server (HSS, cf. Section 2.1.6.2) subscription data as the default APN for the USIM-INI subscription. This APN will be used by the RN in Phase 1 to access the OAM system. For the USIM-RN subscription used in Phase 2, a different APN has to be used that will be selected by the PDN Gateway (P-GW, cf. Section 2.1.6.4) co-existing with the DeNB.
2. There is a P-GW hosting the APN-OAM, so that the P-GW is configured to be able to route traffic to the OAM subnet. The configuration of an APN at a P-GW is a standard procedure, only a new APN number has to be selected, which can be operator-specific. For Phase 2, the special P-GW serving the RN is the one co-located with the DeNB according to the standardised relay architecture.
3. The P-GW utilised in Phase 1 must have DHCP relay (IETF RFC 2131) functionality using USIM-INI to start the auto-connectivity process.

During the UE Attach procedure in Phase 1, the UE function of the RN indicates within the standard Protocol Configuration Options (PCO) of the Initial Attach message element that it prefers to obtain the IP address later with DHCP. Thus, the P-GW performs no automatic address allocation as part of the bearer establishment, as it would do without such an indication. The MME will use the APN-OAM retrieved from the HSS to connect the RN to the right P-GW. At this point, the RN has physical (Layer 2) connectivity to the OAM subnet, but does not have IP (Layer 3) connectivity yet, so the next step is IP connectivity establishment.

In order to establish IP connectivity with the OAM system, the RN now follows the conventional DHCP protocol procedures (IETF RFC 2131), similarly to the procedure described in Section 4.1.2 (cf. Figure 4.9).

The DHCP client on the RN sends a broadcast DHCP DISCOVER over the radio link to the eNB. At the CeNB, the DHCP DISCOVER message is sent over the established GTP tunnel to the S-GW, which forwards it further to the P-GW. At the P-GW, the DHCP DISCOVER message is received by the DHCP relay process which sends the message to its configured DHCP server in the OAM subnet. At the DHCP server in the OAM subnet, the message is processed. In the successful case that an IP address can be offered to the RN (e.g. there are enough free addresses in the address pool), a DHCP OFFER is sent back from the DHCP server to the DHCP relay which in turn sends it back to the RN (together with the ACS, CA server and SEG IP addresses as described in Section 4.1.2).

At the RN, the received IP address is configured and used further for OAM-related traffic (Figure 4.9). The layer 'IP-m' is referring to connectivity (IP addresses, routing entries, etc.) with the OAM subnet. Note that the depicted 'OAM' layer is the OAM protocol stack above IP

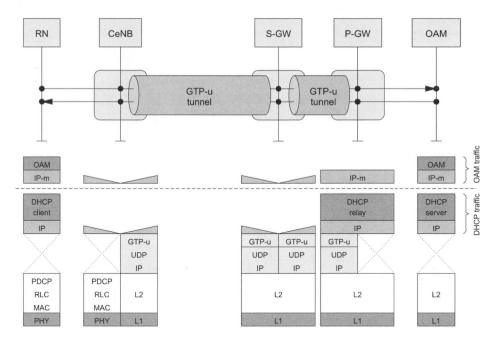

Figure 4.9   DHCP and OAM protocol stack (Szilágyi and Sanneck, 2011). Reproduced with permission of © 2011 IEEE.

for the southbound management interface but any IP traffic could be exchanged over the established connection of course. Additional information received in the DHCP OFFER message is processed and stored in the RN for further usage.

Figure 4.10 shows an overview of the LTE backhaul configuration depicting the relevant nodes and connections from RN auto-connectivity point of view. The figure portrays the common subnet separation scheme that puts the OAM system, the control plane nodes and the user plane nodes into separate IP subnets. The solid thick line indicates the first part of the OAM access path, from the RN to the P-GW. This path (which is the same for any ordinary Uu user plane traffic), apart from the initial RN-CeNB hop (which is the in-band wireless relay backhaul), goes inside the user-plane GTP tunnel that has been established between the CeNB and the P-GW as part of the UE Attach procedure (see Figure 4.9). For the path from the P-GW to the OAM system, three alternatives are shown, denoted by (a), (b1) and (b2). Alternative (a) is accessing the OAM subnet as an external PDN (just like e.g. an IP Multimedia Subsystem (IMS) would be accessed). Alternatives (b1) and (b2) are accessing the OAM subnet through the same provider edge routers that are used to maintain the subnet separation. The difference between these two alternatives is that (b1) does not go through the SEG once more while (b2) does.

The choice between the alternatives (a) and (b1) or (b2) can be a matter of preference or policy from the network operator's side but in fact (a) should work regardless of the network configuration: since all nodes, including P-GWs, need to access the OAM subnet for their own OAM purposes, a P-GW can also provide access to the OAM subnet for UEs (provided that the proper APN-OAM has been configured at the P-GW).

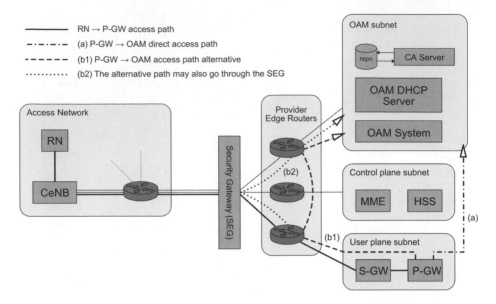

Figure 4.10 LTE backhaul subnet configuration (Szilágyi and Sanneck, 2011). Reproduced with permission of © 2011 IEEE.

4.1.3.4 Secure Connection Setup

After the initial connection has been set up, the RN has IP connectivity to the OAM subnet and the next step is to establish a secure connection to the OAM nodes involved in the rest of the Phase 1 self-configuration process, that is, downloading the initial parameters including a list of DeNBs. For this setup, the RN needs a vendor NE certificate (for mutual authentication with the OAM) and the operator root certificate. The RN contacts the CA and performs certificate enrolment similarly to regular eNBs as described in Section 4.1.2. The RN then establishes a mutually authenticated TLS session with the OAM system.

There are two options in how to do this step: using IPsec and TLS or using TLS only. In the first option, the connection is secured in the IP layer as well by using IPsec tunnel between the RN and the security gateway. It means that in Figure 4.10, the P-GW → OAM path should take the (b2) alternative and even go through the SEG. It should be noted that in this case the overall RN → OAM path goes through the SEG twice: first in the RN → P-GW part, second in the P-GW → OAM part. In the first part, the RN → SEG IPsec tunnel goes through the SEG transparently and it terminates in the SEG only the second time when it turns back and reaches the SEG from within the user plane subnet. This is because the first time all packets (including IPsec and upper layer data) originating from the RN are passed through the transport network as a payload and the SEG at this part only acts as a normal IP router. In other words, the IPsec tunnel is inside the user plane GTP tunnel that spans from the CeNB to the P-GW. However, from the IPsec point of view, the secured connection is between the RN and the SEG, regardless of the transport path of the IPsec packets.

Figure 4.11 shows the corresponding message flow for the authentication and secure link establishment. Alternatively, the IPsec connection to the SEG can be omitted as the RN has in fact already access to the trusted domain by connecting to the user plane subnet and access to

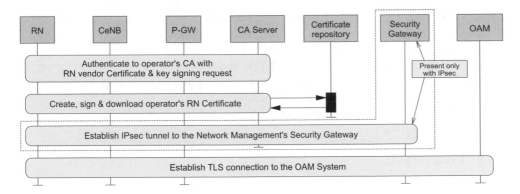

Figure 4.11 Authentication and secure connectivity (Szilágyi and Sanneck, 2011). Reproduced with permission of © 2011 IEEE.

the OAM system is authenticated via TLS handshake. Figure 4.11 shows this alternative procedure as well. The only difference is that this time the SEG is not actively involved since no IP layer security is used. Through the established secure OAM link, the RN can retrieve the initial configuration parameters, such as the list of DeNBs.

4.1.3.5 Switch-Over to the DeNB

After the RN has downloaded the initial DeNB list through the configurator eNB, the RN detaches from the CeNB and enters Phase 2. The RN selects a DeNB from the list it received in Phase 1 and connects to it via the RN Attach procedure that is similar to the UE Attach but with some important differences. In this phase, a secure channel is established between the RN and the USIM-RN with TLS handshake using certificates on both sides; the RN may use the certificate enroled in Phase 1. The RN uses USIM-RN also for attaching to the network and establishing Non-Access Stratum (NAS) security with the MME. The P-GW for the RN this time is the special P-GW co-located with the DeNB. The RN already has an IP address routable in the OAM subnet, which can be reused (after all, only the Layer-2 connectivity is changing) and thus the DHCP part is skipped in Phase 2. The OAM connection is used by the RN to continue with the self-configuration (cf. Section 4.1.2).

Another connection established between the RN and the DeNB is the radio resource control connection used for resource partitioning between the DeNB and the RN. This protocol is similar to the RRC protocol (3GPP TS36.331, 2010) used between an eNB and a UE but it controls the relay link resources. In general, the RN also has to maintain S1 signalling connections towards the SAE-GW and X2 connections to neighbours (either RNs or eNBs whose cells overlap with the RN's own cell). The exact number of signalling connections may vary across the different relay architecture alternatives (3GPP TR36.806, 2010); in the standardised proxy S1/X2 architecture, the RN only has to maintain one S1/X2 interface to the DeNB.

After the OAM connection is in place and the resource partitioning of the relay link is completed, the RN can enter the operational phase, powering its cell and serving incoming UEs. During the operational phase, there are connections which are established on a per need basis, most importantly user plane EPS bearers, created for carrying UE traffic when UEs are attached to the RN and deleted when UEs are handed over to other cells. The Automated Neighbour Relation

(ANR) SON function can create X2 connections between two RNs or between a RN and an eNB as part of establishing neighbour relations between adjacent cells (cf. Section 4.2.3.6).

The discussed auto-connection method (Szilágyi and Sanneck, 2011) has been accepted in 3GPP as the standard way for LTE-Advanced relay nodes to obtain initial connectivity with the OAM node via the Configurator eNB and then switch over to the Donor eNB. In 3GPP, this is referred to as the 'two-phase' configuration for RNs (3GPP TS36.300, 2010), Section 4.7.6.3; the original TDoc reference is (3GPP R3102370).

4.1.4 Conclusions

Auto-connectivity can be seen as the most critical part of the self-configuration process as any further configuration step depends on the success of this phase. Furthermore, it contains the setup of the NE's security environment affecting the security of the entire NE lifecycle. Auto-commissioning prepares the NE during the installation to deliver a network function specific to a planned site. Thus it is possible to deliver the NE only with an 'off-the-shelf' software and configuration installation thereby facilitating the manufacturing process. NE vendor and network operator activities are shifted to a preparation (rather than the actual rollout) phase thereby eliminating the interaction between manufacturer and operator, as well as on-site activities by commissioning personnel as much as possible.

A generic framework for auto-connectivity and -commissioning has been presented containing all relevant constituent parts. As many parts of process are highly vendor-specific by nature (e.g. SW management) and only some aspects are standardised to allow for vendor differentiation, the details of actual auto-connectivity and -commissioning implementations will vary quite a lot (in terms of the degree of self-configuration envisaged, which functionality is associated to which network entity (NE, ACS/different OAM system servers), the exact sequence of steps, etc.).

The presented framework can be applied similarly to 3G, yet observing the RAT specific differences: before executing Node B auto-connectivity, the RNC as an own NE requires (self-) configuration. With regards to auto-connectivity, other transport network technologies than in LTE may be employed (like ATM) which may require either own auto-connectivity procedures or some additional pre-configuration. Auto-commissioning of a new Node B includes also the corresponding configurations in the RNC (cells, adjacencies, transport configuration).

4.2 Dynamic Radio Configuration

The radio configuration of legacy base stations is completely based on a careful planning which needs to be conducted by the network operator prior to the installation, using a planning toolset. When incrementally building up the network without self-configuration functions one can either perform a labour intensive radio planning step every time a new network element is inserted, or perform periodic radio planning updates anticipating the new eNodeB insertions within the next time frame. In the former case the insertion order of the base stations must proceed exactly as planned. In the latter case one needs to accept that the planning is not representative for the operational network at a certain point in time. Without a representative planning, the newly inserted base station and its neighbours will not have a correct configuration matching the current network topology. When using the 'Dynamic Radio

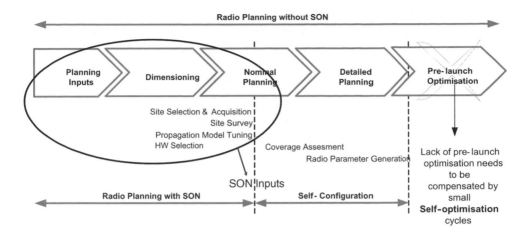

Figure 4.12   Shift of functionality from radio planning to Dynamic Radio Configuration.

Configuration' (DRC) this trade-off disappears, because DRC is adaptive to the current network topology context. The DRC will configure the new base station/cell and its neighbours in such a way that the key radio configuration parameters (Physical Cell ID, Neighbour Relationships, initial power and tilt settings) are correctly configured on the fly, thus enabling deployment in a much more ad hoc manner. It is anticipated that such flexible deployments are required in the future deployment scenarios described in Section 2.2. In general this flexibility enables operators to react more quickly to changed customer requirements, competitor situation and own business priorities.

    While it is still needed to perform a dimensioning and capacity planning for selecting the sites and deciding on the HW resources to be installed, the detailed radio planning can be omitted since the assignment of the corresponding parameters is shifted to the DRC (Figure 4.12, cf. also Figure 4.1). Where off-line radio planning tools completely rely on input provided by the operator (e.g. expected geographic traffic distribution), DRC algorithms can make use of measurements and already optimised parameters of operational neighbours.

    Radio parameters are classified into two major groups (3GPP S5-091879, n.d.):

- *Class A* parameters do not have an influence on the configuration of the adjacent cells or vice versa. Since their assignment is independent from the existing network topology, they are out of scope of the DRC.
- To configure *Class B* parameters correctly, the knowledge of the relationships and the configuration of adjacent cells is needed (including direct neighbours, neighbours of the neighbours or even all cells in a large geographical area). Since these parameter values are dependent on the current network context, it may be necessary to assign them dynamically. Class B parameters are further divided into subclasses defined by Table 4.2.

The activity diagram (Figure 4.13) illustrates the individual algorithms executed within the DRC, their intermediate outputs and their interdependencies. The algorithms build further on (intermediate) results of previous algorithms. Intermediate results are not directly configuration parameters for the eNodeB, but essential inputs for self-configuration algorithms. All activities in the diagram map to a single self-configuration algorithm.

Table 4.2 Classification of radio parameters

Subclass	Definition	Description
A	Single cell scope parameters	
A1	Parameters of class A1 can have a fixed value or are dependent on dynamic properties of the same cell, like the number of users in the cell.	Initially either an arbitrary, default or a traffic modelling derived value can be configured. The value can be dependent on the cell type, urban, rural, highway, indoors, and so on. The value can also be learned from similar already optimised cells in the network. If needed the parameter can be self-optimised in short optimisation cycles when the cell is operational.
		These parameters are out of scope of DRC. Around 80% of the radio parameters belong to this class.
		Examples are RRC, RRM, QCI, scheduler and HARQ parameters.
A2	Parameters of class A2 are dependent on semi-static properties of the same cell.	These parameters could be configured with a rule or policy. A typical example here is the cyclic prefix length. The cyclic prefix length needs to be configured with the value 'extended' when the cell has a radius larger than 1.4 km.
	The dependency is static.	Only a limited number of radio parameters are of this type. The reason is that such dependencies between configuration parameters of the own cell, are already taken into account in the information modelling. For this reasons this type of parameter is not further considered.
B	Multiple cell scope parameters	
B1	Parameters of class B1 need to be configured with the same value in all cells of a large part or even the complete network.	These parameters are read-create and cannot be optimised.
		The value of these parameters cannot be dynamically assigned, and hence need to be assigned by the mobile network operator. These values can be used as input for the online database preparation. Hence this class is out of scope of DRC.
		Examples are PLMN ID, frequency band, EARFCN, #PRB.

B2	The value of class B2 parameters needs to be unique in scope of the complete network.		These parameters are read-create and cannot be optimised. These values are assigned by DRC. The algorithms are considered quite straightforward, although sometimes network wide inputs are needed. Examples are site ID, EGCI and eNB name.
B3	The value of class B3 parameters need to be configured collision free. This means that the same parameter of a direct neighbour cell cannot be configured with the same value.	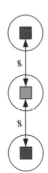	The algorithms to configure the different class B3 parameters will be quite similar to each other. Often additional constraints on these parameters do apply. Not all class B3 parameters need to be initially configured. Frequency sub-band configuration for inter cell interference coordination needs to be only configured when the inter cell interference coordination feature is switched on. This feature is never switched on in new cells. Examples in scope of DRC are *PCI*, PRACH root-sequence (RS) index, cell specific RS cyclic shift (RS CS).
B4	The value of a class B4 parameter needs to be aligned with the configuration of an adjacent cell.	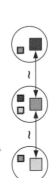	As with the class B3 parameters not all class B4 parameters are in scope of DRC. Some can be configured a default value.
B4.1	The value of a B4.1 class parameter needs to be aligned with exactly one neighbouring cell. For such a parameter a one-to-one relationship between cells can be defined.		Examples in scope of the DRC are neighbour relationship (*NR*) and quality offset for cell reselection.
B4.2	The value of a B4.2 class parameter needs to be aligned in a small cluster of cells. For such a parameter a one-to-many relationship between cells can be defined.		Examples in scope of DRC are tracking area (TA), *Tx power and antenna tilt.*

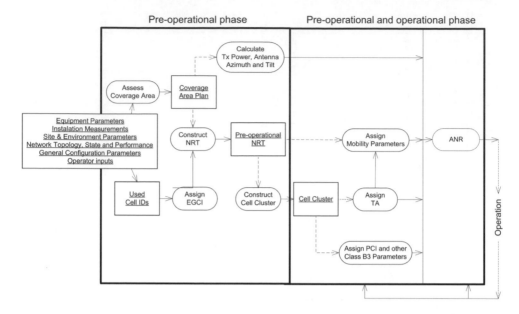

Figure 4.13 DRC process and algorithms.

At the left hand side of the diagram the inputs to the DRC are listed. A high level grouping of these parameters can be made as follows:

- **Equipment Parameters (Inventory Management):** This group of parameters collects equipment properties of the installed eNodeB, like the HW configuration, the type and manufacturer of the different HW components and related properties like the antenna gain, or power amplifier limitations. The HW configuration to be installed on a site is an outcome of the network dimensioning planning activity.
- **Installation Measurements:** Measurements performed while installing the eNodeB, either automatically or manually by the installer. The measurements can be either stored in the eNodeB or in some central repository. Typical examples are the geo-location and the feeder loss.
- **Site and Environment Parameters:** These parameters comprise properties of the site where the eNodeB is installed. Examples are the site ID and the clutter type. Some of these parameters may be assigned automatically (e.g. extract the clutter type from a digital surface map), but often these are configured manually.
- **Network Topology, State and Performance:** The geographic locations, operational states and performances of the existing cells in the network. These inputs are representing the context of the network and are essential in the incremental network growth scenario.
- **General Configuration Parameters:** These include Class A configuration parameters of the new cell, but also of already operational cells in the network. For the new cell(s) these are typically configuration parameters which are derived from the planning during the auto-commissioning phase, like the frequency band, channel bandwidth, and so on.
- **Operator Inputs:** Inputs provided by the remote commissioner in order to control the self-configuration process (e.g. strategies on how to compute the PCIs, cf. Section 4.2.2).

Algorithms for the assignment of the following intermediate outputs and parameters are included in the process:

1. **Coverage area:** Includes the calculation of the coverage of the newly inserted cells as well as the reassessment of the coverage area of already operational cells. The coverage area calculation can be based on matching the link budget with the path loss estimated by a radio propagation model, like applied for the detailed radio planning by off-line planning tools. The algorithm should also calculate the related downlink transmission power, Remote Electrical Tilt (RET) and, if supported by the eNodeB, Remote Azimuth Steering (RAS). They need to be defined in such a way that an optimum overlap with the neighbouring cells is achieved at minimum power radiation. When the dimensioning planning shows that the coverage of the cell should be capacity limited this shall be handled as an extra constraint for the algorithm.

   In case cells are inserted between already existing cells, the coverage area of these neighbours is reduced, their radio parameters are recomputed and these eNodeBs are reconfigured accordingly.

2. **Pre-operational Neighbour Relation Table (NRT):** The pre-operational NRT contains the potential neighbour cells of the inserted cells. DRC needs to construct a NRT in pre-operational phase because this is an essential input for further self-configuration algorithms and for the inter-eNodeB connection setup (3GPP X2 interface). This function is complementary to the Automated Neighbour Relationship function (ANR), cf. Section 4.2.3.

   The algorithm for assembling the pre-operational NRT's will rely on the coverage area assessment of the new cell and all cells within an arbitrary geographic distance of that cell. A neighbour relationship is then identified based on the cell coverage area overlap between two cells.

3. **Cell cluster:** A cell cluster (Figure 4.14) consists of the newly inserted cell, all neighbours of this cell (first tier neighbours) and all neighbours of the neighbours (second tier neighbours). All cells in the cell cluster are identified by the EGCI and PCI. Each cell is given an additional qualifier to identify it as a first or second tier cell.

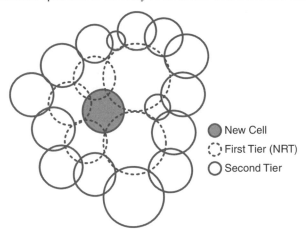

Figure 4.14 Cell Cluster example (Sanneck *et al.*, 2010a). Reproduced with permission of © 2010 IEEE.

The cell cluster is a typical intermediate result, as the eNodeB will not be configured with the cell cluster, but it is an essential input for other self-configuration algorithms requiring information from remote cells, like the PCI assignment algorithm. Depending on the implementation of those algorithms even the nth tier neighbours (with $n > 2$) may be added. The cell cluster is constructed starting from the NRT of the inserted cell (containing the first tier neighbour cells) and going through the NRTs of the neighbour cells until the expected level of neighbours has been collected.

4. **Class B3 parameters:** Class B3 parameters have a restricted value range. Therefore the same value needs to be assigned to multiple cells throughout the network. All class B3 parameters need to be configured collision free (see Table 4.2), which means that the configured value of the parameter needs to be different from the values configured in all the neighbouring cells.

5. **Tracking Area:** In order to maintain the location of mobiles for paging purposes the E-UTRAN is divided in contiguous Tracking Areas (TA). Every cell must be assigned a Tracking Area Code (TAC). Usually tracking areas are assigned manually by planning experts, based on traffic behaviour knowledge in the area.

The size of a TA is constrained by a trade-off between the paging signalling load and location update load. Location area planning can be performed automatically by means of mobility and traffic prediction models. These models aim to compute cell border traversal frequency and paging load to assign location areas for optimum signalling cost. Typical inputs are a geographic map of cells and high capacity routes, the population distribution, user mobility behaviour, paging traffic model and relative resource cost of paging and location updates.

Alternatively a provisional assignment algorithm could be used. For example, a new cell could be allocated to a TA of a neighbour cell while maintaining a TA size limit. Information on TAs (self-)configured on the eNB are sent to the MME in the S1 self-configuration procedure (cf. Section 8.2.1.1).

Because an incremental assignment may create a sub-optimal location of tracking area borders, a periodic TA self-optimisation should be foreseen (cf. Section 8.2.4.2). The optimisation should also take care of reconfiguring other affected cells, since a TA re-configuration may propagate through a large part of the network.

As shown in Figure 4.13, steps (1) and (2) belong only to the pre-operational phase whereas steps (3) to (5) will be executed in both the pre-operational and the operational phase. Also, it is anticipated that for LTE networks step (1) will usually still be executed in a static way, that is, pre-computed at network planning time rather than executed dynamically at the time of base station insertion.

Following the sequence as outlined in Figure 4.13, the next sections present the automatic allocation of the power and antenna setting, the PCIs and neighbourships (ANR) respectively. Section 4.2.4 then summarises considerations on the DRC architecture.

4.2.1 Generation of Initial Transmission Parameters

The radio transmission parameters to be assigned to a newly introduced base station handled in this section include Maximum Transmission Power, Electrical Antenna Tilt and Electrical Antenna Azimuth. According to the parameter type definitions in Table 4.2, all parameters fall into category B4.2 and therefore need to be aligned within a cluster of cells in a one-to-many relationship way. The cluster size depends on the network layout and may include several tiers of neighbouring cells as well as cells on a different layer (macro, micro, pico, femto) than the newly introduced base station.

There are three major options for the configuration of these parameters:

1. Complete pre-configuration of the parameters with values determined by network planning prior to or during installation on site.
2. Pre-configuration of the parameters with some default values; the base station- and site-specific values are then determined in the operational phase using Coverage and Capacity Optimisation, cf. Section 5.4. The default values may be some standard or best-practice values that depend, for example, on the base station type (macro, micro, pico, femto), the location (urban, suburban, rural) and the manufacturer.
3. No pre-configuration of the parameters, the base station and site specific values are automatically determined during self-configuration. The base station is delivered to the site completely without any configuration for the mentioned parameters.

The *first option* represents the traditional way of parameter calculation and assignment using off-line tools. The *second option* avoids this by defining a general default value (which may be relatively far away from the optimum) shifting the site-specific adaptation to the operational (self-optimisation) phase. On one hand this is attractive as simply the self-optimisation functions are re-used for the initial setting. On the other hand it may take a long time to converge to a 'good' parameter setting.

Only the *third option* falls into the DRC category. This means that parameter allocation algorithms are executed in an 'online' way taking into account only a cell cluster rather than an entire network domain. This allows for fast, dynamic parameter allocation, however, with a potentially less accurate result than for a full 'offline' computation for the entire domain. Yet, the goal is that a 'good enough' configuration can be created on which results the self-optimisation algorithms can continue effectively, requiring less time for the adaptation than the second option.

The settings for transmission power, tilt and azimuth highly depend on each other, balancing coverage and capacity. Furthermore, these parameters have also to be adjusted for surrounding base stations. Hence, for the third option, there exist many potential strategies of how to determine these set of parameters. In the following we present an example Strategy A where all three parameters are adjusted. A simplified strategy is Strategy B where the transmission power is set to a pre-calculated value (e.g. by means of network planning) whereas tilt and azimuth are adjusted until the targets for coverage and interference are reached.

4.2.1.1 Strategy A: Complete Determination of Transmission Parameters

In this approach, the values for the transmission parameters are calculated by a centralised algorithm after the commissioning of the new base station (software installation, S1 interface connection setup, OAM connection setup) has been completed, but prior to switching on the radio equipment. The algorithm uses all available information for the new base station, including for example, site ID and sectorisation, frequency band, channel bandwidth, radio propagation model and so on (see also Figure 4.15), as well as corresponding information from

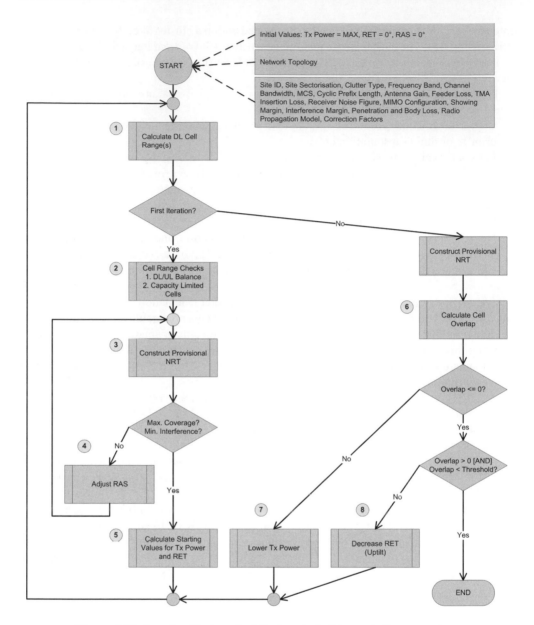

Figure 4.15 Insertion Strategy A: Adjustment of all transmission parameters.

the neighbouring cells. Thus, in fact Strategy A can be described as an approach to move corresponding algorithms from offline network planning (see Option 1 previously) to the point in time just before the radio equipment becomes operational, with the advantage that all input data to the algorithm (i.e. the current network configuration) is up-to-date.

To calculate the values a two-step approach is followed. In the first step the transmission power is calculated, in the second step the antenna tilt and azimuth are calculated. In fact, the values for the new base station start with the largest possible coverage area which is then reduced until the overlap with existing/neighbouring cells is as low as possible.

The algorithm starts with maximum possible cell range:

- Transmission Power (Tx Power) = MAX Note: the algorithm starting value 'MAX' might be lower than the actually possible equipment maximum, for example, for capacity limited cells.
- Remote Electrical Antenna Tilt (RET) = $0°$.
- Remote Azimuth Steering (RAS) = $0°$.
- [OPTIONAL: Position cell → RAS:
  – The azimuth steering of all cells of the eNodeB can run coordinated]
- If overlap between two cells is too large → shrink cell(s) of new base station by adjusting the Tx power and the antenna tilt stepwise:
  – Either newly inserted cell, neighbour cell or both;
  – Focus on inter-eNodeB cell relationships.

Both the Tx power and the antenna tilt calculation are based on the evaluation on inter-eNodeB cell overlaps. The iterations to calculate the optimum Tx power are started with a high Tx power value which is lowered with each iteration until there is no more overlap. Subsequently, to compensate for this loss of coverage, the antenna tilt is adjusted. The starting value for the antenna tilt is the maximal possible downtilt. For each iteration of the algorithm, the antenna tilt value will be increased (uptilted), until there is a minimal cell overlap with inter-eNodeB neighbouring cells. The algorithm is illustrated in Figure 4.15 and explained in detail next.

1. For the new cells calculate the DL maximum possible cell range, R. This is done by setting Tx power to the maximum possible value the PA can deliver and the antenna tilt and antenna azimuth to $0°$. Doing so only the mechanical azimuth and tilt are taken into account.
2. With this maximum cell range two checks are performed. First the balance between the DL and UL needs to be checked, to ensure that both coverage areas match each other. Secondly the maximum calculated cell range is checked against an operator defined maximum cell range. The latter is done when the cell needs to be capacity limited. If one or both checks fail, the cell range needs to be decreased. This is done by decreasing the Tx power value. The antenna parameters are untouched. Once an appropriate value for the Tx power is found which fits both conditions, these checks will not need to be done in any subsequent step in the algorithm. This is because the algorithm will further bring down the initial cell range value.
3. With the calculated cell range in Step 1 and 2, a provisional NRT is constructed. One can exclude intra-eNodeB cell relationships from the table, because these will not be further evaluated.
4. Determination of the RAS value. This is an optional step in the algorithm depending on the fact that the eNodeB supports RAS or not. If RAS is not supported this step can easily be skipped and the mechanical installed azimuth is used in the next steps. The azimuth values of the new eNodeB are adjusted in order to maximise the coverage and minimise the interference. Currently two assumptions are made which need to be verified further with simulations. First a coordinated azimuth steering is assumed. This means that the azimuth of all cells included in an eNodeB are steered in a coordinated way, so the interdependency of

intra-eNodeB cells is more or less fixed. If the azimuth of one cell is adjusted, the azimuths of the other cells belonging to the same eNodeB are adjusted in the same way. Second only the azimuths of the newly inserted cells are adjusted. The latter is because when adjusting the azimuth of already operational cells the propagation effect through the network would be considerable. The exact calculation details are for further specification. With every iteration the NRT should be re-evaluated.

5. With the calculated cell ranges and azimuth values, the overlap between the newly inserted cells and already operational cells can be calculated. From this overlap the cell insertion scenario can be extracted. How this is exactly handled is for further specification. For example, if the overlap $> 90\%$ of the cell's coverage area, it is clearly a cell splitting scenario, if the overlap $< 10\%$, one has a clear coverage driven scenario. Of course this is not in all cases that obvious. Nevertheless it is believed that the cell insertion scenario can be determined quite distinctive. Depending on the cell insertion scenario starting values for the algorithm are determined. This includes starting values for the new cells as well as the already operational cells. Although in a pure coverage driven scenario (e.g. overlap $< 10\%$) the starting values for the algorithm for the already operational cells can be the same as the current configured values. In a cell splitting scenario the Tx power can be averaged between the new inserted cells and already operational cells. The maximum possible Tx power value is determined in Step 1 and 2. The algorithm will start with a high Tx value and iteratively bring it down. For the tilt values the maximum possible downtilt will be used as starting value. The algorithm will iteratively increase the value (uptilt).

6. With the starting values for the Tx power and the antenna tilt, the cell coverage for all cells in the NRT are calculated. The overlap between all cells is calculated.

7. The next step is to calculate the Tx power. This is done by bringing the value gradually down, until there is no overlap between the newly inserted cells and the already operational cells. The Tx power step size used between the iterations is for further specification. Again the scope of this calculation can be only the newly inserted cells or also the already operational cells can be included, this is dependent upon the cell insertion scenario.

8. Finally after the azimuth and the Tx power calculation, the antenna tilt will be calculated. This is also done by iteratively and gradually adjusting the tilt value upwards, until the overlap of the new cells and already operational cells is bigger than zero and lower than an operator defined threshold. So doing the tilt configuration will compensate the loss of coverage cause by the low Tx power value.

4.2.1.2   Strategy B: Pre-Calculated Transmission Power

Within the SOCRATES project (Eisenblätter *et al.*, 2011) a 'Soft Integration' concept has been proposed, which aims at a gradual integration of new base stations into an operational network with little disruption. Also here, the algorithm starts after the initial commissioning of the new base station has been completed, but in contrast to Strategy A the algorithm runs after the radio equipment of the new base station has already been switched on. Starting with pre-calculated values from network planning (that are, however, considered imperfect), Soft Integration uses information from the network or UEs to determine the 'final' configuration of the new base station. This is achieved by stepwise increasing radio parameter values of the new base station, allowing surrounding base stations to adapt their transmission parameters by means of self-optimisation. The simulations conducted used the maximum transmit power as driving radio parameter, and electrical antenna tilt as the parameter to be optimised. In fact, Soft Integration operates at the transition between self-configuration and self-optimisation as the mechanisms used to determine the 'final' configuration are strongly related to corresponding self-optimisation mechanisms. The approach is most sensible for a capacity driven insertion as, in case of a coverage driven scenario, there may not be sufficient measurements available from neighbouring base stations or UEs to verify new settings.

- The algorithm starts with lowest possible cell range:
  - Transmission Power (Tx Power) = MAX Note: the algorithm starting value 'MAX' may depend on the base station type, the existing environment or on pre-calculated values coming from network planning;
  - Remote Electrical Antenna Tilt (RET) = MAX (maximum possible downtilt);
  - Remote Azimuth Steering (RAS) = 0°.
- [Position cell → adjust RAS: this has not been simulated within SOCRATES but would be a logical step].
- Check/simulate area in coverage with current settings, check interference with existing installation;
  - If coverage insufficient increase RET by one step;
  - Adjust Tx Power/RET of neighbours (depending on algorithm only direct neighbours or several tiers of neighbours).
- Return to previous step until target coverage is reached.

4.2.2 Physical Cell-ID Allocation

A fundamental parameter for the LTE radio configuration is the LTE reference signal sequence. It is called the Physical Cell ID (PCI) and used as regionally unique identifier on the physical layer. There are two reasons why these reference signal sequences are used as identifiers. First, they can be read within a very short timeframe (5 ms) but more importantly they are constructed in a very robust way against interference which is important for a reliable identification. The automated configuration of physical cell identities is one of the key SON use cases defined by the 3GPP (cf. Section 3.2.3).

The starting point for the construction of PCIs are 168 pseudo random sequences; the 'cell identity groups'. For each of the cell identity groups three orthogonal sequences are constructed which results in a total of 504 useable PCIs 3GPP TS36.300 (2010). In an LTE network domain there are obviously more than 504 cells, however, which necessitates the reuse of PCIs. PCIs are B3-category parameters (cf. Table 4.2) which therefore need to be configured in a 'collision-free' way, that is, 'neighbouring cells' need to be assigned different PCIs. Neighbouring cells are defined here as set of cells which can be received by an UE at a specific location, that is, there is some overlap of coverage within this set of cells. But since the neighbour relationship tables which are used for handover management are based on the PCIs, an additional requirement needs to be introduced for a proper assignment – 'confusion free' assignment. An assignment is confusion free if there is no cell in the network that has two or more neighbouring cells with identical PCIs. A UE requests a handover to a cell identified by a particular PCI. In case of a confusion, the responsible eNodeB would not be able to reliably identify the target cell and a handover would fail. Figure 4.16 shows both, a confused and a confusion-free assignment.

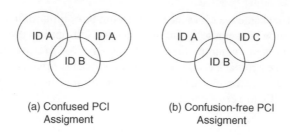

(a) Confused PCI
Assigment

(b) Confusion-free PCI
Assigment

Figure 4.16 Confused and Confusion Free PCI Assignment.

Operators can have different views on how to assign PCIs in the network. In order to carefully deal with a 'limited resource' a high 'reuse-rate' could be applied. In this case as few PCIs as possible are assigned and the same PCIs are repeatedly configured. There are also other possible assignment strategies like assigning PCIs in a way that two cells with the same ID have a maximal distance from each other.

Although, the complexity of a collision and confusion free PCI assignment seems to be quite reasonable with the availability of 504 IDs the impact of a PCI assignment should be considered: every reconfiguration of a cell's PCI may require a restart of the cell or even the responsible eNodeB, which causes a service interruption. Therefore it is important to understand the reasons that require a PCI reconfiguration. On the one hand to perform a reliable reconfiguration and on the other hand to provide an assignment that proactively omits these reasons and reduces the number of required reconfigurations.

4.2.2.1 Related Radio Parameters

The configuration of other radio parameters is directly dependent on the configuration of the PCI. The assigned PCI directly influences the structure of the Uplink and Downlink Reference Signals of a cell (3GPP TS36.211, 2011). Especially if collaborative multipoint transmission with joint beamforming is used the *downlink reference signals* of neighbouring cells should be orthogonal to each other. Since there are six sub-carrier groups, an assignment N_{ID} *mod 6* should be used, where N_{ID} denotes the numerical representation of the assigned PCI. For the *uplink reference signals*, 30 sequence groups are defined. Neighbouring cells should not be assigned identical sequence groups, which implies that neighbouring cells should have different PCIs according to N_{ID} *mod 30*. This is an important example how other configuration parameters which are derived from the PCI configuration impose requirements on the assignment of PCIs in the network.

4.2.2.2 Changes of Neighbourships

For an initial assignment in a non-operational network there is plenty of time to evaluate the planned but theoretical neighbour structure of each cell and assign the IDs in a collision and confusion free way. But there is not only a high probability that in reality not all neighbourships are as planned but they can also easily change due to variations in the radio propagation properties, which is a typical phenomenon observed between the seasons or if buildings are modified or newly constructed. Such deviations from the planned network layout and the involved mal-configurations will only be detectable in the operational phase of the network.

Whenever such an erroneous configuration is detected one or more cells have to be reconfigured.

4.2.2.3 Network Evolution

Apart from the difficulties to determine the actual network layout in a pre-operational phase the layout of the cell structure changes as the network evolves over time. New cells are added to either close coverage holes or to provide more capacity at load hotspots. Cells are replaced by a set of smaller cells or removed when the coverage is provided through larger cells.

Another important fact that has to be kept in mind is that especially in the beginning of LTE network rollouts there will only be LTE island deployments, typically in areas where high capacity is required, for example, at airports, trade fair areas or within large cities. These islands are then step-wise extended until a full coverage is reached. To configure the PCIs of the cells added to such an island properly, the PCI assignment of the neighbouring cells has to be analysed. The results of the analysis are used to assure that the PCI of the newly introduced cell does not collide with any of the neighbouring cells. But especially when new cells are added to an existing cluster of cells, the newly added cell is potentially already confused by two or more neighbouring cells which use identical PCIs as shown in Figure 4.17 (a) and (b). During the evolutionary growth of the network it is important to detect and resolve such mis-configurations as soon as possible.

The confusion probability increases when the coverage areas of multiple deployment islands start to overlap. Each of the islands has been configured collision and confusion free, but if the same PCIs have been used in all islands the probability for the cells 'connecting' islands to be confused is rather high (in Figure 4.17 the new cell with PCI B 'connects' two previously unconnected islands each with a cell having PCI A).

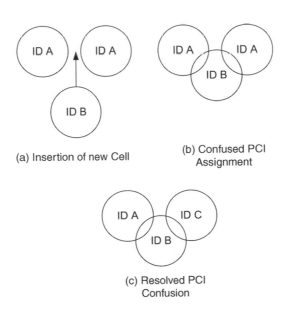

(a) Insertion of new Cell

(b) Confused PCI Assignment

(c) Resolved PCI Confusion

Figure 4.17 Confusion of a newly inserted cell (and its resolution).

Another source of collision and confusion is whenever a high capacity island deployment which consists of multiple small cells is complemented by larger macro cells (cf. Section 2.1.8.5 and Chapter 10 on Heterogeneous Networks). Those macro cells have a common coverage area with several of the small cells. In case the PCIs within the island deployment are assigned using a high reuse rate, the new macro cell will be confused by a larger set of cells.

Independent of the reason for the confusion of the new cell, all but one of the cells that cause the confusion have to be assigned a new PCI. Figure 4.17 shows the process from the introduction of a new cell until the fully resolved confusion.

4.2.2.4   Reduction of Available PCIs through Partitioning of the ID Space

Instead of operating with the full set of available PCIs there are many reasons why an assignment is only done with a subgroup of the full PCI set. In order to proactively avoid repetitive reconfigurations (and the service interruptions they cause), network operators can divide the available ID space into separate blocks and use them for assignment. Sets of IDs could be reserved for different cell types, like a block for macro cells, one for pico cells and so on. Especially for the usage with closed subscriber groups (CSG) femto cells, the 3GPP has defined the possibility to exclusively reserve and signal a group of PCIs (3GPP TR36.902,2011). This is of special interest as a dedicated subgroup of PCIs for femto cells helps to decrease the power consumption of the UEs. Femto cells broadcast additional information to allow the UEs the detection of potential target cells and avoid attempts to camp on a CSG where a UE has no access rights. The broadcasted information contains a CSG ID to identify a particular CSG and an access bit. The access bit allows differentiating between different access schemes like CSG, open or mixed. This information can be obtained by the UEs through additional data acquisition. To avoid those kinds of measurements, (Wu et al., 2010) propose the usage of groups of PCIs to identify femto cells and the access scheme they use.

If in a network that makes use of NE from multiple vendors a single Network Management (NM) system for all NEs is available (cf. Section 4.2.4), the PCI allocation can be coordinated at the NM level; PCIs for the respective vendor domains are then configured using the 'Automatic Radio Configuration Function' (ARCF) being part of the Self-Configuration Management IRP defined in (3GPP TS32.502, 2010) (cf. Figure 4.6). However, groups of IDs could also be assigned to different network domains. In case several domain management systems are used, each of the systems that are responsible for the so called vendor clouds could be assigned a separate group of IDs. This distribution of ID ranges thus avoids the problem of the coordination of PCI configuration across vendor domain borders.

Another very important reason for the usage of PCI groups is that cells located at spectrum license border locations. In these areas a collision free PCI assignment with respect to a neighbouring network operating on the same frequency has to be found. The mobile network operators are not able to influence the ID assignments of the respective neighbouring networks. Therefore they could agree on which operator configures which PCI group to the cells in the border area, so that no further coordination is required.

As it has been shown there are many reasons for separation of the available PCIs into subsets. Finding a way to provide a set of reasonable subgroups of PCIs is a complex task. Up to now there is no final solution to the problem and research in this area is still going on. Especially with the introduction of femto cells a static assignment of PCIs will not be sufficient. Research tries to find ways for dynamic sub-partitioning that meets the requirements of the networks. Lee *et al.* (2009) give an example how PCI groups could be dynamically reserved for the usage with a varying number of femto cells. Dividing the available IDs into smaller groups of IDs leads to the conclusion that it is important to think about the way the IDs are assigned. It requires a strategy that performs an assignment with small number of IDs which still can handle changes in the cell layout, caused especially by the introduction of new cells, to a large degree without requiring any reconfiguration of the already active cells.

The following paragraph summarises the assignment criteria which have been derived from all the factors that have an influence on the assignment and shows examples of different assignment approaches that can be used for SON.

4.2.2.5 Criteria for PCI Assignment

To reach a well defined assignment of PCIs which is in addition also very robust against changes of the cell layout, several requirements have to be satisfied.

- **Collision- and confusion-free assignment:** This is the most obvious requirement on the assignment of the PCIs, which has already been emphasised.
- **Avoid reconfigurations:** The introduction of new cells or the removal of existing cells should not cause reconfiguration of other cells. In case reconfiguration is inevitable the number of reconfigurations should be minimised.
- **Adaptability to different network deployments:** There is no one-fits-all assignment scheme. The used scheme must be adaptable to take the actual context of the new cell and the surrounding network into account.
- **Applicability to initial and evolutionary deployment scenarios:** There should be no major difference if PCIs are assigned to a newly deployed network or to new cells during the network evolution of an operational network. The used approach should be the same or at least be based on the same principles in order to give similar results.
- **Specific physical** (e.g. N_{ID} mod 6, N_{ID} mod 30) **and operator policy constraints**.

4.2.2.6 Assignment Approaches

In the following, an introduction to different approaches to provide a PCI assignment solution that can be used within SON enabled networks is given. Each approach focuses on a different aspect of the PCI assignment.

Graph Colouring-Based, Centralised Approach
An approach for PCI assignment that satisfies the presented requirement has been introduced in (Bandh *et al.*, 2009) and (Bandh *et al.*, 2010).

The main idea of this approach is to map the given problem of PCI assignment to a well know problem from the world of mathematics and computer science. This allows assessment of the characteristics of the PCI assignment even before it is done. For example, a worst case assumption about the required number of IDs can be given but also a first assessment whether an assignment is possible or not. In addition the implementation effort can be reduced as algorithms and methods that where developed for this type of problem are applicable.

As a first step the given cell layout with its neighbours has to be transferred to a graph that reflects the collision and confusion free requirements. Figure 4.18 shows an example of a

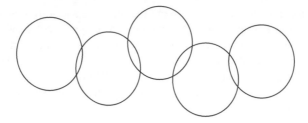

Figure 4.18 Example network layout.

network with five cells. Only those cells that cover an overlapping area are considered to be neighbours.

The considered neighbourships of the cells are assessed from the coverage area assessment in network planning and can be improved through measurements (acquired, e.g. using ANR, Section 4.2.3 or drive tests).

- **Collision Free:** All cells are depicted as nodes in a graph that reflects the cell neighbourships.
- **Confusion Free:** For a confusion free assignment the graph has to be extended with additional edges between the nodes and all neighbours of their neighbours. Figure 4.19 shows the graph for the example cell layout from Figure 4.18.

A generic graph colouring algorithm is applied to the graph which results in a collision and confusion free PCI assignment shown in Figure 4.20.

This approach is especially suitable for an initial assignment of PCIs to a non-operational network but not necessarily to assign PCIs to cells that are added during the operational phase of the network. This is the case since an assignment recomputation could cause a major reconfiguration with a growing probability for service outages. This topic is addressed by

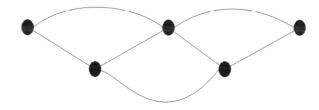

Figure 4.19 Graph with edges to neighbours of neighbours of the cell layout.

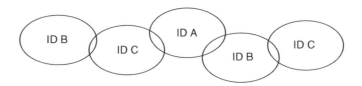

Figure 4.20 Collision and confusion free assignment.

proposing an algorithm that retains the characteristics of graph-based approach while mini-mising the area of potential reconfigurations.

The remaining reconfigurations are based on the fact that independent from the way a PCI is assigned to a new cell, this new cell can already be confused. Those reconfigurations are needed to resolve such confusions. The graph based approach gives a guarantee that the area where reconfigurations may be needed is restricted to the direct neighbours of a cell.

In (Bandh *et al.*, 2010) the approach is extended to avoid reconfigurations that are caused by new neighbourships which can be established through structural changes, like modification of buildings which influence the radio propagation of the cells.

Such new neighbourships cause reconfigurations if a very dense distribution of PCIs has been applied. Instead of applying a maximally spaced PCI assignment, the introduction of 'safety margins' proactively counteracts such reconfigurations while still retaining an eco-nomic use of PCIs. From the research of Welsh and Powell (Welsh, 1967) it is known that the maximal number of colours required for a given graph is directly dependent on the maximal number of neighbours a node in the graph has. Safety margins increase the number of neighbours. Through the selection of a safety margin, the network operator has a direct influence on the spatial distribution of PCIs but also on the maximal number of PCIs required for proactively protected assignment.

Decentralised Graph Based Approach

The approach of (Ahmed *et al.*, 2010) also uses graph colouring as a foundation for PCI assignment. But they propose to use decentralised graph colouring algorithms that run directly in the eNodeBs. To identify neighbourships two phases are used. In a first phase the eNodeB identifies the PCIs of all receivable cells. Based on this acquired information a new PCI is selected and the cell is put into operation. Because the number of receivable cells can be lower than the actual number of neighbouring cells a second phase with UE-based measurements is used to identify additional neighbours. The authors compare a set of potentially useable distributed graph colouring algorithms. For the comparison several metrics are used (like the number of required PCI re-assignments compared to the number of available PCIs and the total number of assigned PCIs) to assess the quality of the used algorithms.

Distributed RSRP Measurement Based Approach

Using UE based measurements to discover neighbours and neighbour-related configuration faults is a commonly used method. It is also used in the heuristic approach presented by (Amirijoo *et al.*, 2008). To gather information about the neighbourships, RSRP measurement reports from the served UEs are evaluated. Those reports already contain the PCIs of handover candidate cells. In case the PCI is not yet listed for a known neighbour, the serving eNodeB requests the UE to read the E-UTRAN Cell Global Identification (ECGI). The new neighbour-ship is then established via an information exchange over the core network. The serving eNodeB informs the target eNodeB about the new neighbourship by sending details like the own PCI and ECGI.

In case of confusion the NRT of the target eNodeB already contains another cell with an identical PCI. This confusion is subsequently resolved by reconfiguring one of the cells.

The serving cell could also be already confused. When it receives a RSRP measurement report from a UE this contains the PCI which is already part of the NRT in this case. The serving cell will prepare for a handover to this cell, which fails because the actual target cell is not the one listed in the NRT. Such confusions can be detected by monitoring the handover success respectively the failure rates. If the handovers from one to another cell often fail, the serving eNodeB will request the UE to report the ECGI of the target cell whenever it receives a request for this particular PCI. This allows the detection of several neighbouring cells using the same PCI.

For measurement based PCI assignment schemes a potentially high number of reconfiguration for already deployed cells could be required until a full collision and confusion free assignment is reached. If a graph based foundation is used guarantees can be given that a stable configuration can be found. A common drawback is the increasing number of measurement reports the UE have to generate, which increases the energy consumption and decreases the run-time.

Summary

Due to the complexity of future heterogeneous networks (cf. Chapter 10), automation of Physical Cell ID assignment is an important requirement. It is crucial that the general requirements as collision and confusion free assignment but also the constraints of closely related parameters like uplink and downlink reference signals are satisfied. To reach the required reliability it is beneficial to base the approach on known problems with known properties, because only then it is possible to give the required guarantees. Multiple approaches have already been proposed for PCI assignment each focusing on different parts of the problem which results in solutions with different characteristics. This shows one of the benefits of SON: mobile network operators have the freedom to choose a solution as a building block that satisfies their needs and requirements and can integrate it together with other building blocks to form a SON-enabled system.

4.2.3 Automatic Neighbour Relationship Setup (ANR)

A labour-intensive activity for mobile network operators is the creation and update of neighbour cell relationships for handover purposes, commonly denoted as Neighbour Relations (NR). To have the right neighbours in place is very important, since a missing neighbour is a common reason for call drops due to failed handovers.

In 3GPP RATs the handover is network controlled, meaning that the radio access network, 2G, 3G or LTE, decides when to make the handover and what the target cell is.

Handover decisions are based on UE measurements which are controlled by parameters given by the specific RAT like, for example, LTE. In 2G and 3G systems up to 3GPP Release 10 the UE needs to maintain a Neighbour Cell List (NCL), meaning that the UE is instructed which frequencies/cells it needs to listen to. An example is shown in Figure 4.21, where Cell B is in the NCL of Cell A, but Cell C is not in the NCL of Cell A, which causes a handover failure for the UE moving from Cell A to Cell C. In addition 3G already supports intra-frequency *detected cell reporting* (DSR) for DSR-capable UEs, which allows the mobile to detect and report cells on the same frequency even if not included in the neighbour cell list.

While in 2G the mobile periodically reports related cells' measurements to the network, in 3G and LTE the network instructs the UE to send an indication only when certain conditions are met; for example, in case of Event A3 (see measurement event description in Section 5.1.3) in LTE, the UE will send a measurement report to the eNB in case a neighbour cell becomes better than the serving cell. The final decision to perform a handover of the UE to the new target cell is executed by the network.

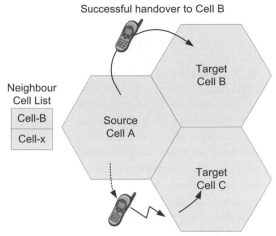

Figure 4.21 Missing neighbours in Neighbour Cell List.

In 2G and 3G systems, neighbours of a given cell are chosen by means of cell planning tools which make use of coverage predictions. Predictions are subject to errors since maps are imperfect and the real environment may change due to for example, new buildings or changed weather conditions. As a consequence the operators need to (re-)check the situation by means of driving tests, coverage and handover regions.

This is today a cumbersome and expensive job for operators which typically have large networks and it becomes even more complicated when another access network (LTE) or another network layer such as micro cells are added on top of the deployed 2G and 3G networks. In addition to the initial network rollout phase, neighbour (re)definitions are required also during the operational phase, since conditions may change; this situation can even get more demanding with future massive deployment of small(er) cells such as micro and pico eNBs, and HeNBs. Cells will pop up and disappear quite frequently and drive tests will become more challenging. Thus an automated setup of neighbour relations may become not only a useful but an imperative feature.

Automatic Neighbour Relations (ANR) allows automating neighbour cell configuration to a great extent. Network performance will also benefit from such optimised and up-to-date lists, for example, proper setup of neighbourships will increase the number of successful handovers and minimise the number of call drops due to missing neighbour relations. ANR is particularly significant for LTE since this is a new technology/network to be deployed. With its flat network architecture without central nodes like the Radio Network Controllers (RNC), the operator would need to configure a huge amount of neighbours' parameters in all LTE eNBs manually if there is no ANR function available. Therefore, ANR helps reducing OPEX by relieving the human operator.

4.2.3.1 Pre-operational Neighbour Relations

In network planning the initial neighbour relations are based on static assumptions; the cell planning tools calculate neighbours based on geographical proximity and direction of antennas.

As said, coverage prediction is subject to errors, for example, due to imperfections in the geographical and existing building development data. As a consequence the operators need to complement the planning phase with drive tests, to verify coverage datasets and identify all handover regions. In addition to that, the above process needs to be repeated since the conditions, like new cells, new building construction, or even seasonal vegetation, change quite often thereby influencing interference and coverage conditions. A conservative approach can consider all neighbourships, even those where the probability for an actual NR is very low. In this case the result is that the NR list could be rather large with a lot of unused NRs. Due to the semantics of NR in LTE such unused NR table entries are less problematic than in 3G, yet a tool chain to come from the pre-operational NRs to distributed operational NRs is required. Operators furthermore may use additional means to optimise the set of neighbourships with tools where neighbours are created or deleted, depending on defined criteria, including inter-cell distance. For example the setting for 3G intra system adjacencies are created based on daily or weekly Detected Set Report (DSR) measurements and KPIs based on Received Signal Code Power (RSCP) and/or Energy Of Carrier Over Noise (Ec/No) values.

Note that in any case (ANR employed or not), pre-operational NRs have to be computed (cf. Figure 4.13). The difference between ANR and the traditional NR determination method is that a more detailed NR planning step and the subsequent distribution of the NR configuration to the NEs can be omitted.

4.2.3.2   ANR Basics

The SON use case 'Automatic Neighbour Cell Configuration and X2 Setup' is defined in the 3GPP Release 8 standards and has been inspired by NGMN recommendations (cf. Section 3.2.2). 3GPP TS36.300 (2010) describes the ANR procedures and relation to its environment.

ANR enables *Intra-LTE* neighbour detection for both, intra- and inter-frequency, and *Inter-RAT* neighbour detection. For the other direction that is, from 3G or 2G systems towards LTE, a standardised ANR functionality is missing; in this case vendor-specific solutions can be applied, for example, to detect the neighbour from LTE towards 3G/2G with UE support and build the neighbourships in the other direction with the help of optimisation tools.

The ANR function relies on the UE to report the Physical Cell IDs (PCI) of detected neighbour cells without any advertised neighbour lists (see Figure 4.22). This capability is possible because of frequency reuse of one in WCDMA and in LTE. The PCI reported by this UE must be mapped to the E-UTRAN Cell Global Identifier (ECGI) before the handover can be executed. The PCI - ECGI mapping can be done in two different ways:

- For *UE-based ANR*, the mapping is done by the UE which decodes the target cell's ECGI from the broadcast channel (Section 4.2.3.3). The capability for decoding the ECGI is an optional UE feature. When one UE has done the ECGI decoding, the information can be stored in the eNB and there is no need for other UEs to do the decoding anymore.
- *UE-triggered ANR with OAM support* (Section 4.2.3.4) does not demand additional and expensive decoding of BCCH system information from the UE, but relies on the network by using the capabilities and information (location) in the OAM system (see Figure 4.22).

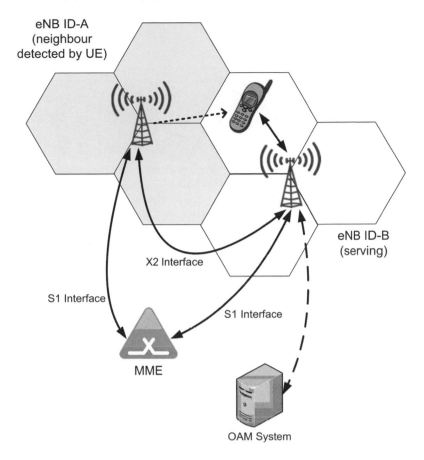

Figure 4.22 Basic ANR principles.

Finally the EGCI needs to be resolved into an IP address of a neighbouring eNB. This is accomplished via the core network (MME), cf. Section 8.2.2, which needs to be pre-configured accordingly, or via the OAM system. Hence, ANR can be seen as a hybrid SON function (cf. Section 3.4) having both a distributed and a centralised component.

4.2.3.3 UE-Based ANR

According to 3GPP the UE measures and reports the following types of cells:

- The serving cell.
- Neighbour cells as indicated by eNB.
- Detected cells, that is, cells which are not indicated by E-UTRAN but detected by UE. The E-UTRAN indicates the carrier frequency.

A detected cell can be an LTE cell which uses *the same frequency of the serving cell*, or *another LTE cell using a different frequency* or the cell *may belong to another RAT*. In the two latter cases the UE needs to be instructed to perform measurements on that frequency.

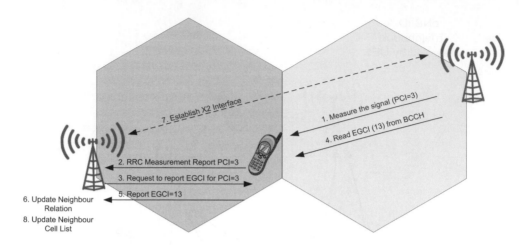

Figure 4.23 ANR in LTE (Holma and Toskala, 2011). Reproduced with permission from John Wiley & Sons, Ltd.

The Automatic Neighbours Relation function covers three steps:

1. Neighbour cell discovery.
2. Neighbour site's X2 transport configuration discovery and X2 connection set-up (Intra LTE) – this may be a logical X2 connection via the MME in case of HeNB or small (micro, pico) eNBs.
3. Neighbour cell configuration update.

A separate but related function is the NR Optimisation (NRO).

Intra LTE/Intra-Frequency ANR
Step 1, the Neighbour cell discovery procedure, is described in Figure 4.23.

1. When the UE changes from idle to RRC_CONNECTED state, it is instructed via Measurement Configuration, to report the detected strongest cells even for PCIs currently unknown (the request is not sent if the PCI is blacklisted in the radio resource control blacklist).
2. The UE measures the PCI of a cell that it has detected (but which is not in a predefined neighbour list) and reports it to eNodeB.
3. If the eNodeB receives a report by the UE that indicates an unknown PCI, it orders the UE to read the ECGI from the broadcast channel of the detected cell.

The network operator can configure 'blacklisted' cells for ANR at the NM level (3GPP TR 32.511, 2011); *blacklisted* cells are broadcast in System Information Blocks 4 and 5 (SIB4 and SIB5), cf. (3GPP TS 36.331, 2010). A blacklisted cell will never be subject to ANR, even if reported as a strong neighbour by a UE. The operator may want to block certain handover candidates, for example, from indoor cells to outdoor cells in high rise buildings or for cells verging to another network without roaming agreement ('no HO' attribute). It is also possible to

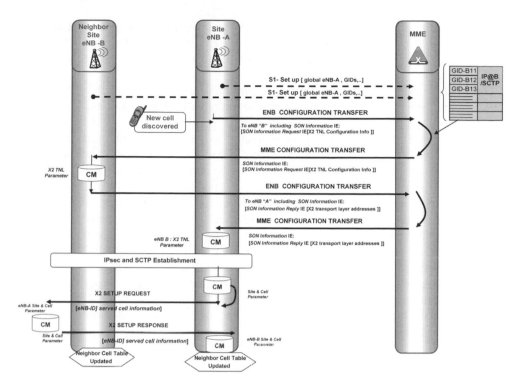

Figure 4.24 Transport Layer Setup.

only blacklist the actual X2 setup, for example, to avoid numerous X2 connections from HeNB terminating at a macro eNB. 'Whitelisting' is also possible, where NRs are explicitly set at the NM level (and can only be removed from there).

Step 2 and Step 3 define the *Transport Layer Setup* and they are described in detail in Figure 4.24.

The ECGI cannot be used to directly address neighbour sites on IP Transport level. Therefore the ECGI needs to be resolved to the transport network control layer configuration which is needed to establish the IP/IPsec connection with SCTP on top of it.

3GPP Working Group RAN 3 defined to use the Configuration Transfer procedure (3GPP TS 36.413, 2011) to exchange the SCTP configuration respectively IP addresses between two neighbour sites (cf. Section 8.2.2). The eNB requests the IP address corresponding to an ECGI from the MME, update its neighbour list, establish an X2 connection with the peer eNB and exchange configuration data over X2. An evaluation of the intra-LTE UE-based ANR can be found in (Dahlén *et al.*, 2011).

NR Optimisation (NRO) is a vendor-specific SON function. The neighbour removal function removes outdated neighbour cell relations. For example, a time stamp may be added to every added neighbour relation; if the cell is not reported for a given period of time, the corresponding relation is deleted.

Inter-RAT ANR

In addition to LTE intra-frequency cases, 3GPP specifies ANR for LTE inter-frequency and inter-RAT. In this case the eNodeB can instruct a UE to perform measurements and detect cells on other frequencies as well as on other RATs. The inter-frequency and inter-RAT measurements need scheduling of the measurement gaps for UEs. For inter-RAT and inter-Frequency ANR, each cell contains an inter-frequency search list. This list contains all frequencies that shall be searched.

The ANR mechanism is very similar to the intra-frequency one, but now the eNB instructs the UE to perform measurements and detect cells on other RATs/frequencies.

1. During connected mode, the UE is instructed to perform measurements and report neighbours on other RATs/frequencies. To do so the eNB may need to schedule appropriate idle periods to allow the UE to scan all cells in the target RATs/frequencies.
2. The UE measures the PCI of a cell that it has detected in the target RAT/frequencies and reports it to its serving eNB. The PCI is defined by the carrier frequency and the Primary Scrambling Code (PSC) in case of a UTRAN FDD cell, by the carrier frequency and the Cell Parameter ID in case of a UTRAN TDD cell, by the Band Indicator + Base Station Identity Code (BSIC) + BCCH Absolute Radio Frequency Channel Number (ARFCN) in case of a GERAN cell and by the PseudoNoise Offset in case of a CDMA2000 cell.
3. If the UE reports an unknown PCI, the eNB instructs the UE to read the neighbours' Cell Global Identifier (CGI) and the Routing Area Code (RAC) in case of GERAN detected cells, CGI + Location Area Code (LAC) + RAC in case of UTRAN detected cells, and CGI in case of CDMA2000 detected cells. For the inter-frequency case, the eNB instructs the UE to read the ECGI, Tracking Area Code (TAC) and all available PLMN ID(s) of the inter-frequency detected cell. This case is more demanding than the intra-frequency one since the eNB may need to schedule appropriate idle periods to allow the UE to synchronise on the indicated frequencies/RATs and read the requested information from the broadcast channel of the detected cell.
4. The UE reports the detected CGI and RAC (in case of GERAN detected cells) or CGI, LAC and RAC (in case of UTRAN detected cells) or CGI (in case of CDMA2000 detected cells) to the serving cell eNB. In the inter-frequency case, the UE reports the ECGI, the TAC and all PLMN-ID(s) that have been found for the detected cell. In Figure 4.25, the Automatic Neighbour Relation principle in case of a UTRAN detected cell is shown.
5. The eNB updates its inter-RAT/inter-frequency Neighbour Relation Table (NRT).

4.2.3.4 UE-Triggered ANR with OAM Support

Another way of implementing automated detection and NR configuration of unknown cells respectively eNBs, is with the help of the OAM system. Since ANR-specific measurement support at the UEs is optional, this mode leverages UEs which do not support ECGI resolution. This is particular relevant for initial LTE deployments where no or only few UEs supporting the ANR-specific measurement may be available. Thus, UE-triggered ANR with OAM support can be regarded as a complementary function to UE-based ANR. It is also a 'hybrid' SON function having distributed and centralised components (cf. Section 3.4). UE-triggered ANR with OAM support is not standardised (i.e. the centralised component is located at the DM level). A standardised solution (at NM level) would be technically possible, though.

The feature includes a preparation phase, where the OAM system creates (and maintains) a list of potential neighbour cells with the related IP connectivity information (PCI – ECGI – IP address).

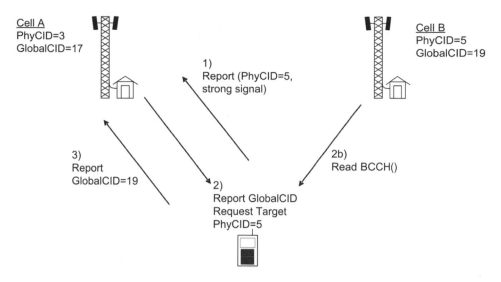

Figure 4.25 Automatic Neighbour Relation Function in case of UTRAN detected cell (3GPP TS36.300, March 2011). Reproduced with permission from 3GPP.

UE-triggered ANR with OAM support comprises the following steps (see also Figure 4.26):

- At the time of the eNB registration to the OAM system, the creation of a table is triggered comprising the PCI/ECGI/IP address data of the current neighbours of the new eNB. This table is downloaded to the new eNB; the tables of neighbouring eNBs of that new eNB are also updated by the OAM system. Note that, given PCI uniqueness in the relevant neighbour area (cf. Section 4.2.2), a PCI can be directly mapped to an IP address. Thus, when creating the table based on a new eNB's registration actual PCI collision/confusion can be detected.

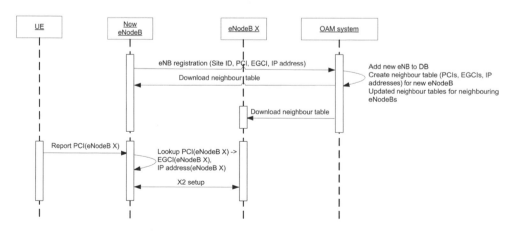

Figure 4.26 UE-triggered ANR with OAM support.

Table 4.3 Comparison of ANR UE-based versus UE-triggered with OAM support

	ANR type	UE based	UE triggered with OAM support
Required functionality			
UE	Standardised PCI report	X (distributed)	X (distributed)
	Standardised EGCI report	X (distributed)	
eNB	Standardised X2 connectivity setup	X	X
	Standardised IP address resolution via S1	X (centralised)	
Domain Manager (DM)	Proprietary address resolution via Itf-S		X (centralised)

- Neighbour cell discovery (radio part): similarly to the UE-based ANR, the UE measures the signal of a new cell and reports the measurement. The UE is then, however, not instructed to read the ECGI from detected cell neighbours' system information, as the neighbour information is already known from the step above.
- X2 transport configuration 'discovery': the IP address of a neighbour eNB hosting the newly discovered cell can be derived also from the mapping table (instead of the MME-based address resolution via the S1 interface).
- The X2 connectivity is then set up and the eNB exchanges information of served cells with the new neighbour as defined by (3GPP TS 36.423, 2011). This step is identical to the step in UE-based ANR.

Table 4.3 shows a comparison between UE- and OAM-based ANR with regard to the required functionality and its standardisation status at the UE, eNB and DM levels.

4.2.3.5 3G Automatic Neighbour Relations

As mentioned in the introduction, Automatic Neighbour Relations is a SON feature defined in 3GPP mainly for LTE. This comes from the fact that operators with existing deployed networks want to reduce to a large extent the effort in network planning and configuration changes due to adaptation of adjacencies when introducing the new LTE layer. Those operators are usually not very interested in extending ANR to mature technologies, such as WCDMA and GSM, since for these technologies neighbours are already configured and stable for years.

Nonetheless, green field operators or operators that intend to expand their 3G network for capacity needs and/or because they do not own a LTE licence, expressed the need to extend ANR to WCDMA as well. This is the reason why 3GPP started discussion for extending this functionality to WCDMA: a UE camping on a 3G cell will be able to report UTRAN, LTE and GERAN neighbour cells even though not advertised in a neighbour list. ANR for UTRAN is specified in (3GPP TS 25.484, 2011).

As described in Section 4.2.3.3, the LTE ANR function is a two-step approach located in the eNB; it is based on detected cell reporting by the UE after which the UE is instructed to obtain System Information (SI) for a specific detected cell. This mechanism leverages UE in active mode; intra-frequency neighbours are simply detected during UE autonomous gaps (DRX), while inter-frequency and inter-RAT require the UE to suspend its transmission or reception to switch to the other frequency (as instructed by the network) to perform cell measurements.

ANR for UTRAN is based on a different mechanism and it builds on UEs in idle mode rather than in active mode. Actually the CELL_DCH/SI acquisition, similar to the LTE method, was originally proposed in 3GPP. In such a case the network would need to configure compressed mode or other means to allow gaps ('idle periods') for the measurements and target cell SI acquisition. This solution, however, was considered too expensive for a single-RX UE and cause significant interruption.

Another alternative proposed in 3GPP was the detected cell re-selection. Here the UE re-selects a detected cell, acquires SI and stores relevant neighbour information for later ANR reporting. The neighbour cell SI acquisition comes 'for free' when the UE re-selects the detected cell. The UE reports to the network at a suitable time, say, at a time when UE has a UL transmission. Although attractive due to its simplicity, this solution was considered not sufficient (e.g. not perfectly working in case of good intra-frequency coverage).

3GPP decided to go for a third solution, called *UE Logging Mode*. Here the currently specified logged Minimisation of Drive Test (MDT, see Chapter 7) methodology is re-used for UMTS ANR needs. The UE performs ANR while in idle mode, CELL_PCH and URA_PCH states.

The steps for the procedure are described in Figure 4.27 (left side): The UE is configured with ANR measurement and parameters. Once the UE is in idle mode, CELL_PCH or URA_PCH has detected a new cell (i.e. a cell that is not in the NCL), it stores the ID of the serving cell and the detected cell plus related SI in a 'logbook'. If the UE can detect cells in another frequency than the serving cell, it can be done only in the frequencies of the cells listed in NCL (maximum two). At the moment where the UE establishes an UL connection, it indicates the logbook availability to the current serving cell, which can then retrieve the logbook from the UE. The receiving RNC uses RRC signalling to retrieve the ANR report over the Uu interface after the

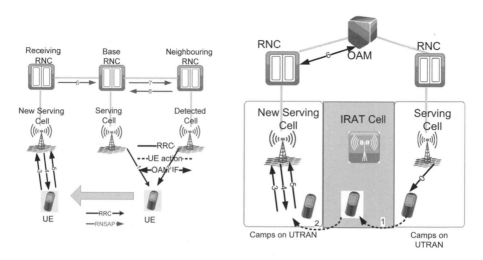

Figure 4.27 3G ANR as in 3GPP Release 10 (intra-3G: left side, inter-RAT: right side).

UE has indicated that it has a logged ANR report available. The receiving RNC handles the ANR report and may forward the ANR report to different RNCs.

The inter-RAT case works quite similar (see Figure 4.27 right side). The UE does not detect new neighbouring GSM or LTE cells when being served by a UTRAN cell, but the UE logbook reporting is done only in case the UE goes back to a UTRAN cell and if the previously used cell in LTE or GSM was not signalled in UTRAN SI blocks. The UE signals the logbook availability to the network when selecting the UTRAN cell and the RNC may retrieve additional cell attributes from OAM.

Note that Closed Subscriber Group (CSG, see Chapter 10) cells are excluded from 3GPP Release 10 ANR.

4.2.3.6 ANR in Relay Deployment

The deployment of Relay Nodes (RNs) is one of the LTE-Advanced scenarios described in Chapter 2; their auto-connectivity is presented in Section 4.1.3.

Relay nodes basically use the previously described UE-based ANR mechanism with minor adaptations due to the specifics of the relay architecture featuring the Donor eNB (DeNB) and the RN itself.

To explain ANR an example of a potential RN deployment is given; Figure 4.28 shows the neighbour relations of the (D)eNB$_1$ without and with RNs deployed in the area. The left side of Figure 4.28 shows eNB$_1$ with only eNB$_2$ and eNB$_4$ as neighbours. The coverage hole between eNB$_1$ and eNB$_3$ does not allow the UE to measure the related neighbour and no X2 interface is in place between eNB$_1$ and eNB$_3$.

To manage the coverage hole the operator may wish to update eNB$_1$ with relay functionalities; eNB$_1$ takes then the role of the Donor eNB. DeNB$_1$ and eNB$_3$ have now a neighbour relation via the RN cell and the X2 interface needs to be established (Figure 4.28, right hand side).

The ANR procedure starts at the moment the RN discovers, via UE measurements, the eNB$_3$ cell. The RN requests the newly detected cell X2 Transport Network Layer (TNL) address to DeNB$_1$ via *eNB Configuration Transfer* (cf. Section 8.2.2). In case no X2 interface between DeNB$_1$ and the newly detected cell is available, the procedure for setting up X2 is executed as depicted in Figure 4.29. The RN does not need to establish a direct X2 interface with the newly detected cell since neighbours are seen through its DeNB: the *MME*

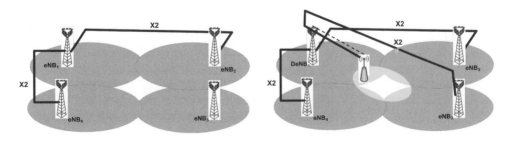

Figure 4.28 Neighbour relation of DeNB$_1$ without (a) and with relay deployment (b).

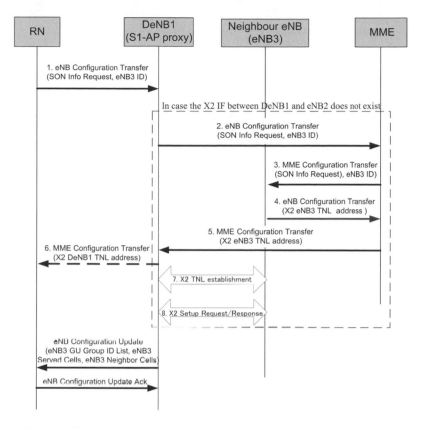

Figure 4.29 X2 setup between RN and neighbour eNB initiated by the RN.

Configuration Transfer message sent from the $DeNB_1$ can contain any value in the *Transport Layer Address* field or even be omitted. In Figure 4.29, the X2 TNL address of the $DeNB_1$ included as example.

After the X2 TNL establishment, the *X2 Setup procedure* between eNB_3 and $DeNB_1$ is executed. The $DeNB_1$ builds the *eNB Configuration Update* message as it was created by the eNB_3 and sends it to the RN (as shown in Figure 4.29). When the RN receives this *eNB Configuration Update* message, it will know that the message is referring to the newly detected neighbour eNB, even though the TNL address of the message is the one of $DeNB_1$. The *eNB Configuration Update* message contains the Globally Unique (GU) Group ID List supported by the neighbour eNB, this allows the RN to decide whether the X2 handover can be applied or not, that is, whether the UE that has to be handed over to the neighbour eNB is served by an MME pool that is supported by the neighbour eNB or not.

If instead the $DeNB_1$ has an already established X2 interface with eNB_3, it can directly send the up to date information regarding neighbour eNB cells forwarding to the RN the latest *eNB Configuration Update* message received from the neighbour eNB. This is the case of eNB_2 and eNB_4 in the example of Figure 4.28, left), where the $DeNB_1$ has already an X2 interface established with them.

If one of the eNB$_1$ neighbours' updates its configuration (e.g. served cells and/or the supported MME pools), it sends an *eNB Configuration Update* to eNB$_1$, which, acting as DeNB, in turns updates the RN.

The availability of X2 interface between the RN and a neighbour eNB depends on the availability of the X2 interface between the DeNB and the neighbour eNB, while the availability of the X2 handover between RN and neighbour eNB is configured by the RN's OAM system (similarly to the case for X2 handover between two eNBs).

4.2.4 DRC Architecture

There are several alternatives (see Section 3.4) for locating the DRC SON functions in the OAM architecture (see Section 2.1.9).

4.2.4.1 Network Management-Centric

In this architectural option the DRC function execution and coordination is located at the network management level, that is, algorithms are executed at a central entity. The main benefit is that the impact on interfaces and standardisation are relatively limited, since the existing interfaces and configuration management functions of the layers underneath can be re-used. The solution can also be deployed on an installed base of eNodeBs without any retrofitting. The main technical challenge of this architecture is the integration of the data from the different new input sources to fulfil the self-configuration process. Also, the abstraction level is rather high, that is, the computation of parameters with a rather high dependency on vendor-specific algorithms is difficult.

4.2.4.2 Domain Management-Centric

To allow for some alignment of parameters between domains, the NM level would act as a relay and proxy function in this variant. A relay function is required in case reconfiguration needs to be triggered and a proxy function is needed in case of data exchange. The relay point on the NM level introduces additional complexity, for example in terms of coordination and exception handling.

4.2.4.3 NE-Centric

The function execution is located at the eNB ('distributed SON', cf. Section 3.4). At the NM level similar impacts as for the above 'DM-centric' solution are expected. In addition the southbound interface needs to support monitoring and control and input data provisioning for the distributed algorithms. For fundamental configuration parameters like PCI, an eNB would need to be able to autonomously discover all neighbouring eNBs (cf. Section 4.2.2). Distributed functions also require extensive data exchange and coordination over the X2 interface. It must be assured at all times that the input parameters are synchronised between the eNBs and aligned with the corresponding centralised management functions. All existing eNBs must have been upgraded to support the required functionality, which may not always be easy to realise in a multi-vendor network.

4.2.4.4 Hybrid Solutions

In hybrid solutions the functions are partitioned such that one part is done at one level and another part at another level. Such a partitioning can be implemented amongst the NM and DM levels, the NM and NE levels (where the DM act as mediator), the NE and DM levels, or even amongst all three levels in the reference architecture. The main pitfall of hybrid solutions is that a careful coordination and synchronisation between the different levels is required (cf. Chapter 9).

4.2.4.5 Preferable Architecture

In principle the preferred architecture may be different for each parameter to be configured. Table 4.4 lists which location in the architecture is recommended to be used for the computation of the respective parameter. For the simplicity of the coordination function, the larger the geographical scope of the input parameters and of the existing cells that may need to be reconfigured actually is, the higher the management level should be where the DRC SON functions and the coordination function are located. Additionally, the dependency of parameters on vendor-specific algorithms must be taken into account.

The algorithms for most of the parameters should therefore reside at network management level. The parameters in brackets indicate a secondary, less preferable location.

In general following characteristics of the DRC functions can be observed:

- **Input parameters:** Required with a multiple cell scope and even a multi-vendor domain scope; need to be collected from different sources and to be integrated.
- **Coordination:** In a chain of sequential functions; between multiple NE: reconfiguration of already operational cells and the move to the operational state of new cells.
- **Limited scalability requirement:** The algorithms are executed at NE or cell insertion time (but for thousands of nodes) and only relatively rarely re-executed for operational nodes.
- The performance with respect to **response time** is not that critical, due to the execution done mainly during pre-operational (build) phase.

Within the scope of the Self-configuration Management IRP (cf. Section 4.1.2), 3GPP has standardised an 'Automatic Radio Configuration Function' (ARCF; 3GPP TS32.502, 2010) where the parameters listed in Table 4.4 for 'NM' are in fact computed at the NM level and then conveyed over Itf-N to the different DMs. For (UE-based) ANR it should be noted that, as

Table 4.4 Preferable location of DRC functions (B3/B4 parameters)

Level	NE	DM	NM
Configuration parameters	NR	(NR)	Antenna tilt
		(PCI)	Tx power
		(PRACH RS index)	TA
		(RS CS)	PCI
			PRACH RS index
			RS CS

outlined in Section 4.2.3, the key component of UE-based ANR is at the eNB, which is, however, complemented by components in the core network (IP address resolution) and the OAM system (for white-/blacklisting).

In summary, the overall solution is hybrid, however, with a strong emphasis on the NM-level computation of radio parameters.

4.2.5 Conclusions

Dynamic Radio Configuration of base stations allows for ad hoc deployments in cellular networks with minimal planning overhead. These functions primarily consist of a set of algorithms to compute the radio parameters automatically and supporting coordination functions to collect the necessary input parameters, to configure the new cells and to reconfigure neighbour cells. Altogether these functions are activated in a (usually centralised) process (cf. Chapter 9) according to a predefined order while taking into account operator policies. Algorithms can be derived from those available in the conventional network planning and configuration tool chain, yet have to be adapted to the dynamic (incremental) way of execution.

The human operator will take a different role than with the traditional base station deployment, where often a (remote) commissioner is involved to provide the planned radio configuration to the access nodes. With self-configuration, the remote commissioner primarily monitors progress of base station rollouts and the performance of the dynamic configuration algorithms on the basis of event notifications, respectively KPIs gathered after entering operational state.

The operator should also be able to control the DRC through policies. This can include the selection of a self-configuration algorithm, fine-tuning of algorithm parameters, defining constraints on parameter values, managing reconfiguration schedules, failure handling and so on. Only in exceptional cases the human operator should be alarmed and take over control to initiate corrective actions at the input data or the policies of the DRC.

The framework enables a transition path from a 'static' infrastructure network where all parameters are pre-planned before deployment to an entirely 'dynamic' infrastructure network where only the site of deployment for a network element is pre-planned. While the outlined DRC vision contains such 'instantaneous' reconfiguration, in a real system there may be constraints: a reconfiguration of existing cells triggered by the above algorithms may need to be postponed until low traffic hours. For example, reconfiguring tracking areas may cause location update storms and temporary paging failures in the area. The modification of an operational radio configuration might also induce a reboot of HW modules, resulting in temporary service outages. In principle, a new cell should not be brought into operation until all adjacent cells affected by the insertion are adapted (however, this may be subject to the operator's policies). Therefore a coordination function is required to control reconfigurations of existing cells and the operational state of new cells (cf. Chapter 9 on SON operation). For the E-UTRAN in the focus of this book, the solution will still not be fully dynamic, however, with the presented framework it is now possible to continue step-by-step towards that scenario. Solutions for 3G will typically be at the same level of dynamicity. The presented concepts are applicable in a similar way with the given RAT-specific constraints (WCDMA scrambling code instead of PCI allocation, cf. Section 4.2.2, NCL instead of neighbour relationship list configuration, cf. Section 4.2.3.5).

References

3GPP TR36.902 (2011) Technical Specification, Technical Specification Group Radio Access Network, Evolved Universal Terrestrial Radio Access Network (E-UTRAN); Self-Configuring and Self-Optimizing Network (SON) Use Cases and Solutions, ver.9.3.1., Release 9, 7 April 2011. Available from http://www.3gpp.org/ftp/Specs/archive/36_series/36.902/36902-931.zip [accessed 30 June 2011].

3GPP R3-102370 (n.d.) *Nokia Siemens Networks, Solution for RN configuration*, TSG-RAN3 contribution, http://www.3gpp.org/ftp/tsg_ran/WG3_Iu/TSGR3_69/Docs/R3-102370.zip. [accessed 19 August 2010].

3GPP S5-091879 (n.d.) *Nokia Siemens Networks, Starting Material for Automatic Radio Network Configuration Data Preparation*, TSG-SA5 contribution. http://www.3gpp.org/ftp/tsg_sa/WG5_TM/TSGS5_64/Docs/S5-091879.zip [accessed 20 March 2009].

3GPP TS36.331 (January 2010) *Evolved Universal Terrestrial Radio Access (E-UTRA); Radio Resource Control (RRC); Protocol specification*, 3rd Generation Partnership Project (3GPP).

3GPP TS23.401 (March 2010) *General Packet Radio Service (GPRS) enhancements for Evolved Universal Terrestrial Radio Access Network (E-UTRAN) access*, 3rd Generation Partnership Project (3GPP).

3GPP TS32.501 (March 2010) *Self-configuration of network elements; Concepts and Requirements*, 3rd Generation Partnership Project (3GPP), (Release 9).

3GPP TR36.806 (March 2010) *Relay architectures for E-UTRA (LTE-Advanced)*, 3rd Generation Partnership Project (3GPP).

3GPP TS32.502 (June 2010) *Self-Configuration of Network Elements Integration Reference Point (IRP); Information Service (IS)*, 3rd Generation Partnership Project (3GPP), (Release 9).

3GPP TS32.532 (June 2010) *Software Management Integration Reference Point (IRP); Information Service (IS)*, 3rd Generation Partnership Project (3GPP), (Release 9).

3GPP TS33.310 (December 2010) *Network Domain Security (NDS); Authentication Framework (AF)*, 3rd Generation Partnership Project (3GPP), (Release 10).

3GPP TS36.300 (March 2011) *Evolved Universal Terrestrial Radio Access (E-UTRA) and Evolved Universal Terrestrial Radio Access Network (E-UTRAN), Overall Description, Stage 2*, 3rd Generation Partnership Project (3GPP), (Release 10).

3GPP TS36.423 (March 2011) *Evolved Universal Terrestrial Radio Access Network (E-UTRAN); X2 application protocol (X2AP)*, 3rd Generation Partnership Project (3GPP), (Release 10).

3GPP TS36.413 (March 2011) *Evolved Universal Terrestrial Radio Access Network (E-UTRAN); S1 Application Protocol (S1AP)*, 3rd Generation Partnership Project (3GPP), (Release 10).

3GPP TS25.484 (June 2011) *Automatic Neighbour Relation (ANR) for UTRAN, Stage 2*, 3rd Generation Partnership Project (3GPP), (Release 10).

3GPP TR32.511 (June 2011) *Automatic Neighbour Relation (ANR) management, Concepts and requirements*, 3rd Generation Partnership Project (3GPP), (Release 11).

3GPP TS36.211 (June 2011) *Evolved Universal Terrestrial Radio Access (E-UTRA); Physical Channels and Modulation*, 3rd Generation Partnership Project (3GPP), (Release 10).

Ahmed, F., Tirkkonen, O., Peltomäki, M. *et al.* (2010) Distributed graph coloring for self-organization in LTE networks. *Journal of Electrical and Computer Engineering*, **2010**. http://www.hindawi.com/journals/jece/2010/402831/ [accessed 20 March 2010].

Amirijoo, M., Frenger, P., Gunnarsson, F. *et al.* (2008) Neighbor cell relation list and physical cell identity self-organization in LTE. IEEE International Conference on Communications Workshops, Beijing, China, 37–41.

Bandh, T., Carle, G. and Sanneck, H. (2009) Graph coloring based physical-cell-ID assignment for LTE networks. 5th International Wireless Communications and Mobile Computing Conference, Leipzig, Germany, June.

Bandh, T., Romeikat, R., Sanneck, H. *et al.* (2010) Optimized network configuration parameter assignment based on graph coloring. IEEE/IFIP Network Operations and Management Symposium, Osaka, Japan, April.

Bandh, T. and Sanneck, H. (2011) Automatic Site Identification and Hardware-to-Site-Mapping for Base Station Self-configuration, IEEE International Workshop on Self-Organizing Networks, Budapest, Hungary, May.

Dahlén, A., Johansson, A., Gunnarsson, F. *et al.* (2011) Evaluations of LTE Automatic Neighbor Relations, IEEE International Workshop on Self-Organizing Networks, Budapest, Hungary, May.

Eisenblätter, A., Türke, U., and Schmelz, C. (2011) Self-Configuration in LTE Radio Networks: Automatic generation of eNodeB parameters, IEEE International Workshop on Self-Organizing Networks, Budapest, Hungary, May.

Holma, H. and Toskala, A. (eds) (2010) *LTE for UMTS: Evolution to LTE-Advanced, Revised Edition*, Chapter on SON, John Wiley & Sons, Ltd, Chichester, UK.

Horn, G., Forsberg, D., Moeller, W.D. and Niemi, V. (2010) Certificate Enrolment for Base Stations, Chapter 8.5, in *LTE Security*, John Wiley & Sons, Ltd, Chichester, UK.

IETF RFC 2131, Droms, R. (1997) *Dynamic Host Configuration Protocol*, IETF, March.

Lang, E., Redana, S. and Raaf, B. (2009) Business impact of relay deployment for coverage extension in 3GPP LTE-advanced. IEEE International Conference on Communications, June.

Lee, P., Jeong, J., Saxena, N. and Shin, J. (2009) Dynamic reservation scheme of physical cell identity for 3GPP LTE femtocell systems. *Journal of Information Processing Systems*, **5**(4), S.207–S.219, http://jips-k.org/dlibrary/JIPS_v05_no4_paper5.pdf.

Sanneck, H., Bouwen, Y. and Troch, E. (2010a) Context based configuration management of Plug & Play LTE base stations. IEEE/IFIP Network Operations and Management Symposium, Osaka, Japan, April.

Sanneck, H., Bouwen, Y. and Troch, E. (2010b) Dynamic radio configuration of self-organizing base stations. 7th International Symposium on Wireless Communication Systems, York, UK, September.

Sanneck, H., Schmelz, C., Baumgarth, T. and Keutner, K. (2007) Network element autoconfiguration in a managed network. IFIP/IEEE Symposium on Integrated Management, Munich, Germany, May.

Sanneck, H., Schmelz, C., Troch, E. and De Bie, L. (2009) Auto-connectivity and security setup for access network elements. IFIP/IEEE Symposium on Integrated Management, New York, NY, June.

Szilágyi, P. and Sanneck, H. (2011) LTE relay self-configuration. IFIP/IEEE Symposium on Integrated Management, Dublin, Ireland, May.

Welsh, D.J.A. (1967) An upper bound for the chromatic number of a graph and its application to timetabling problems. *The Computer Journal*, **10**(1), S.85–S.86.

Wu, Y., Jiang, H., Wu, Y. and Zhang, D. (2010) Physical cell identity self-organization for home eNodeB deployment in LTE. 6th International Conference on Wireless Communications, Networking and Mobile Computing (WiCOM), Chengdu City, China, pp. 1–6, http://ieeexplore.ieee.org/lpdocs/epic03/wrapper.htm?arnumber=5600778.

5

Self-Optimisation

Daniela Laselva, Ingo Viering, Dirk Rose, Jeroen Wigard, Seppo Hämäläinen,
Krzysztof Kordybach, Osman Yilmaz, Jaroslaw Lachowski, Paul Stephens,
Andreas Lobinger, Bernhard Wegmann, Henrik Martikainen and Cinzia Sartori

After having configured the network appropriately as described in the previous chapter, further optimisation steps are necessary during the operation of the network. One reason for this need is the fact that the environment may change, such as:

- Propagation conditions, for example, new buildings or streets, falling leaves in autumn.
- Traffic behaviour, for example, new traffic concentrations.
- Deployment, for example, the insertion of new base stations.

As a consequence, previously configured parameters will become suboptimal. Adaptation of the parameters in order to track those changes can obviously improve the performance of the network.

Another reason for further optimisation steps is the fact that measurements from the operating network under realistic load conditions provide a much more accurate picture of the reality compared with the assumptions which the initial configuration was based on. In many cases, the measurements will reveal problems which could not have been predicted from the initial assumptions on propagation or traffic. Thus, even without the aforementioned environmental changes, the configured parameters should be checked and potentially updated during the operation.

The expression 'self-optimisation' summarises mechanisms which optimise parameters during operation, in particular based on measurements from the network. The demarcation towards 'self-configuration' is not always unique, that is, Dynamic Radio Configuration functions (cf. Section 4.2) like Automatic Neighbour Relation (ANR) are executed during the initial self-configuration phase and then re-executed (when required) as a self-optimising function in the operational phase. The third SON area, 'self-healing' will be treated in the next

LTE Self-Organising Networks (SON): Network Management Automation for Operational Efficiency, First Edition.
Edited by Seppo Hämäläinen, Henning Sanneck and Cinzia Sartori.
© 2012 John Wiley & Sons, Ltd. Published 2012 by John Wiley & Sons, Ltd.

chapter. In this book, the following features are categorised as 'self-optimisation' and discussed in detail in this chapter:

- *Mobility Robustness Optimisation* guarantees proper mobility for the users, that is, proper handovers and re-selection between cells of the same, but also of a different RAT.
- *Mobility Load Balancing* and *Traffic Steering* try to optimally distribute traffic over the cells in particular due to load condition, but also due to other properties such as velocity, QoS or energy consumption. Obviously there is significant interaction with mobility robustness optimisation which is addressed explicitly.
- *Energy Saving* has gained more and more interest than in the past. Savings are achieved on both network and UE side, for example, through switching off inactive network nodes or, for example, reducing transmit power.
- *Coverage and Capacity Optimisation* continuously adapts in particular antenna tilts and transmit powers to maximise coverage, but also to optimise capacity through minimising interference between the cells.
- *RACH Optimisation* finds the best trade-off between performance of the random access and the resources which have to be sacrificed for it.
- Finally there are *parameters for radio resource management* which need adaptive configuration. Many of those parameters can be optimised based on local information inside the eNB and thus are not treated here, for example, configuration of control channels. Others have a clear SON character and will be discussed in detail, namely optimisation of parameters for power control and inter-cell interference coordination.

5.1 Mobility Robustness Optimisation

The general task of Mobility Robustness Optimisation (MRO) is to guarantee proper mobility, that is, proper handovers in connected mode and proper cell re-selection in idle mode. In the following the expression *cell changes* will be used synonymously for both handover and cell re-selection. The explanations will cover the mobility inside LTE (i.e. intra- and inter-frequency mobility), as well as mobility from LTE cells towards other RATs (i.e. inter-RAT mobility). For the sake of simplicity inter-RAT mobility from other RATs towards LTE and mobility inside 2G/3G will be excluded.

5.1.1 Goals of MRO

First the MRO goals will be described in detail. The following list is in the order of importance.

- **Minimise call drops:** In the worst case, a mobility problem leads to a call drop. Call drops obviously lead directly to user unhappiness.
- **Minimise Radio Link Failures (RLF):** After an RLF a connection can be re-established in many cases before the call drops. This special procedure may conceal RLFs for the subscriber; thus pure RLFs are less critical than call drops, though still an indication of a mobility problem. Connection re-establishment is only possible inside LTE, and not if another target RAT is involved.

- **Minimise unnecessary handovers:** The most prominent example for unnecessary hand-overs are ping-pongs, that is, repeated handovers between two cells within a short time. Unnecessary handovers lead to inefficient use of network resources. Furthermore, since user throughput is reduced during handover procedures, those mobility problems may have an impact on user perception as well.
- **Minimise idle mode problems:** Proper re-selection is essential such that a connection can immediately be setup at any time (by paging from network side, or by random access from UE side). Thus, MRO has to guarantee that the user is camping on a proper cell at any time.

5.1.2 Cell Changes and Interference Challenges

Cell changes are initiated due to radio reasons, that is, based on signal strength or signal quality, or due to other policies which are typically summarised under the expression traffic steering. Traffic steering and its interaction with MRO are discussed in the next section. This section will focus only on radio reasons.

First of all the cell changes are categorised into two classes which primarily differ due to the interference situation, namely intra-frequency and inter-frequency cell changes, where the latter includes the inter-RAT case.

For *intra-frequency* mobility, that is, cell changes between two LTE cells using the same carrier frequency, source and target cell interfere with each other. When moving from the serving/selected LTE Cell A to a target LTE Cell B, not only the signal strength of the source Cell A gets smaller, the interference produced by the target Cell B gets larger as well. As a consequence the signal quality drops faster than the signal strength, and a cell change is necessary when the target cell is stronger than the source cell, even if the source cell is still strong. The top plot of Figure 5.1 schematically shows the signal strength (dashed curves) and the SINR (solid curves) of a UE moving from a source Cell A (dark) to a target Cell B (grey). Note that dB/dBm range on both y-axes is identical. Although the signal strength is always high compared with the thermal noise (in the range of -123 dBm), the SINR gets very low as soon as the target Cell B gets better than the source Cell A.

As a consequence, the intra-frequency handover needs to be initiated as soon as a new cell is better than the current cell, irrespective of the absolute signal strength. Figure 5.2 shows an excerpt of a hexagonal network with an inter-site distance of 500 m. Two eNBs are shown and 46 dBm transmit power is used. Dark areas indicate UE locations where there is free choice to connect the terminal either to one cell or another, typically referred to as handover regions. In contrast, bright areas indicate that UEs in those locations can only be connected to one cell, since otherwise the interference from this cell would be too high. Despite the high transmit power (which typically ranges 1 km or longer) and the small inter-site distances, the handover regions are very limited. Consequently the cell changes have to happen in rather precise moments, in particular for fast moving terminals. Note that the figure has been created under the assumption that all cells are fully loaded; the handover regions will slightly grow with decreasing load.

Another important property of intra-frequency mobility is the fact that an LTE terminal permanently measures neighbouring cells on the same carrier frequency. Finally, the pre-paration time for intra-frequency handovers can be rather short through the usage of the X2 interface.

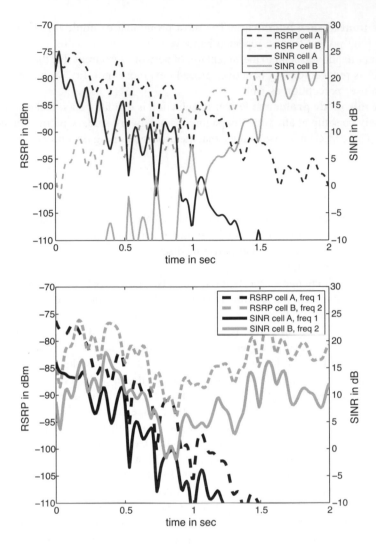

Figure 5.1   Signal Strength (RSRP, dashed) and SINR (solid) towards two cells A (dark) and B (grey) of a terminal moving from A to B. Top: intra-frequency (A and B use same carrier frequency), bottom: inter-frequency (A and B use different carrier frequencies).

In the *inter-frequency case* the source and the target cell use different frequency carriers and thus do not interfere with each other. In terms of interference situation the cell change from one LTE frequency layer to another and the cell change from one RAT to another (e.g. from LTE to 3G) is the same. Signalling is certainly different so later inter-frequency and inter-RAT case will be treated separately. There are some important special cases:

- Co-sited deployment with different frequency bands, for example, co-sited 3G/LTE, or two LTE carriers (if an operator has non-contiguous LTE spectrum): If the frequency bands are not too far apart, the radio conditions will be very similar on both layers. Cell changes due to radio reasons between the layers are not necessary as long as both layers provide full

Figure 5.2 Intra-frequency handover regions (dark grey/black). Light shading indicates areas where the users can only be connected to a single cell.

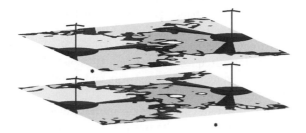

Figure 5.3 Co-located inter-frequency deployment. Users can be connected to both layers at every place.

coverage. The areas where the UEs can be connected to one or the other layer cover actually the whole plane in this case Figure 5.3 gives an example for a co-sited deployment, where the propagation conditions differ very little between the layers (the shape of the cell boundaries is different).

• Coverage layer plus capacity layer, for example, 800 MHz LTE everywhere plus 2.6 GHz LTE in dense areas, or macro layer plus separate frequency layer for picos/femtos: In such a scenario, where (at least) one layer has restricted coverage, UEs must change the cell when they leave coverage of one layer. From a pure MRO perspective, it would be best to keep all UEs in the coverage layer since this would (almost) remove the need to do handovers at all. However, this is obviously not a reasonable solution since traffic must be offloaded to the capacity layer through other mechanisms (traffic steering, cf. Section 5.2) Figure 5.4 shows such a network, where the lower is the coverage layer and the upper layer is just given by some low power hot spot cells. The black area is not covered by the upper layer. UE connected to the upper capacity layer and leaving the bright area have to change the cell, however, should also exploit the capacity layer coverage as much as possible.

Figure 5.4 Deployment with coverage layer (bottom) and capacity layer/hot spots (top).

In general inter-frequency handovers must be done whenever signal strength and/or signal quality of one layer becomes bad and signal strength/quality of another layer is sufficient. The bottom plot in Figure 5.1 schematically shows signal strength and SINR in a scenario, where a UE moves from Cell A on Frequency Layer 1 out of coverage.

In contrast to intra-frequency measurements, inter-frequency (and inter-RAT) measurements are typically not done permanently and by every terminal. The reason is that those measurements require measurement gaps, since the oscillators have to be tuned to the other frequency. The UE cannot be scheduled during those gaps, that is, the data rate drops. As a consequence, the configuration of the measurements has to be done much more economically and plays a much more important role in inter-frequency mobility compared with intra-frequency mobility. Call drops may happen even without the UE being aware of another frequency layer.

Preparation of an inter-frequency handover within LTE can also use the X2 interface and is therefore comparable to the intra-frequency handover. However, the preparation time for an inter-RAT handover has to go through the core network and therefore is longer. This has to be taken into account when initiating those.

Summarising this section, different challenges for intra-frequency and inter-RAT mobility have been observed. Cell changes have to be much more exact in the intra-frequency case, whereas there is much more degree of freedom in the inter-frequency case. However, in the latter case, the availability of measurements is a critical issue since mobility measurements are very expensive there. In the special case of inter-RAT handovers a longer preparation time has to be taken into account.

5.1.3 MRO Relevant Parameters

This section will review the mobility procedures in LTE. Focus will be on the most relevant parameters involved which are crucial for MRO, so this section is not to be understood as a detailed description of LTE mobility. Some of the introduced aspects will be reused in the traffic steering Section 5.2.

In idle mode, the cell re-selection decisions are done by the UEs and the behaviour and the parameters are specified in 3GPP. In connected mode handover decisions are made in the eNBs (or RNC/BSC, respectively) with the help of measurements made by the UEs and the underlying policies and parameters are vendor-specific in principle. However, a typical assumption is that handover decisions directly follow UE measurement reports. The eNBs configure triggers for those reports, such that they are sent when certain conditions are fulfilled. Alternatively the eNBs can also force the UEs to report periodically, but this implies either large overhead (if periodicity is fast) or the risk to miss suitable handover occasions (if periodicity is slow). Therefore, it is typically assumed that the handover behaviour is influenced by changing the configuration of the measurement reporting. Even with periodical measurement reporting and vendor specific handover decisions, it is very likely that the internal rules will follow similar principles as specified for measurement reporting.

This assumption is also followed in this book. Table 5.1 summarises the involved measurement reporting events for the different mobility cases in connected mode. Implementations may vary from this table, but it will be used as baseline for the following description. Recall that only mobility inside LTE and mobility from LTE to other RATs is

Table 5.1   Connected mode parameters and events addressed by MRO

Source	Target	Event	Comment
LTE f1	LTE f1	A3: Neighbour becomes *offset* better than serving	Intra-frequency cell-specific
LTE f1	LTE f2	A5: Serving becomes worse than *threshold1* and neighbour becomes better than *threshold2*	Inter-frequency cell-specific
LTE	2G/3G	B2: Serving becomes worse than *threshold1* and inter-RAT neighbour becomes better than *threshold2*	Inter-RAT not cell-specific

treated here. The relevant parameters are written *italics*. Some of the parameters can be configured in a cell specific way, that is, a cell can configure individual parameters for each of its neighbours so that each cell boundary can be optimised individually. Note that the eNBs can configure further measurement reporting events in addition, even of the same kind, for other purposes, such as MDT or ANR.

For each measurement reporting event there are two further parameters which combat fluctuations of the channel. Note that those are per event and not cell specific, that is, in one cell the same value is used irrespective of the neighbour to which the condition is checked.

- **Time to trigger (TTT):** A report is not sent immediately after the corresponding condition is met. Instead, the condition has to be fulfilled for a certain period indicated by TTT. This avoids that handovers are triggered by measurement outliers. Long TTTs lead to more stable behaviour; however, they may delay the HO decisions unnecessarily which may cause problem for fast UEs.
- **Filter coefficient (FC):** The UE applies filtering on Layer 1 level to average out the fast fading. The exact procedure is not specified in 3GPP, typical averaging lengths are between 50 and 200 ms. However, in addition a UE has to apply a recursive averager to its Layer 1 measurements, and the time constant is configured by the eNB via a filter coefficient index. The relation between this index and the time constant is given in the right plot of Figure 5.5. This stabilises the measurements more directly than TTT since it smoothes the measurements before the condition is checked at all. Similar to TTT, a large time constant also delays the measurements and thereby the handover decisions risking proper mobility in particular for fast users. Note that further averaging can also be done on the network side when periodic reporting is configured, but averaging in the UE lowers the signalling overhead significantly.

Figure 5.5 depicts the standard deviation of the measurement error on the RSRP (reference signal received power) versus the time constant of the Layer 3 filter. Only the impact of the residual fast fading is considered, errors due to noisy estimates are neglected here. Without Layer 3 filtering (time constant 0), the standard deviation is rather large. The standard deviation can be decreased by investing averaging time. This is further illustrated in Figure 5.6 which shows the measurement of the signal strength at several stages. Light grey dashed is the instantaneous signal strength for a 3 km h^{-1} channel. Black dashed is the Layer 1 measurement which still fluctuates a lot leading to unstable handover decisions. The solid curves show the Layer 3 output which is used for the event conditions. Higher filter coefficients smooth the measurements, however, it is obvious that the capability of following the channel is delayed: for

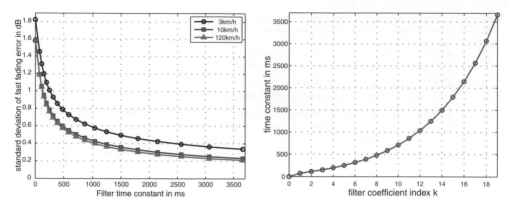

Figure 5.5 Left: Impact of Layer 3 filtering on fluctuation of RSRP measurements (only fast fading impact). Right: Filter time constant versus Layer 3 filter coefficient.

example, for $k = 11$, Figure 5.5 predicts a time constant of $\sim$850 ms, which can be recognised by the delay between the Layer 1 average and the Layer 3 average with $k = 11$. Furthermore, extreme filter coefficients start averaging out slow fading as well, which is not a desired behaviour. In both figures no further measurement errors beyond the impact of fast fading are assumed, for example, due to noisy estimates.

In connected mode, the reporting configuration is sent individually to each UE via RRC signalling after connection setup, after a handover, or whenever a parameter update is intended. So in principle every UE can be configured with individual parameters. However, since the choice of parameters heavily depends on the geographical environment on a cell boundary (shadowing situation, streets, etc.), one cell typically uses the same parameters for the served UEs. Furthermore, UEs may change some properties rather fast, such as velocity, so there is

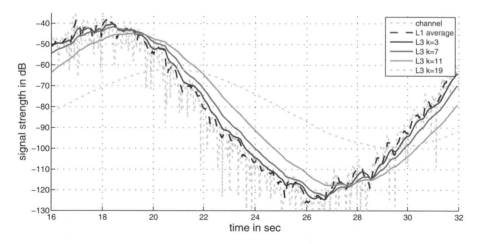

Figure 5.6 RSRP measurements at several stages: plain signal, Layer 1 average and Layer 3 average with different filter coefficients.

Table 5.2 Idle mode parameters and events addressed by MRO

Source	Target	Parameter and Description	Comment
LTE f1	LTE f1 (or f2 with same prio)	Neighbour becomes Q_{offset} better than serving	Intra-frequency, inter-frequency
LTE f1	LTE/3G/2G f2 higher prio	A cell on f2 becomes better than $Thresh_{x,high}$	Inter-frequency, inter-RAT
LTE f1	LTE/3G/2G f2 lower prio	Serving cell becomes worse than $Thresh_{serving,low}$, and a cell on f2 becomes better than $Thresh_{x,low}$	Inter-frequency, inter-RAT

little time for a UE specific optimisation. However, every cell will have different parameters and recall that some parameters are even cell-pair specific.

In idle mode, the re-selection parameters are broadcasted within System Information Block (SIB 3), so those parameters cannot be user specific at all. Recall that the behaviour in idle mode is exactly specified in 3GPP. In addition to the thresholds which are similar to those in connected mode, each frequency layer is assigned a re-selection priority which influences the re-selection behaviour in the inter-frequency case. Table 5.2 summarises the most relevant idle mode parameters.

Equivalent to TTT in connected mode, there are similar timers called *Treselection* in idle mode to avoid re-selection due to measurement outliers. This timer can be different for every frequency layer.

If a UE is camping on a frequency layer with lower priority than another layer, it needs to perform regular inter-frequency or inter-RAT measurements towards the higher layers. Obviously there is no impact on throughput as described for measurements in connected mode. However, too frequent inter-frequency and inter-RAT measurements in idle mode reduce the battery lifetime and thus have to be configured economically as well.

All thresholds discussed for connected and idle mode can be configured either in terms of signal strength (RSRP, RSCP and RXLEV in LTE, 3G and 2G), or signal quality (RSRQ, EcNo, RXQUAL in LTE, 3G and 2G). Signal quality includes information about interference and thus seems to be the more relevant information. However, due to the way it is defined, it also depends on the load in the serving cell. For MRO this dependency is not desired, for example, when looking at the RSRQ towards a target cell. Figure 5.7 illustrates the load-dependent relation between SINR and RSRQ schematically. As a reference the unity slope is added (black dashed). For the same SINR, the RSRQ measurement may differ by up to 8dB depending on the load. For instance, the RSRQ of a UE may drop by 8dB when it starts a download in an empty cell (with 0% load) and is assigned all the resources (100% load), although neither the SINR nor the RSRP have changed. This obviously makes the use of signal quality dangerous in terms of ping-pong behaviour, in particular in low loaded cells. Furthermore, this complicates the configuration of reasonable thresholds in terms of RSRQ.

The rest of this MRO chapter will concentrate on the connected mode. It has already been discussed that idle mode mobility problems are less critical. Furthermore, the MRO work in 3GPP has focused on connected mode as well with the reasoning that idle mode configuration can be derived from the optimised connected mode configuration. Idle mode mobility procedures are further investigated in the traffic steering Section 5.2.

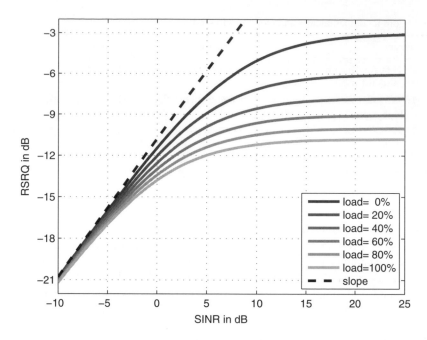

Figure 5.7   Load-dependency of RSRQ measurements (with unity slope for comparison).

5.1.4   Causes for Mobility Problems

This section will describe handover problems in detail. The same classification as done in 3GPP is made. Too late handovers, too early handovers, handovers to wrong cells and unnecessary handovers will be distinguished. Each category is discussed separately for the most challenging case of intra-frequency handovers and comments are given on how this can be applied to the inter-frequency and the inter-RAT case.

The left plot in Figure 5.8 illustrates a *too late handover*. A UE is connected to the left Cell A and moves towards the right Cell B. Cell A initiates the handover too late, or even not at all, so that it suffers an RLF in Cell A. The aforementioned connection re-establishment procedure in LTE will be applied here. After an RLF an LTE UE tries to re-establish the connection

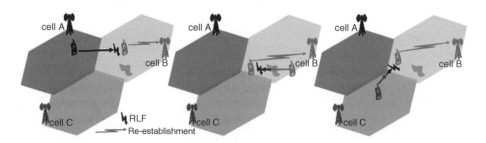

Figure 5.8   RLF-related mobility problems too late handover (left), too early handover (middle) and handover to wrong cell (right).

in the strongest LTE cell which it may find, that is, Cell B. More precisely in the case of the too late handover the UE would send a re-establishment request to the right Cell B. Note that for the description in this section it is irrelevant whether the re-establishment procedure is successful or not.

Whereas too late handovers are the most obvious mobility problem handovers can also be initiated *too early*. In general *too early handovers* happen if a handover is initiated towards a cell whose connection is not stable enough yet. An example is given in the middle plot of Figure 5.8; a UE is connected to the right Cell B, and it moves through a small coverage island of Cell A. Those islands are typically a consequence of coverage overshoots, or of strong shadowing. Cell B might initiate a handover to Cell A. In some cases, the handover to such an unstable cell will not be successful and the UE would simply stay with Cell B without RLF. In other cases, the handover may be successful and there is still enough time to handover back to Cell B. However, in many cases an RLF happens in the left Cell A before this back handover is executed, that is, shortly after the successful handover from B to A. Exactly as in the too late handover case, the UE would send a re-establishment request to the right Cell B. It is important to mention, that from Cell B's perspective, the too early and too late handover are identical since it will not recognise the UE. In both cases it will receive a re-establishment request from an unknown UE. A concrete and illustrative example for a too early handover is discussed later in the simulation section (cf. Figure 5.16).

The last class with involved RLF are *handovers to wrong cell*. An example is given in the third part of Figure 5.8. A UE is connected to the bottom Cell C and moves towards the right Cell B. For some reason Cell C initiates a handover to the left Cell A although the handover should have been to Cell B. This could again be a consequence of a coverage island of Cell A as in the previous case, or to another challenging shadowing situation. After the handover has succeeded, the UE may suffer an RLF in Cell A if the signal degrades significantly before a proper handover to Cell C is executed. The UE will send a re-establishment request towards the right Cell B, just as it did in the case of too early and too late handover.

The previous explanations were based on the assumption that all cells are LTE cells of the same frequency. If the LTE cells are of different frequencies, the problems are actually the same; the only difference is that the underlying radio root cause is typically not interference but probably loss of coverage as discussed in the previous section. Also in the inter-RAT case the problems are very similar. The most significant difference is that re-establishment requests are not sent, RLF will always lead to a call drop if no LTE cell is found after RLF. Instead, higher layers will initiate an immediate re-connection to an available cell which acts similar as the re-establishment.

There are also mobility problems without radio link failure called *unnecessary handovers*. As already mentioned at the beginning, this class of mobility problems is less critical since the impact on user perception is limited. However, they produce unnecessary overhead for the network and thus should be avoided as well. Several cases will be distinguished:

- **Ping-pongs:** Cell A handovers a user to Cell B and Cell B handovers the same user back to Cell A shortly after. In most of the cases, those two successful handovers can be avoided, for example, by a larger handover hysteresis. The left plot in Figure 5.9 shows two ping-pong situations.
- **Short stays/rapid handovers:** Cell A handovers a user to Cell B, and Cell B handovers the same user to another Cell C shortly after. In some cases, one of the two handovers can

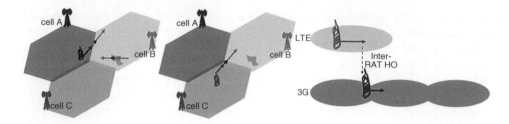

Figure 5.9 Unnecessary handovers ping-pong (left), shortstay/rapid handover (middle) and unnecessary handover to lower priority layer.

be saved by direct handover from Cell A to Cell C. An example is given in the middle plot of Figure 5.9.

- **Unnecessary handover from a high priority frequency layer to a lower priority frequency layer:** In 3GPP terminology unnecessary handovers also cover the situations where one frequency layer (typically with high priority) initiates a handover of a UE to another target layer although the UE could have stayed with the source layer. As a consequence, the coverage of the source layer is not fully exploited. A schematical example is given in the right plot of Figure 5.9 LTE handovers the UE far before coverage ends which is unnecessary.
- **Handover right after connection:** Assume that an idle user is camping on Cell A and connects to the network. The UE will connect to Cell A, but Cell A initiates a handover to another Cell B. This is an indication of a misalignment of idle mode and connected mode parameters. In many cases those handovers can be saved.

Individual cell boundaries are often dominated by one special class of handover problems as described above, in particular if there are streets such that a vast majority of the users cross a cell boundary on the same location, same angle and similar velocity. In those cases, for example, if there are only too late handovers, the potential benefits of MRO are large. Otherwise, if a cell boundary suffers for example, from both too early and too late handovers, improving one class would degrade the other, hence reducing the benefits of MRO. This will be further illustrated by the simulation results later on.

5.1.5 MRO Solutions

It is important to mention that MRO has already been done in the past, but not in an automated way. It was done by human beings who were extensively looking at key performance indicators from the network (such as call drop and RLF statistics), as well as at drive test data. Those statistics were combined with geographical information and with human experience from the past. Changes were proposed for the most critical cell boundaries, in many cases following a trial and error procedure.

- An automatic SON solution for MRO shall not only automate the current manual procedures, but shall also;
- Optimise more/all cell boundaries (not only the most critical ones);
- Optimise more often (permanent capability to react on changes in the network);

- Be based on better statistics (more reliable and less based on human experience); and
- Allow for a de-centralised solution (to keep complexity away from the core network).

The MRO problem can be split into two subtasks, the *root cause identification/evaluation* and the actual *correction of the mobility parameters*. 3GPP has primarily dealt with the former problem, the latter is assumed to be more vendor specific. As in many areas, the actual correction is much less challenging once the exact root cause has been identified.

3GPP has also specified multi-vendor management (i.e. switching the SON function on and off, target setting) and monitoring capabilities for MRO in (3GPP TS32.522, 2011).

5.1.5.1 Root Cause Identification

The general paradigm of the root cause identification is that the 'guilty' cell and the type of a mobility problem have to be identified. The 'guilty' cell is defined as the cell which has caused a specific mobility problem, that is, which should adapt its parameters. If a distributed MRO solution is in the focus the information shall also be brought to the guilty cell, such that this can adapt its parameters. This is inherently accomplished since the knowledge of the guilty cell is needed to uniquely identify the root cause.

The simple example of a too late handover as explained previously will be used to illustrate the challenge of the root cause evaluation. Assume the case of a too late handover from Cell A to Cell B, that is, the handover never happened although handover to Cell B would have been possible. From Cell A's perspective, it simply has lost the UE without initiating a handover, or after initiating the handover. There is no further information at Cell A, although Cell A is the guilty cell here. If Cell B receives the re-establishment request, it knows that the UE had an RLF in another cell, but no further history information. Following the discussion from the previous section, this is not enough to determine the root cause, it could be a too late handover from the previous Cell A, but it could also be a too early handover from Cell B itself (it cannot recognise the UE since it was assigned a new identification by Cell A), or a wrong cell handover from a third cell.

A unique root cause evaluation requires two components:

1. Most of the knowledge about a mobility problem is available at the UE side. So additional reports form the UE have to be considered.
2. Every cell has some knowledge about an RLF. So the cells have to forward their knowledge until the guilty cell has been found.

Both components obviously require support from standardisation. A pure vendor specific solution will not lead to unique root cause evaluation. In the following, we will describe how 3GPP features in different releases help to solve the problem.

The connection re-establishment procedure is already part of *Release 8*. The re-establishment request sent by the UE to the strongest LTE cell already contains useful information about the previous cell, in particular the C-RNTI of the UE in the previous cell (which was the identification of this UE in the previous cell), and the physical cell ID (PCI).

Release 9 has introduced the following features:

- **RLF Report:** Once the re-establishment procedure is successful, the new cell can request the UE to report the last measurements before the RLF.

- **RLF Indication:** The cell receiving the re-establishment request can forward this information to the previous cell via X2 interface. The assumption is certainly that the previous cell is a neighbour, so that the PCI is enough to uniquely identify the old cell. The previous cell will be able to recognise the UE by help of the C-RNTI provided it has stored the UE context. To this end the cells are forced to store the UE context for a certain time after a terminal has run out of sync. The RLF Indication can be sent already after receiving the re-establishment request, or after receiving the RLF Report (with successful re-establishment). Thus, the procedure does not require Release 9 capable terminals.
- **Handover Report:** The previous cell receiving the RLF indication assesses this event by combining the RLF Indication with own information. If it is not guilty for this RLF (i.e. in case of too early HO or HO to wrong cell), it can send a Handover Report to the guilty cell via the X2 interface. Practically it will check whether there was a successful handover of the concerned UE shortly beforehand. In this case it will assume that the root cause was this successful handover (too early handover, handover to wrong cell). The handover report contains in particular the root cause of the mobility problem determined by the previously connected cell. Since the Handover Report typically follows a successful handover, it can be assumed that the cells are neighbours and the X2 interface exists.

Those features enable a unique detection of too early handovers, too late handovers, and handover to wrong cell, provided that the re-establishment request has been received and the features are supported by the involved cells. Furthermore, the procedures guarantee that the guilty cell is informed about the problem and its root cause. The following figures illustrate the identification of too early handovers, too late handovers and handovers to wrong cell by help of message sequence charts.

The first figures (Figure 5.10) describe the detection of the too late handover. The UE moves from Cell A to Cell B, suffers an RLF (1) and sends a re-establishment request ('RER', 2) to Cell B. Cell B may request the RLF report with containing measurements before the RLF (not shown in the figure for the sake of simplicity). Cell B informs Cell A about the detected RLF via RLF Indication (3). Cell A realises that there was no recent handover of this terminal, so Cell A knows that it has created the 'too late handover' problem.

The next Figure 5.11 describes the detection of the too early handover. The UE moves in Cell B through a coverage island of Cell A. It is successfully handovered to Cell A (1), but suffers an

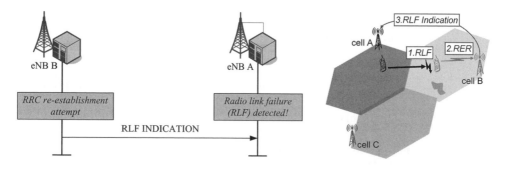

Figure 5.10 Message Sequence Chart of too late handover. (Holma and Toskala, 2011). Reproduced with permission from John Wiley & Sons, Ltd.

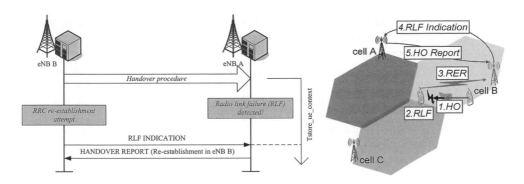

Figure 5.11 Message Sequence Chart of too early handover (Holma and Toskala, 2011). Reproduced with permission from John Wiley & Sons, Ltd.

RLF shortly after (2) and sends the re-establishment request to Cell B (3). Cell B informs Cell A about the RLF via RLF indication (4). Cell A realises that there was a recent handover shortly beforehand such that it had no chance to save the UE. So it will inform Cell B via HO Report (5), that it has created a too early handover.

The final Figure 5.12 describes the detection of the handover to wrong cell. The UE moves from Cell C to Cell B, closely passing Cell A. It is successfully handovered to Cell A (1), but suffers an RLF shortly after (2) and sends the re-establishment request to Cell B (3). Cell B informs Cell A about the RLF via RLF indication (4). Cell A realises that there was a recent handover shortly beforehand such that it had no chance to save the UE. So it will inform Cell C via HO Report, that it has created a handover to wrong cell.

Ping-pongs have not been addressed in Release 9 since they can be read from the UE history which is exchanged along with the UE context during the handover. The UE history also contains the time the UE has stayed in each cell. For a ping-pong: 'Cell A – Cell B – Cell A', the UE history will tell Cell A that the UE was already in Cell A shortly before. Note that Cell A will not recognise the UE, since the previous C-RNTI is lost.

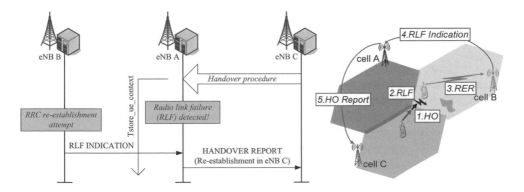

Figure 5.12 Message Sequence Chart of handover to wrong cell (Holma and Toskala, 2011). Reproduced with permission from John Wiley & Sons, Ltd.

Idle mode has not been addressed in Release 9, the assumption is that idle mode parameters can follow the active mode parameters since the underlying radio conditions and user movements are at least very similar and requirements are less strict (idle mode can afford worse radio conditions). This would also avoid unnecessary handovers after connection.

Release 10 will extend the RLF reports. The Release 9 procedure relies on the fact that at least the re-establishment request is successful. Even if the request is successful, but the re-establishment procedure is not successful (and the UE goes to idle), the additional measurements contained in the RLF report are lost. Therefore Release 10 extends the RLF Report such that it can also be sent after the UE went to idle mode.

Whereas Release 9 and 10 SON was focusing on intra-LTE MRO, *Release 11* will introduce inter-RAT features. Similar to the Release 9/10 procedures, suitable UE reports and information exchange between the cells have to be introduced (or existing procedures have to be extended). Note that in 2G and 3G the BSC and RNC are the responsible radio nodes instead of the base stations and as interfaces between eNB, RNC and BSC the RIM procedure can be used as already done for load balancing.

5.1.5.2 Correction of Mobility Parameters

The actual optimisation step, that is, the correction of the mobility parameters, is vendor-specific to a wide extent. Both centralised and distributed solutions are possible, as shown in Figure 5.13. Note that the 3GPP features for root cause evaluation are distributed solutions. In any case, the corrections will typically not be done based on individual events, but on statistics. Every cell will count the identified handover problems caused by itself over a certain period of time. Those *Key Performance Indictors* (KPIs) are typically generated separately for every neighbour cell.

In the *centralised* solution the cells report those KPIs to DM, and let the DM decide parameter changes. The advantage is that the DM can also consider KPIs from other cells and hence can make a more global decision. SA5 has specified the necessary KPIs on the Itf-N interface to enable multi-vendor solution.

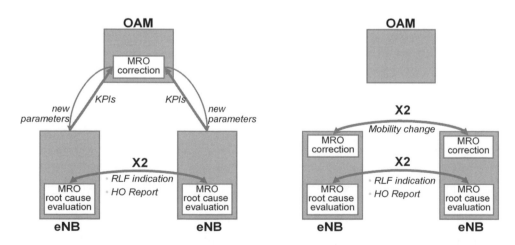

Figure 5.13 MRO architectures: centralised (left) and distributed (right).

Table 5.3 Dominating mobility problems and potential counter measures

Dominating Handover Problem	Action	Parameter Changes
Too early (intra-freq)	Postpone handover	Larger A3 offset, larger TTT/FC
Too early (inter-freq)	Postpone handover	Smaller A5-Thresh1 and/or larger A5-Thresh2, larger TTT/FC
Too late (intra-freq)	Prepone handover	Smaller A3 offset, shorter TTT/FC
Too late (inter-freq)	Prepone handover	Larger A5-Thresh1, smaller A5-Thresh2, Shorter TTT/FC
HO to wrong cell B instead of C (intra-freq)	Postpone handover to B and/or prepone handover to C	Larger A3 offset for B and/or Smaller A3 offset for C Larger TTT/FC
HO to wrong cell B instead of C (inter-freq)	Postpone handover to B and/or prepone handover to C	e.g. larger A5-Thresh2 towards B, Smaller A5-Thresh2 towards C
Ping-pong (intra-Freq)	Postpone handover	Larger A3 offset, larger TTT/FC

In the *distributed* solution every cell will autonomously evaluate the created KPIs, and derive parameter changes from those. The main advantage is that it keeps the MRO complexity in a single node (eNB, RNC or BSC), each eNB is treating its own problems, and DM does not need to be involved. Furthermore complexity in this case will not scale with the size of the network. One could also argue that such a solution allows for more dynamic adaptation than an OAM-based solution; however, the bottleneck might be given by the generation of reliable KPI statistics which will be discussed in Section or the sake of completeness the *mobility change* procedure which has been introduced in Release 9 for mobility load balancing should be mentioned. With this mechanism a cell can propose changes of handover parameter to its neighbour via X2 interface. This negotiation can also be used for further coordination with the neighbouring cells in the distributed case.

Although the adaptation is vendor specific, some examples will be given. Table 5.3 summarises possible actions if a particular mobility problem dominates a cell boundary. The last column contains the concrete parameters changes associated with action. The table focuses on the intra-LTE case, the inter-RAT actions will be similar to the inter-frequency actions however there is no inter-RAT root cause evaluation yet provided by 3GPP.

As already discussed when introducing TTT and FC, those two parameters can also be considered to be UE specific. In general, fast UEs would need short TTT or small FC, and slow UEs would need the opposite. Such UE specific properties are not well-known on the level of the OAM system, so this kind of optimisation can only be done in the eNB. However this approach requires velocity estimates which are not easy to obtain.

5.1.6 MRO Time Scales

As many other self-optimising algorithms, MRO represents a control loop and therefore it must be guaranteed that stability is not endangered. MRO decisions are based on observations from the past, so appropriate observation periods have to be selected such that the user movement after MRO decisions is at least very similar to the one during the observation. For instance, daily MRO decision would inherently assume that users behave similarly on Mondays,

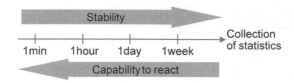

Figure 5.14   Time scale trade-off.

Tuesdays, and so on. In mathematical words, stable MRO requires stationary behaviour to some extent. Furthermore, sufficient statistics, that is, samples of mobility problems must be available to avoid reacting on unusual situations. This implies that the collection of the statistics should cover sufficiently long periods, although especially the distributed MRO implementation would be able to react very fast in principle. Otherwise there is a risk of deteriorating rather than improving mobility.

In very populated areas the statistics should converge after several minutes, however in areas with few users it may take several hours until you have collected enough events. User movement behaviour typically changes in busy hours, between day and night, and it may change between work days and weekends. So the design of MRO requires defining a trade-off between stability and the capability to react on changing UE behaviour. This trade-off is illustrated in Figure 5.14. Note that those stability requirements may limit the dynamics of MRO more than the involved signalling.

Optimisation of TTT and FC could be considered as an exception, if based on UE velocity estimates. Velocity estimates can be obtained not only during handover situations, for example, from Layer 1 Doppler measurements. If those are considered reliable enough, corrections can be applied faster than aforementioned.

As a final remark in particular for inter-RAT MRO, the UE may not re-connect immediately after an RLF. As a consequence, after-idle reporting of a mobility problem might be delayed, so that the MRO observation period should be much larger than those delays.

5.1.7   MRO Performance

Looking again at Figure 5.2 it can be observed that even a single boundary exposes very different shadowing situation, the handover regions can be narrow (which may lead to too late handovers), or wide (which may lead to ping-pongs or too early handovers). If the users cross such a boundary at arbitrary places, in different angles and with different velocities, a mixture of all the explained problems is a consequence. In such a case the potential of MRO will be rather limited. Improving one situation (e.g. too late handover) will degrade others (e.g. too early handovers). However, in reality cell boundaries are very often dominated by a particular user movement. For instance, there might be streets which force a majority of the users to cross a cell boundary on a particular place, in a particular angle and maybe even with a predominant velocity. As another example, a cell boundary may go through a crowded place where users are very static. As a consequence those cell boundaries are also dominated by particular types of mobility problems. And those are the situations where MRO can improve a lot.

As a consequence, MRO simulations cannot be based on random and arbitrary user movement as discussed in the typical 3GPP simulations according to 3GPP TR36.814 (2010). More elaborate mobility models are needed to evaluate the performance of MRO. For the

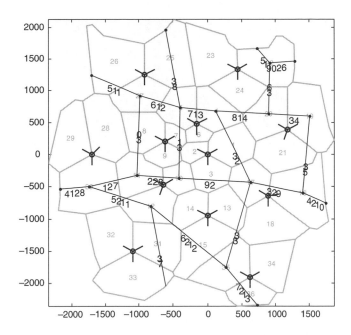

Figure 5.15 Irregular network layout; 'Springwald'.

following performance evaluation an irregular cell layout was chosen as proposed in Turkka and Lobinger (2010). It contains different cell sizes, the ranges being between the 3GPP defined cases with intersite distance of 500m and 1732 m. All users are moving at 30 km/h, following two different mobility models:

1. Users are uniformly distributed in the network, moving arbitrarily and randomly.
2. Users are uniformly distributed in the streets and they are only moving along the streets.

The cell layout including the street layout is given in Figure 5.15. Wrap around is used to reduce border effects. Location-dependent shadowing is used, and the shadowing maps are the same for all investigations, that is, no re-dropping of the shadowing map. This is essential for MRO as explained above. In total 900 users are moving at 30 km/h. Since in the simulation it is guaranteed that the general user behaviour remains constant (stationary), stable statistics are obtained after a rather short time. Here, 90 s are used, although the collection period should be significantly larger in reality, as discussed in the previous section. Long periods in the range of a day would have made the simulation time infeasible. After every collection period of 90 s MRO decides for every cell boundary whether the corresponding cell individual offset shall be changed. Corrections of 0 dB, 0.5 dB or 1 dB are applied. After that, KPI statistics are reset and after further 90 s the next decision is made. Every cell uses an additional handover margin of 3 dB, sometime referred to as 'hysteresis'. That is, in the default setting UEs are executing a handover if a neighbour is 3 dB stronger than the serving cell. Time to trigger is not changed here; it is set to 200 ms.

For such a simulation setup some results will be presented which illustrate MRO properties as discussed in this chapter. Only the intra-frequency mobility is assumed, that is, all cells

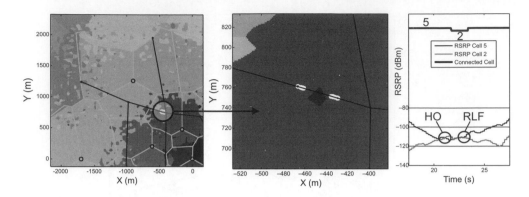

Figure 5.16 Concrete example for too early handover. Left: excerpt of the network; middle: zoom into the circled area; right RSRPs and connected cell.

operate on the same carrier frequency. This has been identified to be the most challenging case in terms of interference.

Before discussing statistical performance results a concrete and illustrative example for a too early handover is taken out of the described simulation setup. The scenario in Figure 5.16 is an excerpt of the irregular layout of Figure 5.15 and it shows a crowded street through a coverage island, so the majority of the users will cross the cell boundary on the street to both directions. The two maps show too early handover events from Cell 5 to Cell 2 (the middle figure zooms into the left figure). The right plot shows the RSRPs of both cells for one UE (bottom) and the cell the UE is connected to (top). The RSRP of Cell 2 exceeds that of Cell 5 only shortly, such that the handover is initiated. However, there is a RLF shortly after the handover. Note that it is a 'normal' coverage island created by the commonly used random shadowing model.

Figure 5.17 shows performance results for the case where all 900 users are moving in the streets. The x-axis basically shows the time axis, one KPI Collection Interval corresponds to 90 s in the simulation, whereas in reality it would tend towards an hour or even a day/week. The y-axis shows the counters for several mobility problems as indicated in the legend. The counter is accumulated over the whole network, that is, all cell boundaries. The left plot shows the

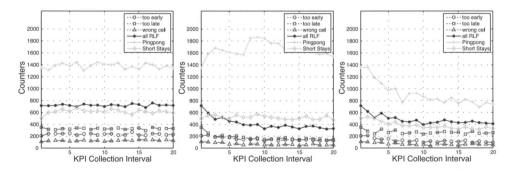

Figure 5.17 Evolution of mobility problems over time, 900 users at 30 km/h in streets. No MRO (left), simple MRO (middle), MRO considering ping-pongs.

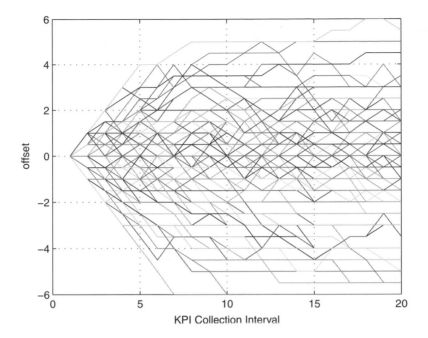

Figure 5.18   Evolution of handover offsets.

reference case where no MRO is used. It can be observed that all events are occurring, 'too late' is the most frequent event associated with RLF and there are even more ping-pongs. Obviously the counters do not change over time (apart from statistical variations). In the middle plot a simple MRO scheme is applied which does not take into account the ping-pongs when correcting the handover offsets. The RLFs are significantly reduced by ~50%. All three classes are improved. Convergence is achieved after 5–10 collection periods, whereas the most significant improvements are already realised after three steps. Unfortunately this seems to be paid by an increasing number of ping-pongs. A more advanced MRO algorithm taking ping-pongs into account as well is applied in the right plot. Now, ping-pongs are reduced as well by ~40%. However, the RLF reduction is slightly impacted, now in the range of 40%.

Figure 5.18 shows how the cell individual offsets behave. Recall that an additional handover offset of 3 dB is used. Stepsizes of 0.5 dB or 1 dB, and the convergence (after 5–10 periods the changes are marginal) can be observed. Furthermore, after convergence handover offsets between −6 dB and +6 dB occur, whereas the negative offsets leading to earlier handovers slightly dominate. This is in agreement with the aforementioned fact that late handovers are more frequent. The figure also demonstrates, that every cell boundary obviously has different optimal offset which emphasises again the need for MRO.

Further clarification is obtained when looking into individual cell boundaries. Three interesting cell boundaries are selected in Figure 5.19. In two of them a single problem is dominating, namely too late handovers for handovers from Cell 29 to 16, and too early handovers from Cell 5 to 2. In both cases the problems can be reduced almost entirely. From Cell 33 to 5, there is a mixture of problems and at a small level. The improvement of the RLF related problems is smaller in this case.

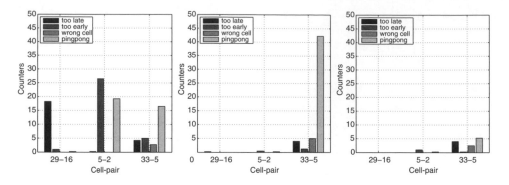

Figure 5.19 Mobility problems on selected cell boundaries.

The next Figure 5.20 is a histogram over the number of RLFs per cell boundary without MRO and with (converged) MRO. Indeed, even without MRO there are many cell boundaries with no or few RLFs. However, there are also some cell boundaries with many RLFs. With MRO, no cell boundary with more than 100 RLFs is left and the number of boundaries with no RLFs is significantly increased. This highlights the fact that the MRO gains can be very large at certain locations, whereas many other locations do not benefit.

Finally, as a comparison Figure 5.21 shows results for the same layout, however with all users moving randomly and arbitrarily ('random walk', instead of moving in the streets).

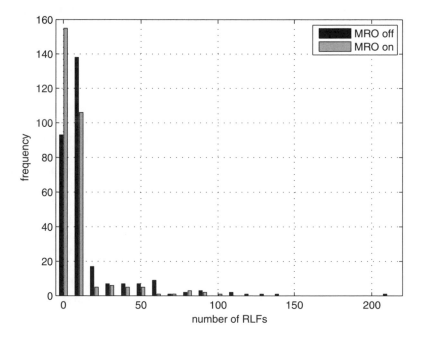

Figure 5.20 Histogram over the number of RLFs per cell boundary within the last ~450 sec of the simulation (i.e. converged results).

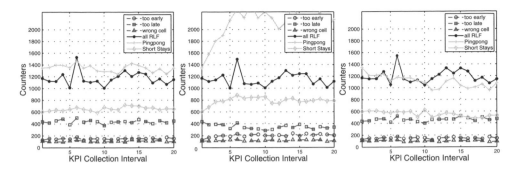

Figure 5.21 Evolution of mobility problems over time, 900 users at 30 km/h with random walk (no preferred directions/HO locations). No MRO (left), simple MRO (middle), MRO considering ping-pongs.

Again the left plot shows the reference without MRO. The total number of RLFs is higher, since users move through coverage holes as well, which leads to further RLFs. In general the fluctuation is a bit larger in this case, that is, the statistics are less stable. Using MRO without ping-pong consideration (middle plot), hardly any improvement is visible. 'Too late' problems are traded against 'too early' problems and ping-pongs are significantly increased. With ping-pong consideration in the right plot, no significant changes are observed.

With the comparison of the two extreme cases, 'all users in a few streets' and 'all users with random walk' it was illustrated that MRO benefits can be large when a cell boundary is dominated by a common behaviour of all users. If a cell boundary is crossed at all places and in different angles, MRO benefits will diminish, and reasonable (symmetric) default values will be sufficient (and not corrected by MRO).

5.2 Mobility Load Balancing and Traffic Steering

5.2.1 Introduction to Traffic Steering

Traffic steering is the ability to control and direct data and voice traffic to the best suitable cell layer and radio technology within any network. It becomes a key feature in LTE networks deployed over multiple layers, such as frequency layers and/or hierarchy layers as macro, pico or femto cells. Those networks make available resources from multiple layers to an end-user in a certain geographical area. Therefore, traffic steering could optimise network capacity and user experience via an efficient utilisation of the whole pool of available resources (e.g. radio, transport, backhaul). Traffic steering can also be used when the LTE network coexists with other radio accesses such as GSM/EDGE, WCDMA/HSPA or WiFi. An example is illustrated in Figure 5.22. Traffic steering solutions may provide OPEX reduction and limit and/or postpone CAPEX expenditure.

The primary challenge for traffic steering is to direct the traffic to the best layer in overlaid cells by coordinating mobility configurations. Yet traffic steering should not compromise robust and optimised mobility performance in terms of for example, handovers, radio link failures and ping-pongs features, as explained in the previous section on MRO. This calls for

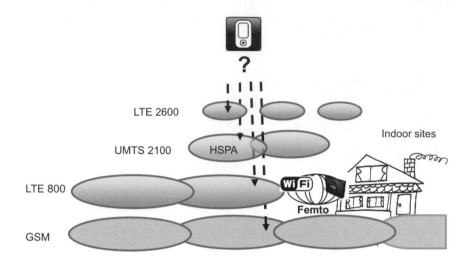

Figure 5.22 Example of the traffic steering decision.

SON features which automatically adjust the mobility configurations on the bases of several factors such as:

- UE capability in terms of RAT, system band or frequency carrier support;
- Network layer capability in terms of transmission power, advanced features support such as advanced antenna systems or beamforming;
- Requested service, user perceived performance, QoS/QoE;
- Enforcement of QoS differentiation;
- Dynamic cell loads and their variations in the available cells;
- Power consumption in UE and BTS;
- UE speed and mobility.

There may be several use cases for traffic steering depending on the network deployment to optimise.

- To balance the load across the network cells referred as mobility load balancing (MLB).The goal of mobility load balancing in general is to move traffic from high loaded cells to less loaded neighbours as far as interference and coverage situation allows. This way better utilisation of cell capacity and larger UE throughputs can be reached.
- To offload macro cells towards low power cells, HeNB or WiFi: The goal of the offloading is to accommodate as large part of the traffic demand, which otherwise would be served by macro cells, over small cells as possible. Techniques such as *range expansion* could be adopted to control the offloading extent.
- To avoid fast mobility UEs to connect to low power cells: The goal of such mobility control relies in avoiding fast mobility UEs from repetitive handovers.
- To minimise eNB power consumption: The goal is to redirect traffic away from those cells which are targeted to be switched off for power savings purpose in the case of low traffic demand. General mechanisms are covered in Section 5.3.

The following sections will mainly focus on mobility load balancing within LTE and between LTE and WCDMA/HSPA. First an overview of the SON policies for mobility load balancing and a theoretical view are presented. Then the standardised features and procedures which enable load balancing are discussed in Section 5.2.4, followed by examples of load balancing in Section 5.2.5. Considerations on the challenges related to the uplink direction are provided in Section 5.2.6. Section 5.2.7 concludes the discussion with the elaboration on the interactions between MRO and MLB.

5.2.2 SON Policies for Mobility Load Balancing

Mobility Load Balancing targets to balance the load amongst the available cells in a certain geographical area by means of controlling mobility parameters and configurations including UE measurement thresholds. The overlaid cells may belong to same or different frequency layers, hierarchy layers, or RATs In the remaining part of the section load balancing and mobility load balancing will be used interchangebly.

Three main phases can be expected in networks for the related SON policies as follows:

- **Phase I:** *LTE introduction*. As typical case of network evolution scenario, the E-UTRAN network will be rolled out as overlay to existing radio access such as 2G/GERAN and/or WCDMA/HSPA systems. Either static configuration of mobility parameters or simple SON policies could adjust automatically mobility configurations such to guarantee service support and basic coverage assurance. That is, push LTE capable to LTE network whenever possible and if poor (lack of) service capability/coverage occurs push LTE capable terminals to another RAT. Mainly idle mobility parameters such as reselection priorities and thresholds could be controlled for such steering strategy. Enabling mechanisms in idle mode are further treated in Section 5.2.4.
- **Phase II:** *Mature phase of LTE networks/high LTE penetration*. The self-organisation property become a key enabler for this phase. More advanced SON policies can adjust mobility parameters and configurations in order to achieve dynamic load balancing between the LTE networks and between LTE and other RATs such as 3G/HSPA. Parameters like handover offsets, measurement thresholds, reselection priorities and thresholds could be controlled. The reader can refer to Section 5.2.4 for further elaboration. Dynamic load balancing between LTE cells/layers and between LTE and 3G/HSPA can provide efficient resource utilisation when the penetration of LTE UEs becomes high enough – above 50%.
- **Phase III:** *LTE deployment* is rather large, LTE UE capability is spread and the LTE provision of voice is fully mature. LTE spectrum most likely will be increased, leading to LTE multiple frequency layers. Enhanced MLB mechanisms will be needed to distribute the traffic optimally over the different bands, while taking advanced features like carrier aggregation into account.

In the following, mechanisms for dynamic mobility load balancing for Phase II are widely discussed.

Load balancing has been done in the past already for instance in 2G or 3G, but in a rather static way. The balancing effect was achieved by considering traffic densities when configuring mobility parameters, either already in the network planning phase, or later on by manual optimisation during operation. In some cases, even semi-automatic mechanisms were implemented in the OAM system which optimise the mobility parameters with little human interaction. However, those approaches have a sluggish nature and cannot react to traffic changes which are by nature rather dynamic. For instance, during events such as a soccer game, the traffic can move within minutes/hours from subway station to the arena and back. Hence, in comparison to many other SON mechanisms, load balancing should be able to react very dynamically. The latter means that fast dynamic load balancing may require, at least partially, a distributed architecture such that decision entities are located at the eNB where knowledge on the traffic changes, radio channel and current user satisfaction are known.

5.2.3 A Theoretical View of Load Balancing

In this section the analytical model of load balancing for a two layer model is presented. Load balancing between two layers can be modelled with a Markov process which is seen in Figure 5.23. Focus here is on an example of a two-layer load balancing system, but the model is general and can be extended for any multi-layer network where the call throughput is known. The states in the model are denoted Px,y, where x and y are the number of users with data in the buffer, connected to the Layer 1 or Layer 2, respectively. $\lambda1$ and $\lambda2$ are the arrival probabilities of new calls in Layer 1 and Layer 2, which are assumed to be Poisson distributed. It is assumed that the arrival probability of new calls is independent of the number of active users in the layer, which is a valid assumption when the population of potential calls in a cell is much larger than the number of active calls. $\mu1$ and $\mu2$ are the probabilities of a call finishing in respectively the Layer 1 and Layer 2. It is assumed that all calls in a layer have the same average length and that the available capacity in each layer is equally shared by the users in that layer, while the maximum number of users in Layer 1 and Layer 2 are N1 and N2.

The probability of a call finishing connected to one of the layers in a state with n users, depends on the size of the call, *callsize* (in bits), the cell throughput C of the layer and the number of users, n; the cell throughput is shared with and can be expressed as:

$$\mu = n \cdot \frac{C}{n \cdot callsize} = \frac{C}{callsize}$$

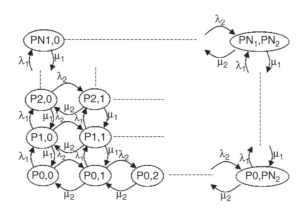

Figure 5.23   Markov process illustrating the two-layer load balancing mechanisms.

It is assumed that the call size is independent of the system but the cell capacities can be different for the different layers.

Assuming an M/M/1/N queue (Poisson arrivals, exponential service time, one server, and a maximum of N users in the system) for both layers the state probability Px,y (the probability of x users in Layer 1 and y users in Layer 2) is calculated (as Schwartz, 1988):

$$Px, y = Px \cdot Py = \frac{(1 - \rho_1)\, \rho_1^x}{1 - \rho_1^{N_{HSPA}+1}} \cdot \frac{(1 - \rho_2)\, \rho_2^y}{1 - \rho_2^{N_{LTE}+1}}$$

where ρ_z equals λ_z/μ_z. The average user throughput on a layer can be calculated as the average throughput per state weighted with the state probability of that layer (Jørgensen et al., 2011, so the throughput for Layer 1 this becomes:

$$TP_{UE} = \sum_{x=1}^{N1} \sum_{y=0}^{N2} P_{x,y} \cdot \frac{C_{1,}}{x}$$

where C1 is the cell throughput of Layer 1. Note that in this model it is assumed that the capacity of a layer is constant and not depending on the number of users in the system, that is, the increased interference or multi-user diversity is not modelled. In order to get the CDF of the throughputs the different bitrates users experience in a cell are taken into account, that is, a cell centre user experiences a much larger throughput than a user at the cell edge. Therefore the probability of a user getting throughput k as $f(k)$ is defined, where it is assumed that the user is the only user in the layer. This function statistically represents the different bitrates a user experiences due to the location in the cell which depends on how the scheduling in the cell is being done. The probability of getting bitrate m in Layer 1 can be approximated as (Jørgensen et al., 2011:

$$P(UE_T P = m) = \sum_{x=1}^{N1} \sum_{y=0}^{N2} P_{x,y} \cdot f(xm)$$

The analytical model can be used to determine the performance of simple load balancing strategies and this will be illustrated in the following example.

A system is considered with one LTE layer and a HSPA layer and focus is on the downlink. A number of traffic steering techniques are considered, which are listed in Table 5.4. Numerical results are shown based on the analytical model explained above to characterise the performance for the case of load balancing the two considered layers. The schemes considered are semi-static load balancing schemes. They range from a random strategy where the assignment of a user terminal to a certain RAT is done purely randomly to a scheme where the instantaneous load distribution of neighbouring cells and UE radio conditions are perfectly known and used as input for the steering. Notice that, without loss of generality, load balancing actions are assumed only during the connection setup phase of each call.

In the next part, the performance of the RA and the PBLA schemes is assessed: Figure 5.24 illustrates the 5% user throughput performance of the RA and the PBLA as a function of the

Table 5.4 Example traffic steering algorithms

	Expression/principle/outcome	Required information	Comments
Random Algorithm (*RA*)	- Steers UEs randomly towards a layer (LTE or HSPA).	No a priori information.	
Push-to-Best Layer Algorithm (*PBLA*)	• A predefined amount of users is steered to each layer (based on the optimal average amount per layer).	Layer capacity in terms of cell throughput per layer.	Will achieve the average optimal distribution of traffic to reflect capacity difference s between the layers
User Load based Algorithm (*ULA*)	• Objective: dynamic load balancing based on time-varying cell load per layer. • Layer decision based on the criterion: $\arg\min_{l\in L}\left(\frac{nUE_l}{C_l}\right)$, where l is the layer index, nUE_l is the number of active users in layer l, C_l is the cell capacity of layer l and L comprises the set of available layers.	- Cell throughput of each layer; - Instantaneous cell load per layer in terms of number of active UEs.	Semi-dynamic. Each UE is steered at connection setup to the layer with the lowest instantaneous cell load.
User Throughput based Algorithm (*UTA*)	• Load balancing criterion: $TP_{u,l} = \frac{SINR2TP(SINR_{u,l})}{nUE_l + 1}$, where $TP_{u,l}$ = average achievable user throughput of user u if served by layer l. $SINR_{u,l}$ is the SINR of user u in layer l, $SINR2TP$ is the assumed function that maps the user SINR to the user throughput for a single user scenario, and $TP_{u,l}$ is then the estimate of the average throughput for user u in layer l.	- Cell throughput per layer; - Instantaneous cell load per layer in terms of number of active UEs. - SINR of every user in every layer. - A priori known SINR to user throughput mapping curve per layer.	Perfect knowledge of throughputs and loads assumed in order to find the upper limit.

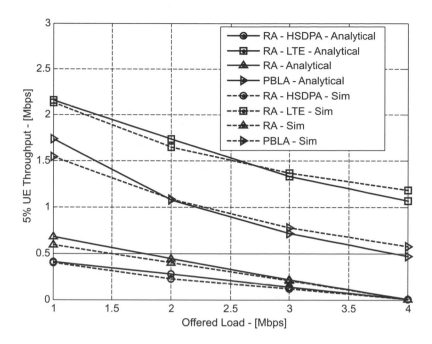

Figure 5.24   5% user throughput results as a function of offered load for the *RA* and the *PBLA* methods via analytical and simulation studies.

offered load in the overall network. The curves named 'HSDPA' and 'LTE' show the 5% user throughput performance for the HSDPA and LTE layer, which are considered here, separately when using the *RA* method. A significant gain is observed when using the *PBLA* over the *RA*. *RA* performance is limited because the HSDPA layer becomes congested as 50% of the users are steered to each layer. The figure shows besides the analytical results also simulation results from Monte Carlo simulations.

Similarly, in Figure 5.25 the average user throughput is shown versus the offered load for the same cases. As expected, the average of the *RA* performance of the HSDPA and LTE layers equals the overall *RA* performance. The trends are the same as for the 5% user throughput. The *PBLA* shows a significant gain over the *RA*. The analytical results match nicely the results from the simulations.

Based on the results in this part it is concluded that the *PBLA* showed the best performance of the two considered algorithms.

Now let us focus on the performance of ULA and UTA compared to PBLA. In Figure 5.26 the 5% and average user throughput is plotted as a function of the offered load. The *PBLA* shows the worst performance and is used as reference in the rest of this section. It is seen from the 5% user throughput that at an offered load of 6 Mbps the network becomes congested. For the reference algorithm starvation of the users at the cell edge happens, while the *ULA* and *UTA* are still able to provide a throughput significantly above zero. When a minimum 5% user throughput of 0.5 Mbps is considered, the gain of the *ULA* and *UTA* is about 20% and 40% respectively in terms of extra capacity they could accommodate.

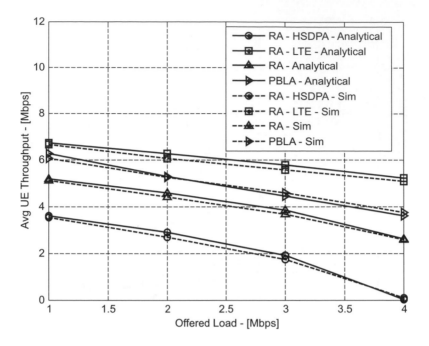

Figure 5.25 Average user throughout comparison for analytic results and simulation results.

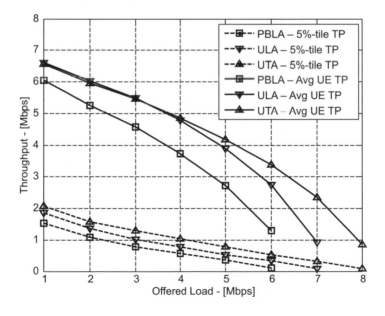

Figure 5.26 Average and 5% user throughput performance for the *PBLA*, *ULA* and *UTA* methods.

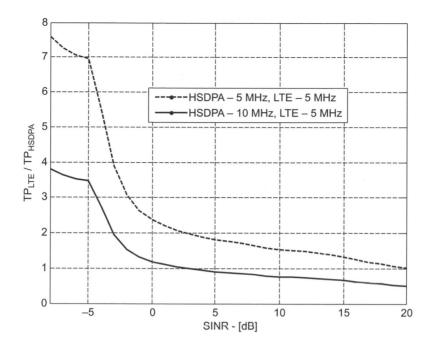

Figure 5.27 LTE throughput mapping gain over the HSDPA throughput mapping.

Also the average throughputs show large gains for the *ULA* and *UTA* over the reference case. For loads up to 4 Mbps the *ULA* and *UTA* perform similar, so it is concluded that no gain is achieved by taking the SINR of the user into account at low loads. For loads over 4 Mbps the *UTA* shows best performance. At a load of 6 Mbps the average user throughput for the reference, *ULA* and *UTA* are 1.3 Mbps, 2.7 Mbps and 3.4 Mbps, respectively. This is equal to gains of 112% and 162%.

These large gains are possible because the reference algorithm does not use any instantaneous information and only relies on the layer capabilities. The two more advanced algorithms on the other hand steer the users based on instantaneous information, for example, the instantaneous load per layer and is thus able to adjust to the dynamics of the system. The additional gain achieved by the *UTA* derives from additionally making use of the instantaneous user SINR information. The user SINR is used to estimate the expected user throughput in each layer at the current load. Figure 5.27 illustrates the LTE-throughput curve divided by the HSDPA-throughput curve versus the SINR. At low SINR the LTE-throughput is up to eight times higher than the HSDPA-throughput and at high SINR the LTE-throughput is only ~1.5 times higher than the HSDPA-throughput. The consequence of this is that the *UTA* steers the low SINR users to the LTE layer and the high SINR user to the HSDPA layer in order to maximise the average user throughputs.

The potential gain from utilising the *UTA* depends on the ratio between the LTE- and HSDPA-throughput. Such ratio depends on system spectral efficiencies, usage of advanced network features such as Multiple Input Multiple Output (MIMO), adopted user receiver type, and so on. The closer the ratio is to 1 within the SINR range, the lower gain can be achieved with the *UTA* over the *ULA*.

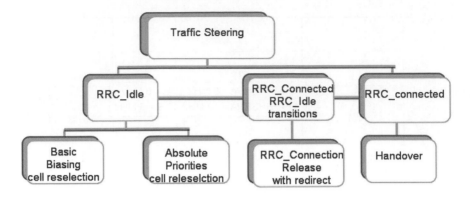

Figure 5.28 Traffic Steering tree of LTE standardised features for directing users.

5.2.4 Standardised Features and Procedures to Direct UEs to the Desired Layer

Load balancing can be achieved by means of controlling standardised mobility parameters and configurations such as handover offsets, reselection priorities and thresholds, and thresholds used in the measurement rules on the base of standardised information exchanged between networks or between the network and the terminals. Based on those enabling features several procedures could be implemented to direct desired users to the more appropriate layer/cell.

In general, the following 3GPP standardised features can be distinguished to direct certain users or user groups to a particular RAT/layer/cell. Those are captured in Figures 5.28 and 5.29. Load balancing in connected mode impact the handover procedures for example, through biased or forced handovers; and

- Load balancing through re-directs: Impact the connection establishment procedure for example, via rejection with redirection; and

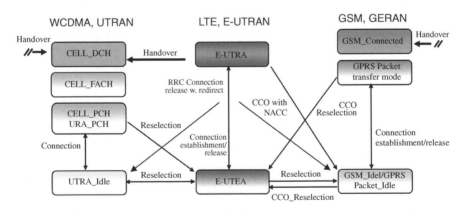

Figure 5.29 Standardised features for directing users from/to LTE to/from WCDMA/UTRAN or GSM/GERAN.

- Load balancing in idle mode: Impact idle mode camping for example, through cell selection/reselection procedure biasing and priorities.

On Load Balancing in Connected Mode

When a change of the state from idle to active (e.g. due to a call or a data transfer) happens, a connection is usually establishing to the cell on which the UE is currently camping. As the camping and cell reselection processes are run autonomously by the UE, it may happen that the current cell has insufficient resources to handle the desired service and is not capable of fulfilling QoS. This can be due to a wrong camping decision, a change in user traffic demands and/or radio conditions. In such a case the UE may be dropped or better directed towards another cell or layer via adjusting handover parameter configurations such as cell specific offset or forced handovers. This involves increased signalling and an increased call setup time. Those solutions have been extensively studied within 3GPP and further details are elaborated in Section where also possible procedures that use handover parameters are proposed.

On Load Balancing Through Redirections

Possible triggers for call redirection in the session setup phase can be based on:

- Load balancing for example, overload; or
- Unavailability of given service in the current network.

The reason for using traffic steering at this stage is to decrease the amount of signalling as opposite to a force handover after connection establishment. The impact on the connection setup can be in the form of a connection rejection with redirection info. Differently from WCDMA, *RRC Connection Reject* message (see Figure 5.30) with redirection towards another frequency or RAT (GSM or E-UTRA) (3GPP TS25.331, 2011) is not supported in LTE. In WCDMA, the information about the frequency or RAT targeted for redirection is provided in IE 'frequency info' or IE 'inter-RAT info' part of the *RRCConnectionReject* message. After receiving such message the UE reselects to the targeted frequency or RAT and disables the reselection to the original frequency/RAT for the time provided in IE 'wait time'. The connection establishment is then re-started on the new frequency or RAT.

It is impossible to redirect a UE in the session setup phase towards another frequency or RAT in the E-UTRA network. However in LTE, dedicated redirection information (3GPP TS36.304, 2011) can be enforced to the UE through the *RRCConnectionRelease* message upon connection release.

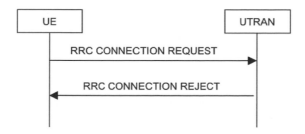

Figure 5.30 RRC Connection Establishment, network rejects RRC connection with redirection information.

On Load Balancing in Idle Mode

Load balancing in idle mode targets to keep idle UEs camped in the most suitable cells such to be optimally provided the service requested when the UE becomes active. Controlling the desired UE distributions across the available network cells when UEs are in idle mode would enable the network to optimally provide the requested services. The benefits of such an approach are the enhanced connection setup times and reduced network signalling. To achieve this, an 'intelligent' network could predict service usage prediction for users' dominant application, for example, by collecting user history profiles, preferences, and so on, in order to be able to steer the given UE towards a cell which can statistically guarantee proper execution of the desired service.

Generally, a cell selection process starts when a UE is switched on for the first time. It scans all radio frequency channels in order to find a cell, which is suitable for camping. The list of RATs that are selected for the scan can be defined by the *Non Access Stratum* (NAS), for example, by indicating RATs associated with the selected PLMN. Restricting the number of RATs can thus speed up the selection process or bias it towards a RAT preferred for a given UE. The cell reselection is instead controlled as follows. Typically a user equipment (UE), while being in idle mode, camps on a given cell, as long as the minimum radio conditions are met (3GPP TS36.133, 2011). The camping process is defined here as synchronising to the cell and listening to its broadcast channel (BCCH). Once the condition of the radio channel towards the camping cell drops below a predefined threshold the UE starts to search for a different cell better suited to camp on. When, a better camping cell candidate is found, the UE synchronises to the new cell (using a cell reselection mechanism) and begins to camp on it while still being in idle mode.

However, UEs perform autonomously cell (re)selection, so the network can only passively steers the UE decision by adjusting cell specific parameters (e.g. reselection offsets, thresholds, priorities). How those features work and how to set and optimise their parameters is described in the following part. The standardised features for controlling camping decisions are described in Section 5.2.4.5.

In principle any combination of the above available standardised mechanisms could be used for directed the traffic. However, whenever possible idle mode steering could be preferred as it is rather ideal for the network and UE side as explained just above. Whenever that is not sufficient, in a more mature phase of LTE deployment, idle mode steering could be complemented with redirection and/or load based handovers. The exact combination will depend on several factors such as deployment scenario, terminal support, traffic type. In case of services which do not tolerate large latency, such as streaming, a redirection based steering solution should be avoided to prevent longer initial buffering time. In conclusion, the enabling features may differ in respect to the following factors which should be then carefully assessed in any load balancing decision:

- Latency, for example, added latency in call setup time;
- User plane interruption, for example, due to an handover execution;
- Control plane overhead;
- UE power consumption; and
- Mobility and QoS/throughput performance.

The majority of available studies and 3GPP discussions focus on load balancing in connected mode, since during that state the eNB has full control over the users. Load balancing in idle mode and through re-directs have not been specifically addressed in 3GPP so far, however, those are possible via the standardised features. Hence, further details and results will be illustrated in the remaining part of the section related to all three load balancing possibilities.

SON based policies for load balancing purposes can adjust the parameters for the features stated above as well as the UE measurement thresholds. Automatic adjustments at a certain node can be based on the knowledge of load conditions and requirements at that node, on information exchanged between cells in terms of for example, cell load and handover configurations. Therefore, the following sections will dive into further details regarding: which UE measurements the above features can be based on, what cell load definitions are available in the standards, other standard possibility for exchanging mobility information between networks as handover negotiation and how it can be useful for load balancing. Procedures for load balancing in connected and idle mode, respectively, are described as well.

5.2.4.1   UE Measurements Relevant for Load Balancing

The majority of above procedures require that terminals perform measurements of the channel quality and/or signal strength based on which the terminals either autonomously perform camping decisions or assist the network to perform handover decisions. For the latter, the terminal can be configured to report the measurement values or specific triggers to the network. Two types of standardised UE measurements are relevant for load balancing. UE signal strength and channel quality: namely, RSRP and RSRQ respectively for LTE cells. An illustration of the two measures versus SINR is provided in Figure 5.31. Similar measurements can be defined for the inter-RAT neighbour cells such as RSCP, EcNo, RSSI for 3G. This was shortly described in Section 5.1.3 on UE measurements. The terminal searches for cells autonomously according to given measurement configurations and within the list of carrier frequencies provided by the network. The network has means to give a so-called black cell list to a UE containing those carriers that should not be measured. Measurements during active mode are performed rather often in the case of intra-frequency neighbours, while transmission gaps are required for performing inter-frequency and inter-RAT measures. In idle mode, the amount of the measurements can be controlled by absolute thresholds relative to the camping cell (*Sintrasearch/Sintersearch*), meaning that whenever camping on a cell which satisfies given thresholds the terminal can avoid performing measurements on other cells, this way saving battery.

5.2.4.2   Load Information Exchange for Load Balancing

Reliable and fast, in the order of minutes, load information from neighbouring cells is very beneficial to track load changes in the network: for instance, to identify those cells which are most likely to accommodate the excess load. In this case handover requests from a cell with heavy load could be directed to those cells which have high probability to fulfil the request. Otherwise, without such load information, a heavily loaded cell can only blindly initiate handovers to arbitrary neighbours. The rejection probability for those requests would obviously be much higher. This creates unnecessary network load, leads to unnecessary delays for offloading and may cause suboptimal load balancing decisions.

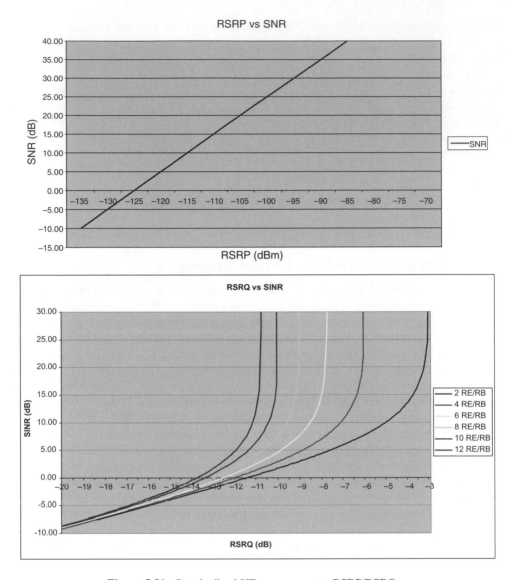

Figure 5.31 Standardised UE measurements RSRP/RSRQ.

In Release 8 the eNBs can already exchange load information via the X2 interface using the Resource Status Update procedure. However, only pure Physical Resource Block (PRB) usage has been specified, separately for Guaranteed Bit Rate (GBR) and non-Guaranteed Bit Rate (non-GBR), and separately for uplink and downlink. The expressiveness of those definitions with respect to how much load a cell would accommodate is limited. Two examples are illustrated in Figure 5.32:

1. A cell has only a single connected user which is running a background download. It may happen that this user will get 100% of the resources. PRB usage would be 100% for

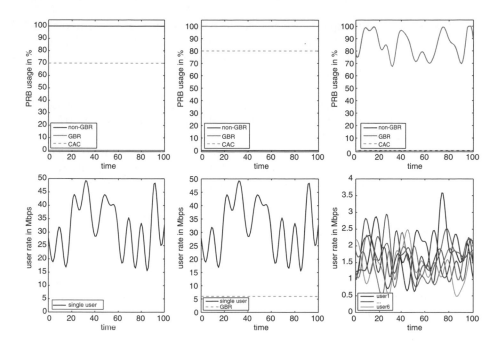

Figure 5.32 Examples for load information: PRB usage and Composite Available Capacity (CAC) for a single best effort user with (left) and without (middle) GBR, and several users with variable bit rate streaming (right).

non-GBR traffic and 0% for GBR traffic. For the neighbours it is not clear how much of the 100% PRBs it can access for load balancing purpose. This depends on the priority of the user. In case of a gold user, the serving cell is probably not willing to compress the resource assignment of this user (and thereby reduce its throughput) significantly. If it is a bronze user, it can compress it. One can conclude that the distinction between GBR and non-GBR traffic is by far too coarse.

2. If the download user in the upper example has a guaranteed bit rate (e.g. 6 Mbps), the serving cell would probably signal 100% GBR traffic and 0% non-GBR traffic. That is, such a cell would look crowded although it is effectively empty.

3. Assume a cell has a lot of gold users with variable bit rate streaming service which currently cause 80% PRB usage on average, that is, it would signal 80% GBR traffic. Nevertheless, this cell may not be willing to accommodate any traffic in order to protect the existing traffic.

Those examples indicate that new information tailored for load balancing would be helpful. Every cell has maximum knowledge about its own traffic including QoS profiles and radio conditions and thus it should be allowed to include all this knowledge into the load information instead of being forced to report pure PRB usage.

Release 9 introduces a simple and pragmatic remedy for the aforementioned problems by introducing new load information called 'Composite Available Capacity' (CAC), which is added to the existing Resource Status Update procedure. Instead of defining an exact method to calculate the CAC, 3GPP states that 'a cell is expected to accept traffic corresponding to the indicated available capacity' (3GPP, TS36.300, 2011). In other words, a cell informs its neighbourhood with a value expressing how much load it is willing to accept. This may include any kind of limitation which the cell is currently subject to, such as transport load, control channels, baseband capacity, and in particular QoS current PRB usage. Revisiting the two examples and Figure 5.32 above, the CAC could be signalled as follows:

1. The serving cell considers the priority of the connected user and decides that the user should not get less than 30% of the resources since it is a gold user. So it would signal 70% available capacity to its neighbours.
2. The serving cell approximates how much resources (including some margin) the user would need for the GBR, for example, 20% (with 100% PRB usage it obviously achieves ~30 Mbps, so 20% will lead to the GBR of 6 Mbps. So it would signal 80%.
3. In this case the serving cell would signal 0% available capacity in order to protect the streaming users. The neighbourhood would be aware that this cell will not accept handover requests due to load balancing.

In principle, a cell is always free to signal 0% available capacity even if it does not suffer from any limitation. However note that every cell is a part of the operator's network and thus should behave cooperatively. Even if it behaved in a selfish manner it would improve its KPIs by accommodating load balancing traffic (e.g. larger throughput, larger number of satisfied users, etc.).

Release 9 has specified the CAC for inter-RAT as well, that is, an LTE cell can also signal the CAC to cells of other RATs (GERAN and UMTS) using the RIM procedure. Release 10 is currently discussing whether the CAC will also be generated in GERAN and UMTS to be part of the load information provided by a GERAN/UMTS cell to an LTE cell. Details on inter-RAT load balancing will be treated in Section 5.2.5.2.

5.2.4.3 Negotiation of Handover Parameters Mobility Change Procedure

The mobility change procedure is a mechanism which guarantees that users which are offloaded to a neighbouring cell will not be sent back due to radio handover. Such procedure is at the moment standardised in LTE only for intra-LTE neighbours; later on it will be obvious that the risk of such a back handover is small in the inter-RAT case and therefore such a mechanism is not needed there. The previous section shows that a cell with heavy load knows which neighbour would accept how much load. The left plot in Figure 5.33 shows

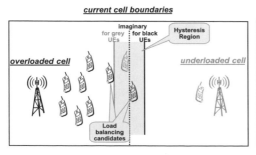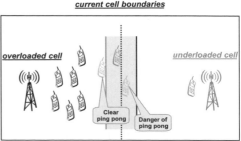

Figure 5.33 Overloaded cell (dark) with underloaded neighbour (grey). Left: only radio handover. Right: forced handover towards underloaded neighbour.

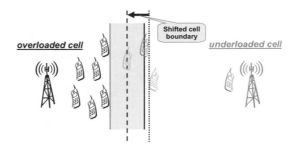

Figure 5.34 Forced handover with subsequent shift of cell boundary.

schematically a situation where an overloaded cell (black eNB, black UEs) has a low-loaded neighbour (grey eNB, grey UEs). The handover parameters are such that black UEs would handover on the black line, the grey UEs would handover on the grey line. The area between those lines is typically called the hysteresis region; users in that area are connected to either of the cells. The hysteresis region is a part of the handover region. The boundary of the hysteresis region indicates where handovers are initiated; the boundaries of the handover region indicate when the handovers must be completed. The two black candidates are candidates for load balancing handovers which are those closest to the grey cell. All other black UEs are assumed to be out of the hysteresis region and thus cannot connect to the grey cell at all. In the right plot the two candidates are handed over to the grey cell.

Now interesting effects can be observed. One of the candidates is in the hysteresis region. On one hand, the UE indeed will be connected to the better (grey) cell, so it may have even better SINR after the handover. On the other hand, the UE has much less ping-pong protection than after normal handovers (executed on the black line), so the risk of ping-pong is rather high for this UE. This is even more extreme for the other candidate. The grey cell will definitely handover this user back as a result of radio conditions and the current setting of the handover parameters.

Release 9 also introduces a mechanism to negotiate new handover parameters between the involved cells, that is, to shift the cell boundary appropriately. This is illustrated in Figure 5.34. With the 'Mobility Change Procedure' the black cell can send a request via the X2 interface, where it proposes to shift the common cell boundary by a certain delta. As already discussed in the MRO section, handover decisions are vendor specific and this is the reason why a delta value is exchanged instead of absolute values. Another advantage of the shifted cell boundary is that future handovers due to radio reasons may be done on the new cell boundaries as well. As long as the traffic situation remains similar, no further load balancing actions are needed.

5.2.4.4 Load Balancing Mechanisms in CONNECTED Mode

In principle, load balancing in connected mode can be done by simply forced handovers; that is handing over users from a heavily loaded cell to less loaded neighbours or adjusting automatically handover parameters such as cell/frequency specific offsets according to the load such that handovers are favoured automatically towards low loaded cells. In the inter-frequency case where the overloaded cell and the offload target use different frequencies there are no mutual interference problems. However, in the intra-frequency case those handovers

Figure 5.35 Scenario with two eNBs serving three sectors each; dark areas indicate locations where users can be connected to either one or another cell, that is, users in those areas are candidates for load balancing.

often need to be forced against radio conditions, that is, radio conditions will effectively limit the area where balancing between two cells in a certain frequency layer is feasible. Figure 5.35 repeats Figure 5.2 from Section where the dark area represents locations where a user can be connected to more than a single cell. Only users in those areas can be candidates for load balancing handovers, and proper mobility must not be endangered. Furthermore, forced handovers are typically associated with a degradation of the SINR since the strongest cell has now become an interferer. This SINR loss has to be compensated, that is, for the same QoS a user will produce more load in the target cell. When evaluating the benefits of intra-frequency load balancing this load increase needs to be considered. Later on it will be discussed how this can be quantified.

The following discussion describes procedures to perform load balancing in CONNECTED mode in the case of the intra-frequency, inter-frequency and inter-RAT case.

Intra-Frequency Load Balancing Procedure
This section discusses how intra-frequency load balancing can be implemented. Figure 5.36 shows a message sequence chart of the whole mechanism. In parallel, Figure 5.37 gives an example for an underlying scenario, where Cell A is overloaded with another overloaded Neighbour B and an underloaded Neighbour C. As a starting point a Cell A has detected an overload, that is, the served users cannot be satisfied due to any limitation. The classical reason is that all PRBs are already used; however, there might be other reasons such as limitation of control channels or transport network.

This cell will request Resource Status Updates from its neighbours via X2, and it collects the CAC values. From those it can determine which cells are willing to accommodate traffic at all. Cell B is heavily loaded as well, so Cell C is the only candidate. Next it will select appropriate candidates amongst the served UEs with the following properties:

- **The UE is close enough to a neighbour with spare capacity (i.e. CAC > 0).** If measurement reports are configured appropriately the UEs will report RSRP/RSRQ measurements to relevant neighbours to the serving cell, so that it can determine close cells for each UE. In the figure those UEs are encircled. All other UEs are no candidates.
- **The UE will be better served by the neighbour.** From the RSRP/RSRQ measurements the serving cell can also approximate the SINR decrease that a UE would suffer in the corresponding target. In combination with the QoS profile of this UE, the cell can also approximate the required load in the target. In the figure the size of the UEs indicates their QoS requirement. UE #1 and #2 have good chances, UE #4 would probably suffer too high

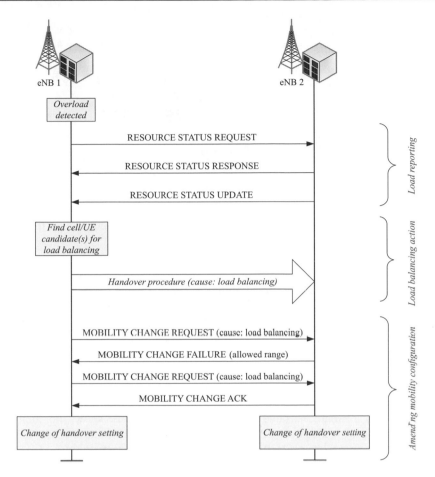

Figure 5.36 Mobility Change procedure in the intra-frequency case (Holma and Toskala, 2011). Reproduced with permission from John Wiley & Sons, Ltd.

SINR degradation and UE #3 would suffer less SINR degradation: however, will produce high load due to high QoS.

• **The UE would offload the cell significantly.** Although a UE with small data rate (e.g. voice) would have good chances to fit into a neighbour, it would not lead to significant offloading. Hence, a cell with a larger rate would load to better offloading. Offloading UE #1 or UE #4 would lead to very little offloading, UE #2 and #3 (if possible) would be better choices.

For those candidates a handover request is directed to the corresponding target. Note that the target cell is still able to reject the request. However, if the load approximation is done properly the probability should be very low.

Finally a new cell boundary is negotiated between the two cells via X2 using the Mobility Change Procedure to avoid ping-pongs as discussed before. Note that it is vendor specific whether this happens before, during or after the handover (or even not at all).

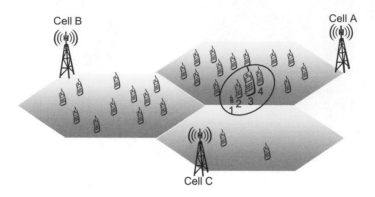

Figure 5.37 Exemplary scenario for MLB between Cell A and Cell C within a single frequency layer. Only cell edge users are MLB candidates.

Inter-Frequency and Inter-RAT Load Balancing Procedure

Although all the overall procedure has been discussed for the challenging intra-frequency case, it is in large part also directly applicable to the inter-frequency and inter-RAT cases. The composite available capacity is also available in the inter-frequency case via X2, and even for the inter-RAT case it is specified using the RIM procedure, but only from LTE towards 2G or 3G. From 2G or 3G, currently only the actual radio load is available instead of the MLB-tailored composite available capacity. The most significant difference is that there is much more degree of freedom when selecting UE candidates for load balancing when two cells do not share the same frequency layer. Figure 5.38 shows the same situation as Figure 5.37 but with a co-located second frequency layer which is rather empty. It is obvious that all users can be handed over to the other frequency layer provided that it has the same coverage.

Figure 5.38 Exemplary scenario for MLB between two different frequency layers/RATs, both with macro coverage. All users are MLB candidates.

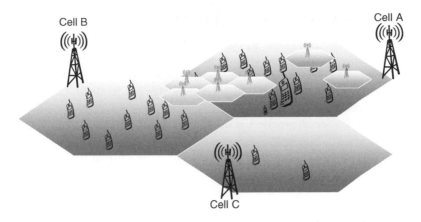

Figure 5.39 Exemplary scenario for MLB between two different frequency layers/RATs, one of them with small cell deployment. All users in the small cell areas are MLB candidates.

Another example is given in Figure 5.39. The previous situation is assumed to be the macro layer of a network, for example, at 800 MHz. In some areas there are cells of a capacity layer, for example, at 2.6 GHz. From the capacity layer point of view, in principle all users are load balancing candidates. From the macro layer point of view, only those users which have coverage of the capacity layer are candidates which should still be a larger number provided a reasonable network planning.

However, in contrast to the intra-frequency case it cannot be assume that RSRP/RSRQ measurements for a different frequency layer or RAT are available. Inter-frequency or inter system measurements require measurement gaps and will therefore be used economically. Hence an important component of inter-frequency and inter-RAT load balancing is the appropriate configuration of UE measurements. Additionally, inter-system handovers are more costly from a signalling overhead point of view compared to intra-system handovers, thus they may be triggered carefully.

From those UEs for which inter-frequency/inter-RAT RSRP/RSRQ measurements are available, the overloaded cell can again select a subset which would fit into the inter-frequency/inter-RAT neighbour would lead to significant offloading. The same procedures as for intra-frequency load balancing can be used, however, as already mentioned, many more candidates will typically be found.

In the MRO section it has been discussed that inter-frequency handovers are typically triggered by the A5 event (or a similar vendor specific event). That is, handovers are initiated if the serving cell becomes worse than a threshold and an inter-frequency neighbour becomes better than another threshold. As a consequence, the risk of ping pongs is much smaller for inter-frequency load balancing. More specific, as long as the target cell is not very weak (which can be read from the measurements) there is will be no ping pong. If a condition similar to A3 is used for inter-frequency handovers, typically a large hysteresis is used. So in general the negotiation of new handover parameters via mobility change procedure seems to be less critical for inter-frequency load balancing. Although the procedure is available proper inter-frequency load balancing might be possible without. Notice that the mobility change procedure is not at all specified across different technologies.

5.2.4.5   Load Balancing Mechanisms in IDLE Mode

Load balancing in connected mode is a powerful solution as the network with full control and knowledge can resolve uneven load conditions, However, it may cause degradation of UE performance due to breaks in the data delivery and signalling overhead at the network side. Therefore, load balancing would be beneficial if applied in idle mode to achieve balancing in the camping user distributions.

Following assertions for idle mode UE distribution can be made:

- In a roll-out phase of LTE camping on LTE should be prioritised (due to lower LTE UE penetration rate);
- A UE should camp on frequency/RAT which can provide most efficiently the expected requested service; and
- The idle UE distribution amongst different frequencies/RATs should follow the inverse of the load experienced in the active state.

As the cell reselection process is run autonomously by a UE, to enable IDLE mode load balancing the network has to impact it indirectly. Currently, there are two standardised methods which could be used to achieve this target:

- Basic biasing (BB) – Adjusting cell-pair offset in order to effectively extend or reduce the range within which a cell is likely to be selected by a terminal.
- Absolute priorities (AP) – Adjusting cell priorities to increase or decrease the likelihood of a cell being selected for camping.

Basic properties of the two methods are described in Figure 5.40. A simple illustration of the methods is shown in Figure 5.41.

The basic biasing method is the main procedure for cell reselection and used in cases of intra-frequency reselections. This method is also applicable for cell reselections between different frequencies (of the same RAT) with equal priorities (3GPP TS36.304).

The Absolute Priority method enables to prioritise given frequency or RAT in the reselection process and is applicable only for inter-frequency or inter-RAT cell reselections.

Since both methods have their own limitations and are used in different scenarios, an effective IDLE traffic steering solution should consist of a both methods. Below the characteristics of the two methods are highlighted.

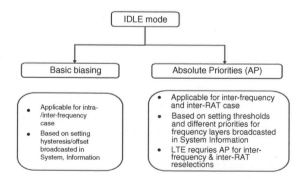

Figure 5.40   Idle mode traffic steering methods.

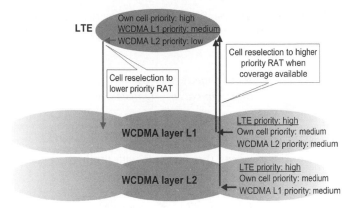

Figure 5.41 Traffic steering between LTE and HSPA via Absolute Priorities reselection or basic biasing mechanisms.

Basic Biasing (BB)

Basic biasing is based on a standard cell selection/reselection procedure as depicted in Figure 5.42.

To determine if a measured cell (3GPP TS36.133, 2011) is suitable for camping, a cell selection criterion known as *criterion S* is used. The criterion is fulfilled when:

$$Srxlev > 0 \, AND \, Squal > 0,$$

where *Srxlev* is derived based on the measured cell received signal level (RSRP for E-UTRAN or RSCP for UMTS) and *Squal* is derived based on the measured cell quality level (RSRQ for E-UTRAN or Ec/No for UMTS) (3GPP TS36.304).

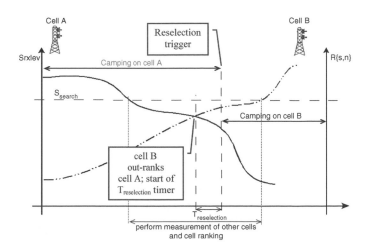

Figure 5.42 Basic cell selection/reselection.

When more cells fulfil the *criterion S*, cell ranking is performed. The ranking algorithm evaluates the *criterion Rs* for the serving cell and *Rn* for neighbouring cells according to the following formulas:

$$Rs = Qmeas, s + Qhyst$$
$$Rn = Qmeas, n - Qoffsets, n,$$

where *Qmeas* is the measured cell received signal level, *Qhyst* is the serving/camping cell hysteresis and *Qoffsets,n,* is the offset between the two cells (s,n).

The cell ranking algorithm selects then the cell with the highest rank according to the formula:

$$Selected\ cell = \max \{Rs, Rn\}.$$

Measurements which are used for cell ranking can be based either on signal strength or on signal quality and are defined by the network in the System Information.

While camping on a cell with signal level below a predefined threshold (S_{Search}), UE will regularly search for a better suited cell, as a reselection candidate, according to the cell reselection criteria (3GPP TS36.304). The UE performs continuous ranking of all cells that fulfilled the cell selection *criterion S* and reselects to the best cell only if the following conditions are met:

- The new cell is better ranked than the serving cell during a time interval $T_{reselection}$, and
- More than 1 s has elapsed since the UE camped on the current serving cell.

Modification of the *Qoffsets,n* values in the ranking equation enables the network to bias the reselection process towards a preferred for example, less loaded, cell, thus in effect steer the idle users. The *Qoffsets,n* values can be either broadcasted by each cell in the System Information or can be set individually for each target (neighbour) cell. However, the application of the offset should be limited, as a high offset can lead to radio link failures due to high interferences in the case of co-channel network deployments.

Absolute Priorities

E-UTRAN supports a priority based cell reselection procedure for inter-frequency and inter-RAT reselections (3GPP TS36.304), (3GPP TS35.304, 2011). Notice that GERAN and UMTS supports it from Release 8 onwards. With this solution absolute priorities of different frequencies or inter-RAT frequencies may be provided to the UE in the system information, in the *RRCConnectionRelease* message, or by inheriting from another RAT at inter-RAT cell reselection. The priorities provided to the UE in dedicated signalling (*RRCConnectionRelease* message) supersede those provided by the system information and are valid till UE re-enters the connected state or the (optional) validity timer (T320) expires.

With the priorities in place UE reselects to a higher priority frequency or RAT if the targeted cell fulfils $Squal > Thresh_{x,highQ}$ (Figure 5.43).

If the serving cell fulfils $Squal < Thresh_{serving,lowQ}$ the UE will reselect to a lower priority cell if $Squal > Thresh_{x,lowQ}$ is met for the targeted cell during the time interval $T_{reselectionRAT}$, as depicted on (Figure 5.43). The quality thresholds $Thresh_{x,highQ}$, $Thresh_{serving,lowQ}$, and

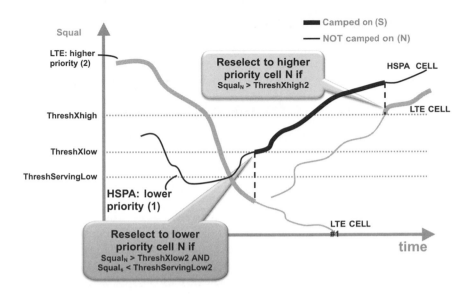

Figure 5.43 Idle mode inter-frequency/inter-RAT reselection algorithm.

$Thresh_{x,lowQ}$ are broadcasted in the System Information (3GPP TS36.331, 2011). In case the System Information does not hold the quality thresholds the $Srxlev$, $Thresh_{x,highP}$, $Thresh_{serving, lowP}$ and $Thresh_{x,lowP}$ are used in above mentioned criterions.

Cell reselection to a higher probability frequency/RAT takes precedence over a lower priority frequency/RAT if multiple cells of different priorities fulfil the reselection criteria. In case multiple cells with the same priority fulfil the reselection criteria cell ranking is used as in basic biasing case.

An example of priority based cell reselection is depicted on Figure 5.43 where the E-UTRAN cells have higher priority than the UMTS cells.

Anticipation of the future UE traffic for example, base on the call history or mobility can be used as basis for an individual (per UE resolution) adjustment of priorities. Here the dedicated priorities and thresholds, conveyed to a given UE through the *RRCConnectionRelease* message upon connection release, as well as the validity timer are the main instrument to be used.

5.2.4.6 Alignment of Load Balancing Procedures in Idle and Connected Mode

Similarly to the avoidance of handover ping-pongs which may occur when a user offloaded to a neighbouring cell is immediately handed back due to radio reasons, it is of key importance to avoid idle-to-connected ping-pongs. A idle-to-connected ping-pong is defined as the situation when a terminal handed over to a certain cell or layer when in connected mode immediately reselects another cell or layer when moving back to idle or vice versa.

Therefore, a procedure to align the mobility adjustments performed in idle mode with the adjustment taken in connected mode should be applied. That is, for instance, if a cell or frequency specific handover offset is applied in connected mode, similar offset should be applied also to the reselection decision.

5.2.5 Exemplary Results of MLB

This section illustrates several numerical results of load balancing, namely load balancing in intra-frequency LTE scenario via handover adjustments; load balancing in inter-RAT LTE-WCDMA/HSPA scenario via redirections at call setup and via priority-based reselection.

5.2.5.1 MLB Applications to the LTE Intra-Frequency Scenario

Before showing numerical results the functionality by a screenshot of an intra-frequency simulation are illustrated. Figure 5.44 shows an excerpt of a network. Every UE is served with a constant bit rate of 300 kbps, that is, it is watching a YouTube video. The light grey bars indicate the load of every cell. Numbers above '1' indicate the degree of overload. The dark grey bars represent the number of unsatisfied users, that is, those who do not get the 300 kbps. The left plot shows the situation before switching on load balancing. There is a user concentration in the middle which creates overload in the adjacent cells, that is, the bars are larger than one and there is a non-zero number of unsatisfied users. The surrounding cells are carrying much lower traffic. Those are willing to accommodate traffic. The right plot shows the same situation with load balancing being switched on. The arrows indicate those users which have been handed over against their radio conditions. As expected, only those users are affected which are close to a low loaded neighbour. Obviously, in those special situations where an overloaded cell serves users on the cell edge towards low-loaded neighbours, large gains can be achieved. In this example the yellow cell can offload enough traffic to fall below 100% load, such that all users are satisfied.

For numerical evaluation of the load balancing gain, it is assumed the same irregular layout as in the MRO section with cell sizes between 500 and 1732 m (Turkka and Lobinger, 2010) and wrap around is used to remove the impact of the network edges. All users are served with constant bit rate. The advantage of CBR services is the clear definition of user satisfaction. Users either get the required data rate, or they are unsatisfied. In particular CBR services will create cell loads between 0 and 100% (in terms of radio resources). In contrast full or finite buffer models typically create either 0% load (if there are no users to be served), or 100% load

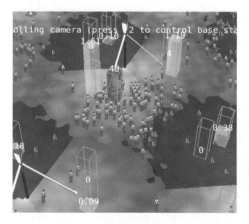

Figure 5.44 Intra-frequency Mobility Load Balancing.

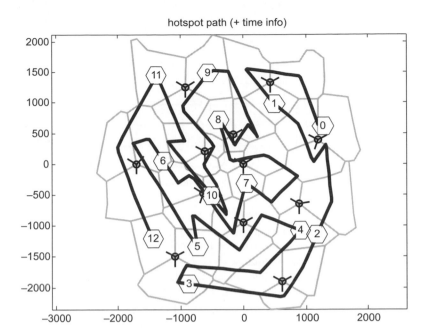

Figure 5.45 Network layout and motion path of hexagonal hot spot. The numbers indicate the time instances (from 0:00 to 12:00).

(already a single user fills the cell). Those CBR properties simplify the evaluation of load balancing.

Every cell serves approximately five moving users which generates a background load for every cell. In addition there is a single hotspot of 40 users moving in a restricted area bounded by a hexagon with the diameter of 333 m. Thus the hotspot creates a capacity demand of 20 Mbps in addition to the background demand of 2.5 Mbps. An LTE downlink with 10 MHz is considered which has a CBR capacity in the range of 10–12 Mbps. The hotspot moves through the network as indicated by the path in Figure 5.45. Every hour the position of the hotspot is highlighted by the bounding hexagon including the timestamp, that is, the hour. The velocity of the users as well as the velocity of the hotspot is 3 km/h so that the total velocity is no larger than 6 km/h.

Figure 5.46 shows the number of unsatisfied users in the whole network over time, that is, those users who do not achieve the CBR of 300 kbps. Note that dissatisfaction can only occur in the area of the hotspots since the background load is very low. Comparing the results with the path and the layout of the network it can be observed that gains can be achieved if the hotspot is close to cell edges but still does not split into two or more cells. If the hotspot is close to an eNB, all users are in good radio conditions and all users can be served properly. Between 5 and 10 additional users can be satisfied applying load balancing (solid black) compared with the reference case (dashed grey) where the users are handed over symmetrically with an offset of 0 dB and a handover hysteresis of 3 dB. This gain may seem moderate when referring to the whole network, however, looking at the restricted area of the hotspot with 40 users it is more

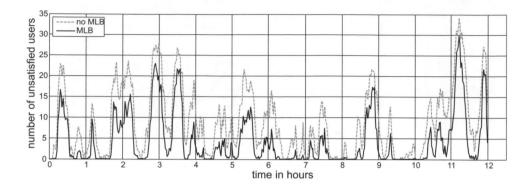

Figure 5.46   Unsatisfied users with and without MLB.

significant. Furthermore the results show that load balancing can dynamically follow spatial changes of the traffic far below an hour.

Concluding the performance of load balancing, local gains can be achieved in special situations where traffic concentrations create high load in a cell on one hand, and are close enough to a less loaded neighbour cell. Furthermore, obviously load balancing can collide with MRO, in particular at higher velocities (the simulation results above the users were moving only at 3 km/h).

5.2.5.2   MLB Applications to the Inter RAT (LTE and HSPA) Scenario

Load Balancing via Redirection Upon Connection Setup
In Section 5.2.3, the Push-to-Best Layer Algorithm (*PBLA*), a static scheme which randomly steer users to any of the layers, was compared against the User Throughput-based Algorithm (*UTA*), which performs the redirection decision based on the UE throughput achievable in the different layers. It was shown how *UTA* outperforms *PBLA* when increasing the offered load level thanks to the knowledge of UE channel quality and cell load. In this section it will be shown how the LTE penetration level affects the performance of the *PBLA* and *UTA* schemes. Two cases with a HSDPA and LTE bandwidth of 5 MHz at the offered cell load of 3 and 5 Mbps are considered. The third case considers an LTE network with 5 MHz system bandwidth and HSDPA network with Dual-Carrier HSDPA (10 MHz bandwidth) when the offered cell load level is kept 5 Mbps. Figure 5.47 shows the gain in average user throughput from using the *UTA* over the *PBLA* scheme for the three cases. The gain is plotted for different LTE penetration levels, ranging from 50% capable LTE terminals to 100% capable LTE terminals.

At 100% LTE penetration the gain of using the *UTA* scheme is larger for higher offered load, as it was already shown in Section 5.2.3 in Figure 5.26. The latter is due to the fact that at low cell load the likelihood of one or no active user per layer is high and in that case there is no difference between the algorithms.

Instead, the *UTA* gain at any LTE penetration is larger than in the first two cases when adopting Dual-Carrier HSDPA (DC-HSDPA). This is due to the fact that in case of DC-HSDPA the throughput of HSDPA is roughly doubled, making HSDPA the preferred layer. *UTA*'s gain is then larger as the scheme is able to exploit the information of the available capacity in contrast to *PBLA*, which selects users randomly.

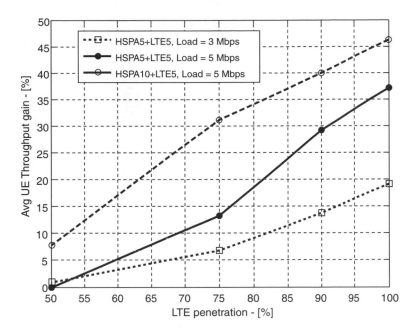

Figure 5.47   Average user throughput gains versus LTE penetration of the *UTA* over the reference *PBLA*.

In any case, the gain of using the *UTA* scheme decreases the lower is the LTE penetration level. This can be explained by the fact that when not all terminals are LTE capable a group of UEs which do not support LTE will be steered to the sub-optimal HSDPA layer. Then, the best decision that the *UTA* scheme can perform is to steer all LTE capable terminals to the LTE layer, just as the *PBLA* algorithm. This way no gain can be expected when using the *UTA* over the *PBLA* scheme.

This investigation shows that when the LTE penetration level is 50% or below, then the simple *PBLA* performs as good as the more advanced load balancing algorithms. When the LTE penetration levels increases to about 75% or higher then it becomes beneficial to use an adaptive load balancing algorithm. The breaking point where it becomes beneficial to utilise a non static algorithm depends on the actual scenario: for example, on one side, if the LTE layer capacity is increased then a static algorithm is sufficient at higher LTE penetration levels. On the contrary, if the HSDPA layer capacity is increased then the advanced algorithm is beneficial at lower LTE penetration levels. Furthermore, it is important to consider also the traffic volume aspects. In the presented numerical study, the traffic volume generated by any user is the same regardless of the terminal capability. Instead, it may be expected that LTE users will generate larger traffic volumes than HSDPA users making dynamic load balancing algorithms beneficial at lower penetration levels.

Load Balancing via Priority Based Reselection Policies
The numerical examples in this section illustrate how different user distributions can be obtained in the inter-RAT co-sited scenario where both LTE and HSPA networks provide the same coverage applying priority-based reselection mechanisms. As related to the

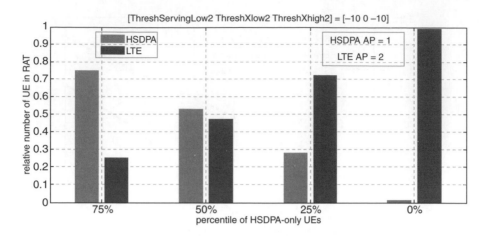

Figure 5.48 Steering of idle mode UEs: different HSPA-only UE penetration rate. LTE Priority 2 (high), HSDPA Priority 1 (low).

absolute priority feature, two sets of parameters can be adjusted: namely, reselection priorities and reselection thresholds.

An example result of idle UEs reselecting towards LTE, that is, the prioritised RAT is depicted in Figure 5.48. There the impact of the reselection priorities, while reselection thresholds are kept constant and rather aggressive, and different UE LTE-capability penetration ratio is shown. As it can be seen that by means of prioritising the reselection in a given RAT (in this case LTE network), it allows to control the camping UE distributions between the RATs.

The above static configuration is rather sufficient as long as LTE penetration is rather modest, below 50%. However, when the LTE UE penetration increases using a configuration being aggressive towards LTE may lead to non-optimal resources utilisation and eventually user starvation in LTE. Therefore, SON policies may be required to adjust more dynamically the reselection thresholds and/or priorities once the ratio of multi-RAT capable UEs, present in the network, increases over a level of expected balance for example, 45/55 in the example that follows. As shown in Figure 5.49 it is possible to balance the idle UEs distribution between different RATs/frequencies solely by changing the AP thresholds. Idle UE distribution amongst different frequencies/RATs should follow the inverse of the load experienced in the active state. The actual threshold values depend on the deployment of the nodes and the values below should be seen only as an example.

Given the assumption that idle and active mode load balancing techniques are fully aligned (see cf. Section 5.2.4.6 for further details) the user distribution per RAT shown in Figure 5.49 translates into mean UE throughput per RAT and in the system as depicted in Figure 5.50. The example assumes packet call generation per user according to an exponential distribution and mean packet size of 400 kb.

It is worth mentioning, that in case of co-sited nodes deployment, the RAT or frequency with the higher priority assembles UEs with good channel conditions as compared to the low priority RAT/frequency. This can be seen in Figure 5.51.

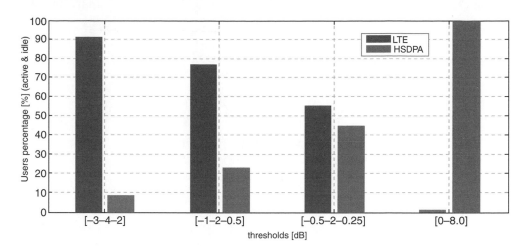

Figure 5.49 Impact of the AP reselection thresholds on idle user distribution. Four sets of thresholds [$Thresh_{x,high}$, $Thresh_{serving,low}$, $Thresh_{x,low}$] are considered. LTE Priority 2 (high), HSDPA Priority 1 (low).

Load Balancing via Connected Mode Mobility

This section illustrates a numerical example when applying dynamic load balancing in connected mode for the inter-RAT scenario of LTE and HSPA networks. As described earlier in Section 5.2.4.4, not only there exists a SINR penalty for a UE whose handover decision is biascd via a ccll bascd offset in the case of intra-frequency handover, but also the interference level will be affected. Instead, biasing the handover towards a cell on a different frequency layer or different RAT will cause no mutual interference problems. Anyhow the SINR can also be worse in the latter case.

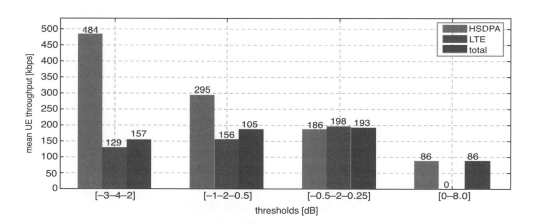

Figure 5.50 Impact of AP reselection thresholds on mean UE throughput shown per RAT (LTE/HSPA) and per system (total).

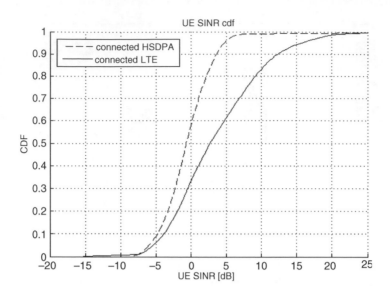

Figure 5.51 Scheduled SINR distribution; LTE Priority 2 (high), HSDPA Priority 1 (low).

An automatic adjustment of the handover parameters (handover offset in the example in Figures 5.52) depending on the cell load difference between the serving cell and a potential target cell. The larger is such load difference the larger the handover offset will be and therefore more handovers will be favoured towards the less loaded cell. Of course, the larger the handover offset the larger is the penalty in average user SINR or path loss which has to be compensated by the load difference. Symmetric adjustment of the handover offset on the cell-pair level is also required in order to avoid handover ping-pongs.

The cell load difference can be mapped to a UE throughput gain in the target cell compared to the throughput achievable in the serving cell. This way clear performance benefit is identified before performing any handover(s) to a less loaded neighbour cell. The following expressions

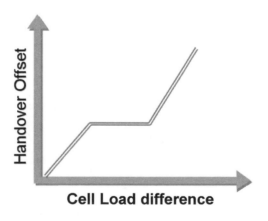

Figure 5.52 Example of automatic adjustment of handover offset versus cell load difference between serving and target cells.

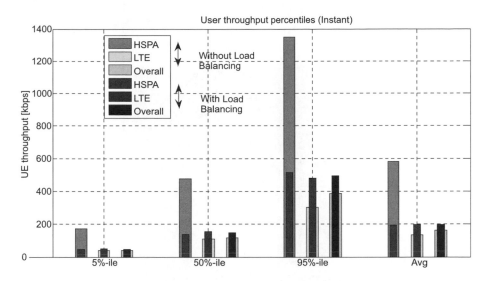

Figure 5.53 Performance of automatic adjustment of handover offset; UE throughput percentiles.

provide an example of the computation of the load difference in terms of normalised UE throughput:

$$Load\ Difference = \frac{TP^n_{NORM} - TP^s_{NORM}}{TP^s_{NORM}},$$

where TPnorm is the average cell throughput for serving cell s and target cell n; *and*

$$TP^i_{NORM} = \frac{TP^i_{UE}}{Used\ Resources},$$

where TP_UE_i is the average user throughput for users in cell i; $UsedResource\_i$ is the amount of utilised radio resources (PRBs) in cell i.

Figure 5.53 shows the performance results of such algorithms in the case of co-sited macro LTE and HSPA deployment, both networks using 5 MHz as system bandwidth. Results are illustrated in terms of UE throughput percentiles with and without dynamic load balancing scheme. Finite buffer users are dropped randomly in the macro cell area according to a Poisson arrival generating a high offered load. Dynamic load balancing, clearly improves all the UE throughput metrics as it allows targeting more even user distributions in the two RATs. The latter can be seen from Figure 5.54.

5.2.6 Uplink Load Balancing

All previous considerations had a strong focus on downlink. In Release 9, the uplink has not explicitly been addressed as well. As long as the downlink is the limiting link in terms of load, this is a valid approach. In fact, most of the current applications such as web browsing, downloads, and so on are strongly downlink-centric. Nevertheless, it should be made sure that the uplink is not impacted, if load balancing is based on the downlink. An uplink user which is

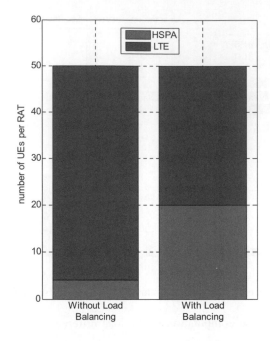

Figure 5.54 Performance of automatic adjustment of handover offset; UE distribution.

handed over against the radio conditions may produce larger interference to the overloaded cell, however this will only affect a small set of PRBs. Furthermore the dynamic range in the target cell might be impacted. However, as long as the uplink capacity limit is not reached, this will not degrade the system. Moreover, a smart power control will react appropriately with small adjustments of power control parameters (cf. Section 1.7.2. Note that in the worst case, a target would reject the handover request if its admission control would detect that the user would not fit into the uplink.

It has been discussed that an overloaded cell can approximate whether a user would fit into the downlink of another target cell with slightly worse radio condition. Although the composite available capacity is defined for the uplink as well, such an approximation would require more information such as interference level and power control parameter in the target cell.

Detailed investigations on uplink load balancing are found in Nihtilä *et al.* (2011). In this publication it has been observed that control channel limitations are more severe than in downlink as a result of the more strict power limitation. That is a cell running out of PDCCHs may benefit from initiating load balancing, even if many PRBs are available.

5.2.7 Interactions Between TS/MLB and MRO

In the last sections it has been discussed that traffic steering/mobility load balancing in general is accomplished by:

- Making 'forced' handover decisions (based on internal/vendor specific parameters).
- Changing cell re-selection priorities.

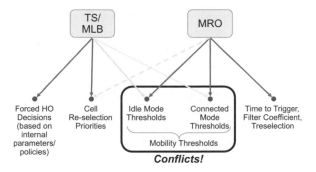

Figure 5.55 Parameters affected by TS/MLB and MRO.

- Adapting re-selection parameters in idle mode, (such as $Qoffset$, $Thresh_{x,high}$, $Thresh_{serving, low}$, $Thresh_{x,low}$).
- Adapting handover parameters in connected mode, such as thresholds for the measurement reporting events A3, A5 or B2.

However, recalling the previous chapter on mobility robustness optimisations, partly the same parameters have to be addressed by MRO to guarantee there are no mobility problems. Namely, MRO is accomplished by:

- Adapting time to trigger, filter coefficient and re-selection.
- Changing the cell re-selection priorities.
- Adapting re-selection parameters in idle mode, (such as $Qoffset$, $Thresh_{x,high}$, $Thresh_{serving, low}$, $Thresh_{x,low}$).
- Adapting handover parameters in connected mode, such as thresholds for the measurement reporting events A3, A5 or B2.

This section addresses the interactions between MRO and TS/MLB. The description is kept general, it holds for the intra-frequency (co-channel) case, the inter-frequency case, as well as the inter-RAT case. In the following the parameters associated with the latter two bullets will be summarised as 'mobility thresholds'.

Figure 5.55 illustrates the TS/MLB and MRO together with the impacted parameters. It is obvious that the mobility thresholds are adapted by both SON functions. Section 9.1 extensively addresses the problems associated with such a parameter conflicts. Note that MRO also has some responsibility for the cell re-selection priorities, however, the task is considered to be straightforward and not part of the conflict since only obvious inconsistency of the priorities can create mobility problems, in particular ping-pongs.

Even if the mobility thresholds are not addressed by TS/MRO avoiding aforementioned conflicts thereby, the decided HOs and the modification of re-selection priorities still may create mobility problems such as ping-pongs or too early handovers. Without any interaction between the two SON functions, there is the risk that MRO detects those problems as 'self-made' and tries to counteract by adjusting the mobility thresholds. In case the problems were actually created by bad TS/MLB decisions and not by bad mobility thresholds, performance could be even degraded.

Solving these problems from a high level with SON coordination (cf. Section 9.2.1) may not be able to exploit the full potential of both functions. A tighter collaboration ('SON function co-design', Section 9.2.1) will definitely have benefits. Certainly, a joint design of TS/MLB and MRO can achieve the best performance, however restricts the options for implementation. For instance, centralised Domain Management (DM)-level implementation of the MRO corrective actions has been discussed in the MRO chapter, whereas TS/MLB is very likely to be implemented at the eNBs. A joint design is obviously difficult in this case. Therefore a solution will be discussed in the following which minimises the interactions between MRO and TS/MLB (and thereby allowing independent implementation) whilst leaving full potential for both functions.

First of all, according to the previous discussion the problem is split into two parts:

1. Double responsibility conflict (both functions access the mobility thresholds).
2. TS/MLB created mobility problems (MRO should not try to correct those problems).

The latter problem obviously requires a distinction between MRO and TS/MLB created mobility problems. Following the structure introduced in the MRO chapter, this task needs to be fulfilled by the root cause evaluation. As far as radio link failures are concerned this is already possible in Release 9, since the *HO report* already contains the *HO cause* of the handover of interest. If the HO cause was load-related, the guilty cell knows that the handover was initiated due to load, that is, by TS/MLB, and so the mobility problem must be corrected by there. Otherwise, if the HO cause was radio-related, the problem has to be corrected by MRO. The Release 11 SON Work Item in 3GPP will investigate the aforementioned distinction and will provide further solutions if necessary.

The former problem of double responsibility raises the question of priority of the two functions. A typical assumption is that connection of stability is one of the highest targets. A subscriber will be much more unsatisfied with connection problems compared with sub-optimal throughput. As a consequence, MRO should have more responsibility on the mobility thresholds than TS/MLB. In other words, MRO must be able to restrict the way that TS/MLB modifies the mobility thresholds. Concrete examples for those constraints are given in the following list:

- TS/MLB is not allowed at all to modify the mobility thresholds. This is certainly the simplest method, which however may cut the potential benefits of TS/MLB significantly. Note that in particular in the inter-frequency and inter-RAT case, where there is no mutual interference between source and target cell, there will still be degree of freedom to steer the traffic via forced handovers and cell re-selection priorities.
- TS/MLB is only allowed to modify the mobility thresholds if the MRO root cause evaluation indicates no mobility problems. The simulation results in the MRO section have already indicated that mobility problems only occur on some cell boundaries. At those locations MRO may create significant improvements, whereas TS/MLB actions would probably degrade mobility further. On the remaining cell boundaries TS/MLB gains can be achieved without negative impact on mobility.
- MRO configures parameter ranges, that is, upper and lower limits for the mobility thresholds, and TS/MLB is free to modify them within the given ranges. Such a solution obviously requires smart mechanisms within the MRO function to adapt the upper and lower limits, however, imposes minimum constraints for MLB whilst not endangering mobility.

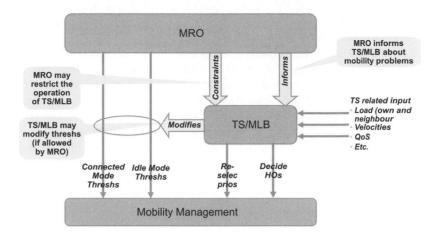

Figure 5.56 Possible solution for proper coexistience of MRO and TS/MLB.

Figure 5.56 illustrates the interaction concept derived from the discussion above. MRO informs TS/MLB about mobility problem created by TS/MLB. This information is primarily used to keep cell-reselection priorities consistent and to adapt the internal policies for forced handover decisions. Furthermore TS/MLB may modify the mobility thresholds; however, MRO is able to restrict this ability via constraints as discussed above. For the sake of completeness the usual input for TS/MLB has also been added to the figure. The actual mobility management is carried out based on the combined input of MRO and TS/MLB.

5.3 Energy Saving

Deployment of Energy Saving (ES) is motivated by both, to reduce CO_2 footprint and to optimise the costs. The RAN and particularly the radio base stations (RBS) have been identified to have the highest share of the mobile networks' overall energy consumption and hence the largest potential for energy saving measures.

5.3.1 Introduction

Within the recent past the mobile communication industry has seen a dramatic increase of awareness regarding energy efficiency.

The reasons are twofold: on one hand, the network traffic is showing very high growth rates and this trend will continue as new technologies, highly demanding applications are made available in short cycles. On the other hand operators are facing higher costs per kWh and, most of all, there is the responsibility of the global society to protect the environment.

As a consequence, mobile network operators are committing themselves to reducing the overall energy consumption by deploying highly energy efficient networks and, at the same time, limiting possible environmental impacts.

Traditionally the main target has been the UE side as to prolong stand-by time and talk time with the focus to ensure subscriber satisfaction. Nowadays energy saving has become a wider scope, the overall energy consumption of the network has become of interest. Furthermore not only the energy in terms of kWh but also the CO_2 emission (kg) is targeted by enabling energy saving SON functionalities. In the context of SON, energy saving is addressed by a functionality to coordinate Network Elements (NE) in order to find and invoke an ES-optimised network configuration. At the same time the user perception should not be compromised. Quality of Service (QoS) and Quality of Experience (QoE) shall stay within the desired target ranges. A large potential for SON-based ES originates from the traffic profile, that is, service demands versus time. Naturally the traffic is low during night and reaches a maximum during a given hour during the day, the so-called Busy Hour (BH). The networks are typically dimensioned to fulfil the traffic demands during the BH whereas the required capacity during off-peak periods reaches only a few percent. Consequently a large amount of the networks capacity is not required during these periods and SON ES functionality aims to provide the most energy efficient configuration. There are several main aspects with regards to energy saving:

- **Inter-RAT ES:** Nowadays several radio technologies are deployed in parallel. In many cases this holds also for a given geographical area. For several reasons and so for SON it is advantageous to have a 'single-network' approach. In this sense the capacity of the different RATs can be shared. Furthermore different technologies often have a common coverage footprint and can backup each other. This allows for a 'sleep mode' of complete cell, sites or even RAT-layers.
- **Inter-eNB ES:** Energy Saving takes place within LTE and could also be called intra-LTE, generally intra-RAT. This scenario can be further subdivided in an eNB overlaid scenario and a capacity-limited scenario.
- **Intra-eNB ES:** From conceptual perspective SON requires the interworking/coordination of two or more network elements. Hence, local, intra-eNB features are not instantaneously within the scope of SON. However, such measures should be considered as input to SON ES in order to find an overall optimised ES configuration. Local ES features could be adaptations of TX power, reduction of the number of TX antennas, reduction of the number of carriers, deployment of MBSFN, DWpTS for ES purposes. Some measures clearly have network impact. This becomes obvious if the cells' coverage is affected and therefore also the neighbours of the considered cells.
- **ES in terms of power-efficient HW/equipment and local deployment options:** As above, HW-/site-/cell-specific features are not really within scope of SON-ES. Energy efficient HW can contribute a great deal. Typically the power consumption versus the air IF load/max RF power is a main efficiency indication. The HW-/site-/cell-specifics may be used as an input in the actual SON-ES procedures. These characteristics behaviours with regards to power consumption as function of Uu load is described by means of an energy model.
- **Type of energy generation:** There are different ways to power a base station. Today, the majority of the sites is connected to the general power grid. Depending on availability it might be necessary to produce energy on-site by means of a generator (diesel-based). The latter is clearly one of the worst scenarios when the CO_2 footprint is considered. Much more favourable is on-site renewable (solar, wind, etc.) energy production. Eventually the network's green performance could be evaluated based on pure energy consumption and additionally based on its CO_2 footprint. Consequently an energy model of a site could be based on CO_2 rather than on pure energy and used in the SON ES process.

5.3.2 Requirements

Main requirements for an acceptable ES algorithm can be concluded as follows. In the sense of constraints ES actions shall:

- **Avoid coverage holes.** In the majority of practical relevant scenarios the existence of overlay networks is a clear advantage to overcome such risks. Furthermore, very often multi-radio base stations are deployed which are co-sited and will provide similar coverage footprint in more than a single technology. For example, even in case there are temporarily coverage holes for a dedicated technology (e.g. LTE): From single network perspective no coverage hole exists if the legacy RAT is providing basic service. A wake-up mechanism can ensure that the LTE service is back in quasi real-time. It is assumed that all UEs are capable of 2G and 3G.
- **Not compromise user perception:** In order to guarantee QoS for the incoming service requests a traffic detection mechanism should exist. The legacy RAT should have the capability to distinguish if it can provide the required service/QoS itself or if it has to trigger an activation of a superior RAT. Naturally the UE capabilities have to be taken into consideration.
- **Maximise the energy saving potential** by taking into account the power consumption and traffic situation.
- **Avoid interference with existing SON features,** but rather take advantage of already defined SON algorithms and definitions.
- **Not lead to instabilities and ambiguous/undefined states.**
- **Do not require extensive (manual) intervention** at the OAM system-level but follow SON principles.

As an example key performance indicators (KPIs) can be used in order to describe the performance of the ES algorithms. The most important KPIs typically are call drop ratio, call blocking, user-throughput, voice quality (MOS: Mean Opinion Score), mobility-related KPIs as HO failure rate. Many variations might be in use, as 'mean time between drops', tenth percentile throughput and so forth. In order to describe ES further KPIs could be defined as Energy/bit, CO_2/bit, Power/Erlang, Power/bps.

A straightforward approach to measure ES efficiency and to incorporate it into the SON context could be to reach or to stay within the target values. The actual targets are either operator specific or might be imposed by regulatory aspects.

5.3.3 Energy Saving Management

From management and architecture perspective ES can be envisaged to be done via 3GPP TR32.826 (2010). This *Technical Report* includes following alternatives and list, for each of those, advantages and disadvantages:

- Distributed approach which does not need to involve OAM.
- Centralised view, that is, the central management system is in charge. It requires necessary inputs, analyses ES options and finally triggers ES related actions.
- Hybrid architecture combines the distributed and centralised aspects.

Key questions are to identify triggers for switch on/off cells and/or to start/end additional ES actions. For instance, candidate inputs in the decision process might be:

- Fix time schedule (e.g. between 01:00–06:00 a cell is switched off).
- Historic traffic data. The advantage is it can be made cell-specific. A drawback is that it is rather static and could hardly cope with dynamical effects as short-term fluctuations. Long-term trends could be analysed by for example, regular measurement campaigns.
- Dynamic load analysis. Based on quasi real-time load the cell could autonomously decide to enter a dormant state. In a centralised way the OAM system would act in a similar fashion. In a full centralised version the eNB would inform the OAM system for example, every 5 min about the current load situation. The final decision for switch off is made by the OAM system. An advantage of the centralised way becomes directly obvious; the OAM system has the full picture of all relevant cells. In a hybrid way the eNB could deploy OAM defined thresholds and act triggered by events to inform the OAM system. Load increase consequently could be used to activate dormant cells again.
- QoS requests. In case a QoS request cannot be handled by the currently active NEs/cells, a wake-up functionality could be utilised.
- Availability and activity of local ES functions.
- Energy models of the involved NEs.
- Equipment alarms, faults, KPI-alarms.

5.3.4 eNB Overlaid Scenario

The scenario is described by a coverage layer and one or more capacity boosting cell(s) within its coverage (3GPP TR36.927, 2011). The coverage layer might be provided by LTE (i.e. an eNB) or a legacy RAT as 2G/3G. The scenario is subdivided into the following use cases: in UC1; intra-LTE inter-frequency, coverage and capacity layers deploy different frequency. Typically, a case is macro-pico cell deployments. UC1a is specifically addressing femto cells such as Home eNB under the coverage of a macro cell layer. UC2 is the above mentioned inter-RAT case. The corresponding network layout is sketched in Figure 5.57.

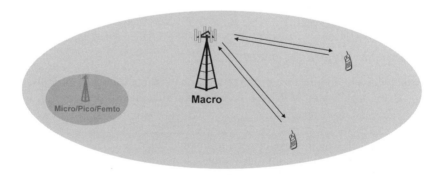

Figure 5.57 eNB overlaid scenario.

A complete switch-off of the hotspot (pico/femto cell) during periods of low load/no load is feasible as coverage is maintained and sufficient capacity can be ensured by the macro cell. It should be noted that the scenario potentially could be extended towards deployment of capacity-enhancing relay nodes. Relay deployment with the purpose to extent coverage might prohibit such an ES scenario.

The eNB overlaid scenario has been further investigated in 3GPP TR36.927 (2011). The scenario is structured in 'inter-RAT scenario 1', the coverage layer is provided by legacy RAT(s), for example, GSM (2G) and/or UMTS (3G). A second scenario is termed 'inter-eNB scenario 1', hence both, the hotspot and the coverage providing cells are LTE-cells.

Main issues and potential solutions are being discussed:

Basic procedure for switch on/off cells
- Via regular Configuration Management.
- Via local policies downloaded from the OAM system. Fix timing is used. Additionally load information may be analysed locally to make an autonomous decision.
- Based on signalling across RATs without OAM system involvement. Furthermore the related (inter-RAT) neighbours shall be informed about the decision and may request a switch on, for example, in case their load motivates such a request.

How to switch on the correct cells
This question arises in a situation where the macro's controlling node detects a need to offload traffic. The user distribution might strongly deviate from uniformity; hence from a coverage perspective the hotspots' ability to contribute in the offload is generally very mixed. The target is to find the most promising candidates, that is, the hotspot(s) which can take the highest part of the load.

There are several candidate solutions:
- Based on pre-defined high load/low load periods for each hotspot (most probably derived from historical traffic statistic data) the hotspot(s) with the highest/high (historical) load is chosen to be switched on.
- All hotspots shall be switched on. Hotspots which experience low/no load may be switched off again afterwards.
- The dormant hotspots keep, or temporarily activate, a listening mode and observe the Interference over Thermal Noise (IoT). The IoT per hotspot now gives a good indication if (or how many) active users are nearby. An example situation shows a resulting IoT for three hotspots under a macro-sector coverage, see Figure 5.58. The users are indicated by dots and are assumed to contribute equally to the macro load. In the majority of the cases the raw IoT gives a good indication which hotspot is a potential target for activation. In the special case that hotspot are heterogeneously located (close/far from the macro) the *Path Loss* (PL) dependency of UL power control can be taken into account and improves the result. In the example (Figure 5.58) Cell 1 is obviously the best cell for activation as it could occupy 50% of the overall load (the load relation is 2:1:1 for Cells 1, 2, 3 respectively).
- The inactive hotspots are going in a probing phase, that is, DL reference signals are transmitted. This signal ('pilot') can be measured by idle and connected UEs and shall be reported to the macro cell's eNB. The latter can make an appropriate decision which cells to activate.
- Correct neighbour(s) to be activated are attempted to be identified based on UE positioning methods and geographical data of the hotspots.

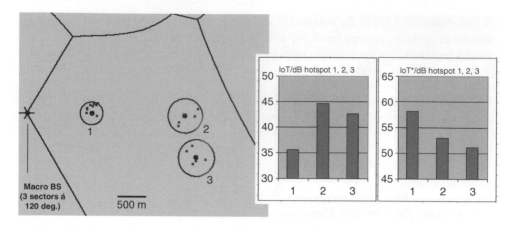

Figure 5.58 IoT samples: raw data and adjusted (IoT* = IoT $-$ PL $+$ Power offset).

How to switch off the cell efficiently

In 3GPP Release 9 switch-off decisions are implementation specific and decisions might be not
coordinated. Two main concerns need to be analysed: (i) how to maximise the overall energy
efficiency and (ii) how to avoid overload caused by local switch off decisions. The solution
proposals are:

- *Solution 1:* The coverage cell observes its own and neighbours load situation and takes
  the decision which hotspot(s) should be switched off.
- *Solution 2:* Hotspots send switch-off requests, the coverage cell makes the decision.

Information exchange for coordination

Inconsistencies of local policies could lead to frequent switch-off/-on ('ping-pong effects'), for
instance: the hotspot switches off, the macro cell request switches on, the hotspot switches
off and so forth. A further problem might be an unnecessary switch-on request, in cases
where the capacity of the macro alone is sufficient. The solution can be to exchange related
ES parameters.

5.3.5 Capacity-Limited Network

The capacity-limited scenario in the context of ES is introduced in 3GPP TR32.826 (2010) and
has been adopted under the term 'inter-eNB scenario 2' (3GPP TR36.927, 2011). The main
principle is sketched in Figure 5.59: in full operational state, during peak-traffic periods, all
cells are in No-ES state and are needed to manage the offered traffic, that is, guarantee the target
Grade of Service (GoS)/QoS. In off-peak traffic situations the number of active cells/sites can
be lowered. ES-compensate cells/sites are deployed to ensure entire coverage. Additionally the
target GoS/QoS is ensured since these ES-compensate sites/cells provide sufficient capacity for
the remaining traffic requests. ESaving cells/sites can be completely switched-off.

From network planning/dimensioning perspective the inter-site distance and hence the
number of sites per unit area is either dictated by wave propagation limits or by the traffic/
capacity demand. In lightly populated, rural areas the former applies: the density of sites needs to
be sufficiently high in order to reach the intended coverage targets ('coverage-limited' scenario).

In urban/suburban areas, however, the density of subscribers and hence the required capacity
is high ('capacity-limited' scenario). The expected BH, traffic demand (Erlang/km^2, Mbps/km^2),

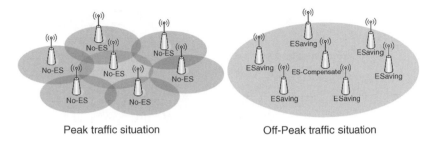

Peak traffic situation   Off-Peak traffic situation

Figure 5.59 Capacity-limited network scenario (3GPP TR36.927, 2011). Reproduced with permission from 3GPP.

defines the average cell area. As this dimensioning is done for BH and due to the great daily-/ weekly-/season-/event- and so on-depending variations of the traffic demands the network is over dimensioned during most of the time. Dynamical adaptation of the capacity enables energy saving opportunities. In detail this adaptation can be done by reducing the cell-/site-density by switching-off/-on complete cells, or even better, complete sites.

It needs to be mentioned that in the capacity-limited case the coverage needs to be controlled. Too large coverage would be counter-productive as it would rather impose interference problems to neighbour cells ('interference-limited' scenario). This is particularly important in the normally deployed 1/1 frequency re-use. In order to limit the described problem coverage-limiting measures should normally be adopted, for example, antenna downtilt, proper azimuth settings, power reduction, low antenna heights.

Considering dynamically changing the networks' cell density it becomes obvious that also these settings should be handled dynamically. Hence, the attempted switch-off/-on actions should be accompanied by further modifications (antenna height, azimuth, downtilt, TX power) in order to get the coverage managed. It should be noted that such actions also influence other planning-, optimisation- and SON-related issues as for example, ANR, intra-/inter-RAT neighbour planning, RACH/PRACH configuration, PUSCH/PUCCH setting/hopping, PCI planning/optimisation, MRO, and so on. This fact may require further attention and should be handled very carefully (cf. Chapter 9 on SON operation). Additionally a close observation of the networks performance, for example, by means of KPI, MDT, is highly recommended. Because the coverage situation is very difficult to predict, the risk of coverage holes, interference, mobility problems is high.

The transition phases seem to be particularly difficult. Connected and idle UEs should be impacted as little as possible. Connected UEs may get handed over during this phase either indirectly as a consequence of the experienced network changes or directly triggered by the algorithm. Idle UEs should be able to access the network but also to be reachable for paging during the whole time. Coverage holes, high interference need to be avoided.

To describe the scenario and related algorithms in a systematic view the following ES states and state transitions (Figure 5.60) are introduced (3GPP TR32.826, 2010). For the NE state:

- **No-ES (notParticipatinginEnergySaving) state:** No ES related measures ongoing.
- **ESaving state:** NE switched off or restricted in resources.
- **ES-Compensate (compensatingForEnergySaving) state:** NE ensures coverage.

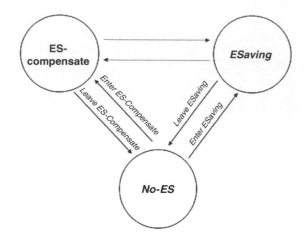

Figure 5.60   Network Element: ES states and state transitions.

5.3.6 Equipment/Local ES

Besides the switch-off options, more specifically the dynamic adaptation of running equipment and further ES measures are to reduce the energy consumption of the equipment by utilisation of state-of-the-art energy efficient hardware. Further attention should be paid to local deployment options and the opportunity to utilise renewable energy. The power model of a site is typically described by energy consumption as function of air-IF load, max RF power. There is a great variety of these models, actual figures depend on the vendor, deployment, temperature, and so on and hence are difficult to standardise.

As a typical model, a linear behaviour is assumed: $P = P_0 + m \cdot P_{RF}$.

where P_0 describes the power consumption of a running RBS without load; by switching off the RBS or HW parts the power consumption P* can be further lowered to values $0 \leq P* \leq P_0$.

The RF TX power depends, not necessarily linear, on the actual load/activity: $P_{RF} = P_{RF}(load)$.

In a detailed view the power consumption needs to be evaluated for each component. More details can be found in EARTH INFSO-ICT-247733 (2010).

A large part may be consumed by cooling, hence the (time-depending) temperature and the location (climate, outdoor/indoor) have large impact. Another main concern is the actual configuration. A distributed deployment with Radio Remote Head (RRH) or remote RF, (an example is shown in Figure 5.61) can contribute considerably to the energy saving budget; typical feeder plus jumper losses are in the order of 2, 4 dB (note, 3 dB being approximately a factor 2) while the jumper losses are typically <1 dB.

To estimate the total energy consumption for a longer period, for example, 24 h, the different time periods can then simply be summed up:

$$E_{total} = P \int_{\Delta T} P(T\{t\}, load\{t\}, ESmeasures \{t, load\}) \, dt.$$

T: Temperature.

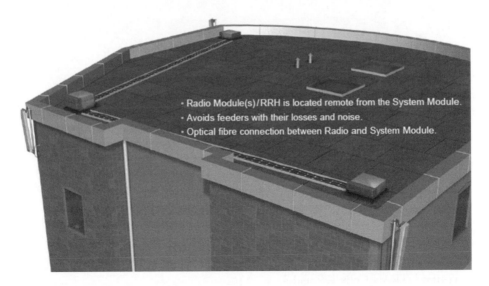

Figure 5.61 Remote option: high ES gains due to feederless deployment and lowered AC needs.

A definition of RBS site energy consumption for test purposes can be found in (ETSI TS102706, 2007). Though the scope of this study is related to GSM/EDGE, WCDMA and WIMAX (Worldwide interoperability for Microwave Access) it can be easily extended in order to include the LTE case. Two main cases are distinguished, concentrated and distributed RBS. For the concentrated case the power consumptions are:

$$P_{site} = PSF \cdot CF \cdot P_{equipment}$$

where Cooling Factor (CF) and Power Supply Factor (PSF) are correction factors. RRH specifics are further considered for a formula describing the distributed case.

Consumption has been highlighted so far from a pure energy perspective. In order to address the CO_2 emission view additionally the CO_2 footprint can be mapped to the electricity emission factor (A WRI/WBCSD GHG Protocol Initiative calculation tool, 2007).

Depending on local electricity mix this factor is for example, 0.502, 0.341, 0.573 kg CO_2/kWh: for the world, EU, USA respectively.

In order to motivate utilisation of renewable energy (solar, wind, hybrid) one could also subtract the appropriate parts from the sites' balance sheet.

5.3.7 Example Scenarios and Expected Gains

To estimate gains of potential solutions a baseline should be defined. Inputs into the analysis have to be defined; traffic model and energy model being the most important. The efficiency of a single ES feature can be quantified by relating for example, a 24 h analysis: (i) ES feature active and (ii) baseline (ES feature inactive).

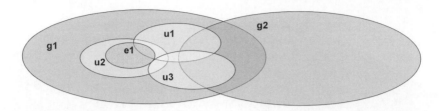

Figure 5.62 Example: inter-RAT overlay, considered is the area covered by g1, u2, e1.

More complex will be cases for which several ES features run in parallel. The single ES gains will add up but also will partly overlap.

An example:

- An inter-RAT overlay scenario is sketched in Figure 5.62.
- The considered ES area is the common coverage area of e1, u2, g1.
- e1: eUTRAN Cell 1.
- u1, u2, u3: UTRAN Cells 1, 2 and 3.
- g1, g2: GSM Cells 1 and 2.

In order to take QoS constrains into account the following assumptions have been made:
    Mix of UE-capabilities:

- 20% type A (2G-only).
- 40% type B (2G, 3G).
- 40% type C (2G, 3G, LTE).

Further input is the daily profile of offered traffic over the day originating from UE types A, B, C (see Figure 5.63).

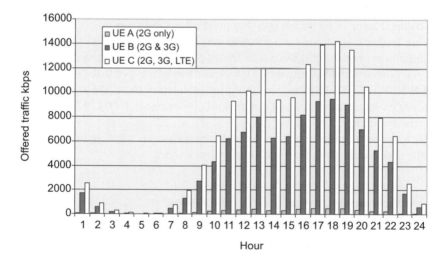

Figure 5.63 Offered traffic [kbps]; daily profile [per hour].

A (simplified) example RBS configuration and power model has been assumed. Note, in order to target the main goal of ES, reduction of CO_2-emission, the power consumption is expressed in 'CO_2-emission units'. The load L is given by the ratio of actual throughput by current capacity. Besides full cell switch off also a local intra-BS ES measure has been taken into account: TRX switch off. The number of the switched on TRX is counted by N_TRX.

2G:

- Configuration: 1..4 TRX (GSM/GPRS).
- CO_2 footprint: Cf = 35 + 20 * N_TRX + 40 * (N_TRX-1) * L.

3G:

- Configuration: 1 or 2 TRX (UMTS/HSPA).
- CO_2 footprint: Cf = 45 + 20 * N_TRX + 40 * N_TRX * L.

LTE:

- Configuration: 1 TRX 10 MHz.
- CO_2 footprint: if L > 0.25 Cf = 20 + 40 * L − 30 else Cf = 0*.

*it is assumed that 30 CO_2 units can be saved by on-site generation of renewable energy.

Estimated gains for the considered example scenario are shown in Figure 5.64. Due to the high efficiency of the LTE eNB in combination with on-site generation of renewable energy a complete switch off of this station is not favourable. In the particular example a switch off of the 3G site during night is the best choice. The traffic can be compensated by 2G (non-LTE capable UEs) and LTE. The example yields an overall ES-gain of 1/5, that is, CO_2 emission is reduced by 20%. The gain strongly depends on the actual input parameters (traffic mix, UE capability, power model of the base stations, actual deployment indoor/outdoor, temperature, on-site generation of green energy etc.). Hence the possible gains may vary to a large extent. Particularly in cases where BS deployments are less efficient the inter-RAT ES gains are significantly higher than 20%.

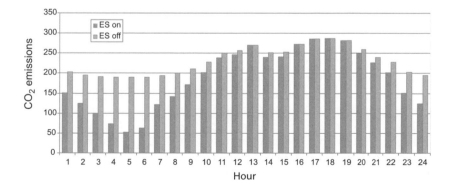

Figure 5.64   Gain estimation. CO_2-emission units versus hour of day.

It should be noted that the example gives an ES-only driven view. Only the total throughput has been checked and further considerations as for example, UE capability limitations, QoS constrains, operator policies, network sharing strategies and so on could inhibit certain scenarios.

5.3.8 Summary

ES has attracted large interest in the recent past due to increase of the perception and responsibility of the global environment, in addition to optimisation of the energy bills. Therefore energy/kWh as well as carbon dioxide emission/kg CO_2 are main performance indicators to describe ES performance. Besides local deployment and efficient equipment solutions also an interworking of several base stations intra-LTE but also inter-RAT-wise is potentially contributing a great deal to the networks overall balance sheet. Finally all ES actions, however, should maintain QoS and QoE, need to guarantee coverage and user accessibility.

At present time it seems that the exploitation of the existing ES-potential has just started and loads of efforts should be put in further investigations.

5.4 Coverage and Capacity Optimisation

To optimise network coverage and capacity manually is a typically very expensive and time consuming task. Therefore Coverage and Capacity Optimisation (CCO) use case has been identified for automatic optimisation of network resources. Capacity and coverage area of a cell may vary due to changed environment. Changes can be diverse, such as changes due to season when trees are leafless during winter and in full leaf during summer, changes in man-made environment due to erected or demolished buildings, insertion or deletion of base stations, or wrong parameter selection in the network planning phase. Changes can be also during the day when different coverage or capacity is needed in different cells during busy hour if compared to quiet periods during nights since traffic distribution might vary. In this case, however, changes inside for example, the busy hour are slow. When the environment changes from the assumptions made when the network was planned and configured, capacity or coverage of the system will be reduced compared to what could be achieved with optimum settings. This sub-optimal capacity and coverage lead to a waste of network resources and lower quality. For example, if cell coverage is such that there is high interference between cells, the overall spectral efficiency becomes lower and more hardware is needed to serve user traffic. Changes in the environment typically happen relatively slowly. As changes are slow, no fast reaction is needed either. Therefore CCO related algorithms are executed over long intervals, only when changes are large enough to have impact on the network performance. CCO can be also done in different parts of network at different times, triggered by values crossing threshold for related KPIs.

In addition to reacting to a changing environment CCO can also help in the planning phase. Due to CCO, accurate network planning is not needed, instead a rough plan is enough and then optimised based on measurements and counters when the base station is activated. For example, it would be enough for an operator to re-use an existing 3G network plan with LTE network and let CCO do needed fine tuning to compensate differences between 3G and LTE plans. This would reduce manual work, save cost and make the planning phase faster. Reusing 3G network plan with LTE would be particularly relevant in the 3G-LTE co-siting case, when minimum additional network planning would be required for installed LTE base stations.

The main objective of CCO use case is to provide sufficient coverage and capacity in the whole network area with minimal radio resources. CCO use case can be further divided to three sub-objectives: maximising the relative coverage in the area so that continuous coverage would be achieved, where the relative coverage can be defined as the probability that the received signal quality is better than the minimum required received signal quality; providing a sufficient received quality in terms of achievable bit rate over the entire area; and to maximise the system capacity in terms of Mbps. The results from CCO are thus improved coverage, cell-edge bit-rate, and cell throughput. Since it can be very expensive to achieve full coverage in terms of radio resources, a trade-off between the objectives is typically needed. The weighting between these objectives should be under the control of the operator.

Targets for CCO can be met in different ways, for example, by adjusting antenna parameters, which is the most apparent way. This would require use of Remote Electrical Tilt (RET) capable antenna or use of active antennas in the base station. Then appropriate signal level could be arranged in different corners of the cells with sufficient coverage by adjusting the amount of antenna tilt, half power beam width or direction antenna is steered. Alternative ways would include adjusting of for example, downlink transmission power, transmitted reference signal power, or optimisation of power control parameters.

As for any SON functionality, the logic of CCO will be defined by the network equipment vendors while, 3GPP standards provide support for CCO functionality. In 3GPP it was seen that the CCO function should typically use a centralised architecture (cf. Section 3.4), that is, it is located in the OAM system (3GPP TS32.522, 2011). In 3GPP's approach actual optimisation would be done in OAM system (at either the Domain Management (DM) or Network Management (NM) level) while detection of coverage and capacity related problems would be done as a part of other SON use cases in the system, namely minimisation of drive test (MDT) functionality and mobility robustness optimisation located in eNodeBs. The problems detected in involved base stations by MRO and MDT are notified to the OAM system. In case for CCO being located at the NM level, this is done via the standardised (not SON-specific) management interface between the DM and NM levels (Itf-N). Therefore most of CCO-related standardisation is related to Itf-N. The configuration parameters to centrally optimise coverage and capacity are overlapping with those employed for the compensation of cells in energy saving (ES, cf. Section 5.3) state or of failed cells (Cell Outage Compensation, COC, cf. Section 6.4). Consequently functions implementing the CCO use case need to be coordinated with regards to the ES and COC activities (cf. Section 9.1).

5.4.1 CCO with Adaptive Antennas

Antenna tilt control has always been an important optimisation tool for all generations of cellular technology. Older deployments mostly used antenna solutions where the value of the downtilt was fixed during the planning phase. This limitation meant that subsequent alterations to the tilt were costly. The introduction of RET enabled antenna optimisation in areas where it would previously not have been cost effective. The usage of RET antennas removes the need for tower climb and base station site visits since tilt angle can be controlled via the O&M

subsystem. Therefore, remote electrical tilt has become an attractive means for the network operators to optimise the capacity and coverage in a self-organised manner.

Antenna downtilt can be adjusted mechanically and/or electrically. In electrical downtilt, phases of antenna elements are adjusted so that the desired tilt angle is achieved by tilting main, side and back lobes uniformly (Lee, 1998) in contrary to mechanical downtilt (Siomina *et al.*, 2006). Hence, higher optimisation may be obtained as it is discussed in (Yilmaz, 2009a). Different techniques for electrical tilt such as Remote Electrical Tilt (RET), Variable Electrical Tilt (VET) and fixed electrical tilt exist. Mechanical tilt can be changed remotely with motors that physically tilt antenna or antenna elements within antenna. RET has become more popular for networks operators for example, adjusting tilt angle when eNodeB insertion or deletion occurs. On the other hand, mechanical downtilt is also needed because electrical tilt range is limited compared to mechanical tilt.

Active Antenna Systems (AAS) are candidate means for the CCO concept as they have potential to provide better system performance and radio network capacity by changing several antenna parameters, such as tilt, azimuth or beam shape. As matter fact, AAS can be seen as an enabler for SON as it can be used for many situations. In addition to coverage and capacity optimisation, AAS can be a powerful tool for cell outage compensation.

A change of tilt has an immediate effect to the coverage boundary. This effect is dependent on the ratio between the height of the mast and the distance to the edge of cell. It is also affected by the intervening topology. It should, however, be noted that a simple change in electrical tilt has a similar effect on the cell boundary across the whole azimuth of the sector. Therefore, wrong antenna tilt can lead to areas with poor coverage or high interference (Yilmaz *et al.*, 2010). Coverage problems may occur when too much antenna tilt is used leading to low call quality at cell borders, and system suffering many call drops and radio link failures or handover failures. From user point of view, this means that there is no continuous service available or service quality is reduced in some parts of the network. When antenna tilt is too little, antenna is directed to area that is meant to be covered by a neighbouring cell and therefore excess interference is created. This is defined as a so-called over-shooting problem. This excess interference makes neighbouring cell users' signal quality worse in the downlink, when that cells' capacity is lower. In the uplink other cell users cause additional interference for the cell where too little downtilt was used. Coverage problem can be alleviated by uptilting antennas in the cells concerned. In a similar way, interference and capacity problem can be reduced by downtilting antennas in the cells concerned.

One of the challenges for CCO is to detect where the problems are and then, on the other hand, how to solve problem (e.g. uptilt/downtilt). Tilt or other antenna parameter optimisations can be done per subnetwork on need basis. Optimisation can be started for a sub-network if certain condition fulfilled for example, too many radio link failures or poor throughput. Downtilting or uptilting need to be done carefully so that coverage problem would not become as interference problem or vice versa. Also, optimisation should be done in only one cell in the area at the time in order to avoid problems due to for example, two neighbours cancelling actions made each other or amplifying action leading to new problems. Actually it is often beneficial to do optimisation jointly for several cells which suggests a centralised approach. Antenna parameter optimisation should not be done too frequently due to the problems cited above. This is also a reason why tilt is not seen as a candidate for cell load balancing, which requires fast reaction to changing load situations.

The impact of antenna parameters including different antenna tilt techniques on the coverage and capacity of LTE networks have been discussed in (Yilmaz *et al.*, 2009/1). It is noteworthy that especially the optimisation of electrical tilt provides significant improvement in coverage and capacity for both interference- and noise-limited simulation scenarios defined by 3GPP (Yilmaz *et al.*, 2009/2). In Section 5.4.2, performance analysis for centralised CCO using AAS is shown based on the work (Yilmaz *et al.*, 2011b).

5.4.1.1 Antenna SON Use Cases

The following use cases are all based on changes to the effective cell boundary and as such, are cases that could be achieved through alteration of the antenna pattern. 3GPP SON solutions achieve this by altering the handover offset parameters rather than by changing the actual coverage of the cell.

COC is perhaps the best candidate for antenna based correction since unlike the other cases extension of a cell's coverage cannot be achieved by changing handover offsets. In this case it is probably reasonable to uptilt the antenna in the direction of the failure cell in order to provide the required additional coverage.

Mobility Load Balancing requires a dynamic adjustment of the boundary between a cell and its neighbours, in response to an overload condition at the cell and available capacity at the neighbour.

Mobility Robustness is an alteration of the handover point to ensure adequate coverage overlap to support handovers but where possible reduce any overlap leading to too many unnecessary handovers.

Energy Saving can be achieved by removing cells from service during off-peak periods where there is sufficient overlap coverage available from neighbouring cells. This is similar to the technique applied in cell outage compensation in that neighbouring cells endeavour to extend their coverage in the area of the cell that is out of service.

5.4.1.2 Advantages

Cell border adjustment for SON using 3GPP approach is based on applying neighbour specific handover offsets. This implies that handovers do not occur at the equal power boundary between the cells, but at a point slightly removed according to offsets and hysteresis. The antenna based SON approach changes the location of the actual cell coverage rather than adjusting offsets, so apart from hysteresis, the handover can take place very close to the equal power boundary.

5.4.1.3 Disadvantages

The major issue with antenna based SON is the limited feedback loop. Coverage changes from antenna adjustments are very difficult to measure. In legacy networks, such changes were measured by drive testing the cell. This is not an option in the SON case as it is contrary to both cost and time requirements. In data-centric networks like LTE, the value of drive testing is significantly reduced since its original goal was to determine street level coverage. Most users are now indoor and potentially distributed across many floors. Antenna adjustment can correctly modify the original handover point but may create either coverage holes or interference in other parts of the cell. The integration of radio planning tools allows some degree of prediction of the effects of antenna adjustment on the cell boundary, but their accuracy tends to be limited especially in predicting the coverage in the high-rise urban environment. Also Minimisation of Drive Test functionality which is a part of SON functionality can be used to detect coverage and interference situation in the network.

Antenna based SON has inherent latency. In the case of motorised tilt, there is a delay in changing to a new tilt angle. This delay is not an issue in RET, however, given the potential negative impacts to the cell edge coverage, it is advisable to make the changes in very small increments, allowing time for sufficient traffic to validate any change before making further adjustment.

5.4.1.4   Costs

Antenna based SON requires some additional hardware which increases the overall cost of deployment. These additional costs may be offset by operability savings when compared to traditional manual optimisation. Additional cost is mainly due to RET antenna systems. Deployment of RET maybe particular cost issue when existing 2G/3G sites are used with already installed antennas.

5.4.2   Performance Analysis for Antenna Parameter Optimisation Based CCO

The proposed evaluation methodology aims to create such a scenario that optimisation may be evaluated from different aspects, meaning that both coverage and capacity problems are present in the system. In performance evaluations, tenth and fiftieth percentiles of throughput and SINR Cumulative Distribution Functions (CDFs) are used as performance measures. The tenth percentile of throughput distribution can be seen as the performance for users in cell border. Therefore by setting a minimum requirement for tenth percentile throughput, coverage of system can be analysed. On the other hand, the fiftieth percentile throughput shows average throughput for all users in the system, therefore showing capacity of the system. Throughput and SINR distributions are generated by using a radio network simulator in which performance measures are collected for multiple eNBs and variable number of mobile users.

5.4.2.1   Scenario

The simulated LTE network planning scenario is originally a 3G WCDMA scenario adopted from (Laiho *et al.*, 2006) and shown in Figure 5.65. A 3G plan is selected for simulations as it leads to suboptimum but still realistic initial situation for an LTE network. As 3G takes benefit from macro diversity, more overlap between cells is desired. This is not optimum for LTE network that relies on hard handover, when excess overlap means just more interference between the cells.

Users are distributed to the simulated area by taking location, for example, buildings, streets and water areas into account. Propagation is modelled in three dimensions by adding the impact of shadow fading, fast fading and antenna configurations summing with the propagation losses, calculated with ray tracing. Fast fading is modelled using a statistical sum-of-sinusoids method as described in (Zheng and Xiao, 2002). Mobility of users is restricted by coastlines. System level throughput is estimated by mapping calculated SINR from system level simulation to

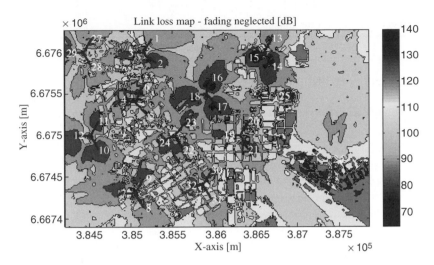

Figure 5.65 Network topology of Helsinki downtown and link loss without no shadow fading.

throughput by using link level simulation results shown in 3GPP, TR36.942 (2010). Link level simulations have been done by assuming a 6-tap Typical Urban (TU) channel and 10 km/h user mobility. System level bandwidth efficiency with respect to the link level efficiency is taken into account to remove the overhead of control and pilot channels in downlink from Physical Downlink Shared Channel (PDSCH) calculations (Mogensen *et al.*, 2007). As the simulated area is finite, inter-cell interference levels in border of the deployment scenario are low. Therefore, the cells close to the network borders are not included in performance evaluations and optimisation to make the simulation scenario more realistic.

Modelling of antenna parameters and main simulation assumptions follow those selected by 3GPP for performance evaluations of LTE and LTE-Advanced technologies (3GPP, TR36.814, 2010) as shown in Tables 5.5 and 5.6.

Adaptive antenna parameters that are optimised in our study can be defined as follows:

- **Antenna azimuth** is defined as the direction, referenced to true north, that an antenna must be pointed.
- **Half-power (3dB) beamwidth** (HPBW) defines the aperture of the antenna. The HPBW is defined by the points in the horizontal and vertical diagram, which show where the radiated power has reached a -3dB level with respect to the main radiation direction. A change in HPBW does not only have impact on the shape of the antenna beam pattern but also on the antenna gain.
- **Antenna tilt** is the angle of the main beam of the antenna below the horizontal plane is called antenna tilt. Positive and negative angles are also referred to as downtilt and uptilt respectively (Siomina *et al.*, 2006).

In electrical downtilt, main, side and back lobes are tilted uniformly, while with mechanical downtilt, antenna main lobe is lowered on one side and the antenna back lobe is raised on the other side. This happens because antenna elements are physically directed towards ground (Siomina *et al.*, 2006) as shown in Figure 5.66.

Table 5.5   Assumptions used for antennas in analysis

Parameter	Modelling and Values
Antenna loss pattern (horizontal)	$A_H(\varphi) = -\min\left[12\left(\dfrac{\varphi}{\varphi_{3dB}}\right)^2, A_m\right]$
Antenna modelling parameters (horizontal)	Default HPBW: $\varphi_{3dB} = 65°$,
	Range of HPBW: [35 135]°
	Range of azimuth steering: [−40 40]°
	Granularity and standard step size of HPBW: 5° and 10°
	Granularity and standard step size of azimuth steering: 5° and 10°
	Maximum attenuation $A_m = 25$ dB
Antenna loss pattern (vertical)	$A_V(\theta) = -\min\left[12\left(\dfrac{\theta - \theta_{etilt}}{\theta_{3dB}}\right)^2, SLA_v\right]$
Antenna modelling parameters (vertical)	HBPW: $\theta_{3dB} = 8.5°$,
	Side Lobe Attenuation: $SLA_v = 20$ dB,
	Range of RET: $\theta_{etilt} = [0° \; 10°]$
	Granularity of RET: 0.1°
Antenna pattern combining in 3D	$A(\varphi, \theta) = -\min\{-[A_H(\varphi) + A_V(\theta)], A_m\}$
Antenna gain	17 dBi
Antenna configuration	MIMO-off – 1 × 2 MRC
Total TX power (One sector)	46 dBm – 10 MHz carrier
TX cable loss	2 dB
RX antenna gain and body loss	0 dB and 0 dB
RX noise figure	9 dB

5.4.2.2   Used Antenna Parameter Optimisation Algorithm

The selected antenna optimisation approach is to collect different Key Performance Indicators (KPIs) formed from measurements and counters from eNBs and make CCO decision in central location, that is, EM or NM. Therefore optimisation is based on KPI distributions in different cells and no specific terminal measurements or reporting are needed. KPIs are a set of selected

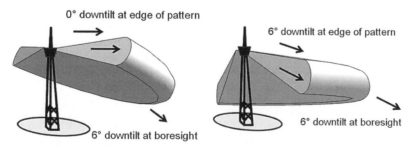

Figure 5.66   Mechanical (left) and electrical (right) downtilt.

Table 5.6 Main assumptions for analysis

Parameter	Modelling and Values
System frequency	2000 MHz
System bandwidth	10 MHz
Number of PRBs	50
Frequency reuse factor	1
Number of total active UEs	1152
Monitored cells	2, 3, 4, 5, 6, 9, 10, 11, 14, 15, 16, 17, 18, 19, 20, 22, 23, 24, 25, 30, 32
Shadowing STD	8 dB
Shadowing decorrelation	0.5 (sites), 1 (sectors)
Shadowing decorr. distance	50 m
Number of total active UEs	1152
UE velocity	10 km/h
Channel model	Typical Urban (TU)
UE height h_{UE}	1.5 m
eNB antenna height h_{eNB}	18.5 m (mean), 16 m (min.), 20 m (max.),
Thermal noise per PRB	−121.4 dBm
Packet scheduler(s)	Resource-fair
Traffic type	Full-buffer

indicators used for measuring the current network performance and trends, for example, radio link failures or call drops statistics, Channel State Indicator (CSI), Channel Quality Indicator (CQI) statistics or handover related statistics. The optimisation algorithm uses KPIs for prioritising the corrective actions. On the one hand, action for optimisation is taken only if there is a significant problem detected, based on statistically reliable KPI data, with respect to the coverage and capacity performance targets.

For deciding how to change antenna configuration, a Case Based Learning (CBL) algorithm is used. CBL algorithm stores decisions from previous cases or instances in memory and applies these cases directly new situations (Uehara, 1998; Langley and Iba, 1993; Aha, 1991). It should be noted that different approaches could be used as well. In recently published works (Islam *et al.*, 2010; Wang *et al.*, 2009), the algorithm of simulated annealing and a Genetic algorithm (GA) are used for the WCDMA and LTE network optimisation respectively. On the other hand, the feasibility of those methods in practice is a question mark due to tight optimisation constraints of SON. For instance, pure meta-heuristic methods like simulated annealing may slow down the optimisation as it has to handle optimisation separately for each cell or randomness in GAs while modifying antenna parameters may result in unexpected consequences. Besides minor dangers and problems, there are major challenges in the SON algorithm development as follows (Dottling and Viering, 2009):

- There is no clear mapping from a large vector of input data to the root cause of a problem, or a promising strategy to improve the situation.
- Incomplete and partly erroneous information may not be sufficient for the decision.
- Full search approaches are not possible due to the large parameter space.
- Trial-and-error in a live network is prohibited due to the risk of negative performance impact.

CBL algorithms are appropriate for the SON use case investigated in this study because of their low complexity in implementation, intelligibility in operational phase by a human operator, as well as high accuracy in case of small number of training examples and limited number of features (Uehara, 1998). The core of the proposed optimisation approach is a CBL algorithm called *k-Nearest Neighbour* (*k-NN*), where k is equal to three (Uehara, 1998; Langley and Iba, 1993; Aha, 1991). The k-NN algorithm can be used when sufficient amount of training samples are stored in the memory. In case there are too few samples in the memory, a temporary predefined decision criterion is used for classification which is used to detect if antenna is cell-edge focus or the opposite. In addition, a meta-heuristic structure for correcting wrong decisions and defining stopping condition is needed as shown in Figure 5.67. After radio

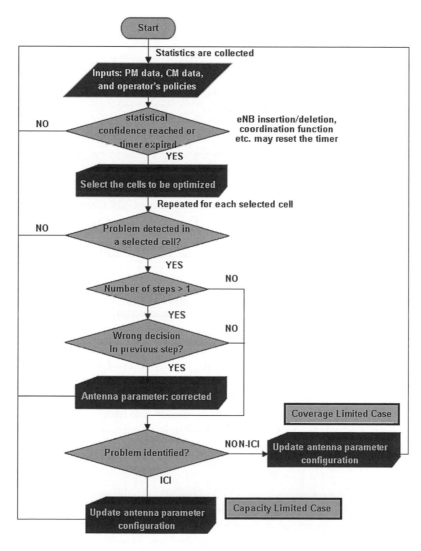

Figure 5.67 Flow-chart of the algorithm.

Table 5.7 Changes in antenna parameters by self-optimisation

Parameter	Number of Cells Where Re-Configuration Made	Mean[a] Parameter Change	Minimum[a] Parameter Change	Maximum[a] Parameter Change
Azimuth	19/21	16.50°	5.00°	40.00°
Remote Electrical Tilt	17/21	2.45°	−3.50°	6.60°
Horizontal HPBW	17/21	−21.76°	−30.00°	20.00°

[a] Statistical figures do not include the information of cells where initial configuration kept.

network performance statistics are collected, all numeric features are standardised so that the biasing due to varying KPI scales is avoided.

Weighting of each KPI feature is recalculated every step of the optimisation by measuring the KPI's contribution to the correct classification (over random guess) using *leave-one-out cross validation technique* (Alpaydin, 2010). The weighting is limited by a minimum weighting value to keep self-optimisation running.

5.4.2.3 Performance Evaluation Results

The amount of how much network performance can be optimised depends on the selected configuration and on how optimum the selected configuration was when optimisation was started. Here same sites and tilt values as with a 3G network are assumed, leading to a realistic but suboptimal network performance. As shown in Table 5.7, most of the cells in the monitored area of the network are downtilted as a result from optimisation. The reason for this is because the initial tilt angles were planned considering soft handover gain (i.e. macro diversity) for the WCDMA network while hard handover is used for LTE instead. That is one of the main reasons why tilt angles are increased and cell overlapping area gets smaller after the self-optimisation of remote electrical tilt.

Figures 5.68 and 5.69 justify the previous studies (Yilmaz, 2009) on antenna parameter optimisation space and remote electrical tilt performance. More significant SINR gains are seen in the capacity performance rather than coverage.

Average improvement in cell capacities are 4%, 11% and 7% respectively for azimuth, tilt and half-power beam width self-optimisation as shown in Figure 5.70. On the other hand, the level of capacity gain might differ from cell to cell due to varying traffic and propagation conditions. Also, it is noteworthy that capacity optimisation is observed by means of the decreased inter-cell interference for the cells in which antenna parameters are not changed.

In Figure 5.71, it is observed that the number of call drops and radio link failures are decreased significantly but not consistently. This is due to large cells which take over the noise limited area from unmonitored cells in the network. Another valuable result observed is that optimisation of horizontal HBPW provides the largest improvement in radio link performance by means of increased antenna gain when the beam is narrowed.

As shown in Figure 5.72, service quality during handover improves in parallel with coverage optimisation. Hence, in the coverage optimised scenario, looser handover hysteresis could be

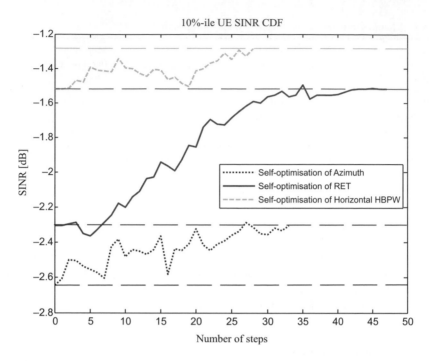

Figure 5.68 The tenth percentile points of SINR CDFs after each optimisation step (Yilmaz *et al.*, 2011b).

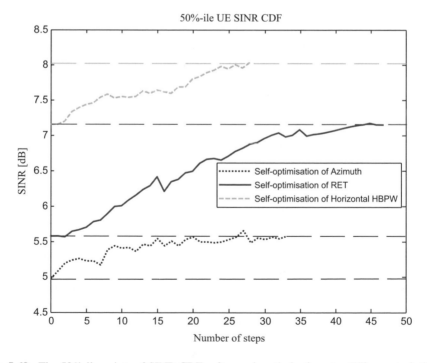

Figure 5.69 The 50%-ile points of SINR CDFs after each optimisation step (Yilmaz *et al.*, 2011b).

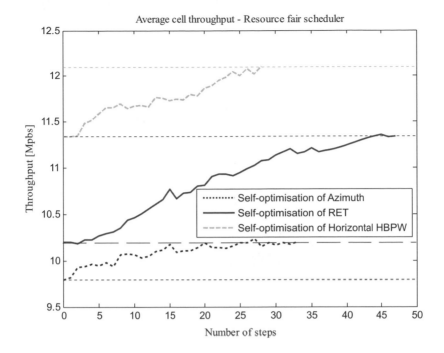

Figure 5.70 Average cell throughput (Yilmaz *et al.*, 2011b).

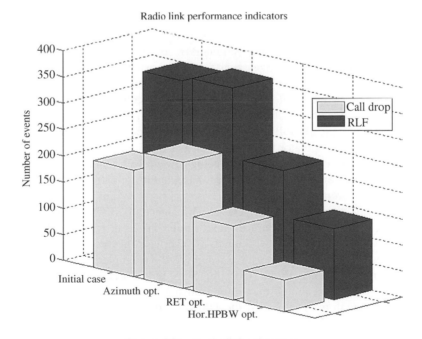

Figure 5.71 Radio link robustness.

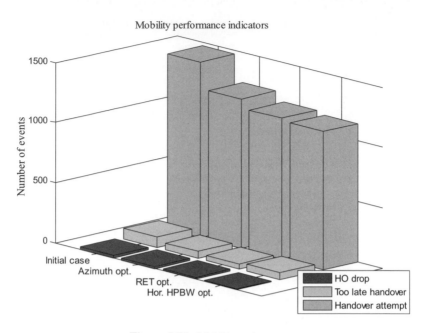

Figure 5.72 Mobility robustness.

used. Further mobility load balancing becomes more robust as further flexibility in cell individual offsets (CIOs) would be possible.

5.4.3 CCO with TX Power

Base station transmission power optimisation is one of the possibilities to get CCO benefits through the avoidance of excessive inter-cell interference in dense networks. However, power optimisation involves certain challenges, such as changes in amplifier long term behaviour and dramatic changes in handover region. Power optimisation maybe problematic for handover as the received power does not change only in a certain area of cell, but also in other parts of cell, actually in entire service area.

In LTE downlink, typically full power is used when opportunity for power optimisation becomes limited. In case operator has used lower than maximum power, power optimisation can be used for coverage optimisation. By reducing power levels, reduction in interference and power consumption for energy saving is aimed. Also, optimisation of reference signal power could be made to clean pilot pollution.

Power optimisation can be made in parallel with antenna optimisation (see Figure 5.73). Then, however, both mechanisms aim to have similar implications. Therefore, power and antenna optimisation need to be coordinated, as otherwise they could interfere each other leading to undesired impact, even negative result (cf. the example in Section 9.2.6).

The Integration Reference Point (IRP) for SON policy defined in (3GPP TS32.522, 2011) allows to convey mean and maximum transmit power. This should enable to either monitor a CCO function executing at the DM-level or to provide input data to a CCO function at the NM-level.

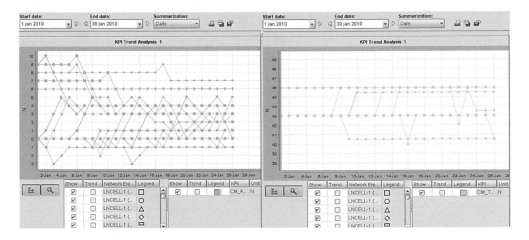

Figure 5.73 Example of RET (left) and TX power (right) optimisation. TX power changed rarely due to coordination action between two algorithms.

5.5 RACH Optimisation

5.5.1 General

The Random Access Channel (RACH) needs to be accurately configured so that it provides sufficient number of random access opportunities to UEs in any of the possible cell sizes. In addition RACH needs to be properly optimised for the cell load and physical environment.

In order to describe RACH optimisation, this chapter gives first a short introduction of the RACH and random access transmission; deep technical inside is given in LTE technology specific books, for example, (Holma and Toskala, 2011) for LTE Radio Access Network.

The random access (RA) procedure is deployed for several purposes (3GPP TS36.300, 2011):

- Initial access from RRC_IDLE;
- RRC Connection Re-establishment procedure;
- UL data arrival during RRC_CONNECTED requiring random access procedure[*];
- Handover;
- DL data arrival during RRC_CONNECTED requiring random access procedure[*];
- For positioning purpose during RRC_CONNECTED requiring random access procedure[*].

[*]Note, a random access procedure is required if the UE is not UL synchronised and/or does not have an SR (scheduling request) assignment in case of UL data arrival.

There are two forms of the procedure which are classified as contention and non-contention based random access procedures as shown in Figure 5.74. For the latter three of the above listed purposes the non-contention based procedure may be used as the eNB is able to signal a dedicated preamble sequence.

In order to start the access procedure the UE synchronises to downlink and reads system information (SIB 2) before preamble transmission. The distance to the eNB or timing advance is not known by the UE at this point. The eNB estimates the required timing advance (TA) based

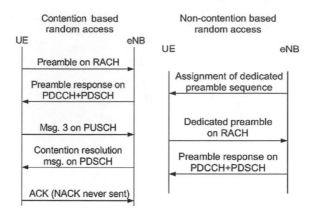

Figure 5.74 Contention and non-contention based random access procedures.

on the received UE signal arrival time. If UE is synchronised the timing advance is valid. Otherwise the RACH-procedure establishes a valid timing advance.

In the contention based case a preamble will be selected by the UE and transmitted in the available subframe(s). Based on correlation the NB may detect the access and furthermore can measure the timing of the UE transmission. The NB answers mirroring the preamble and at this point also timing advance will be fixed. Information on the scheduled resource will be exchanged and a temporary C-RNTI will be assigned. The UE sends its ID within Message 3. The type of ID depends on the state of the UE. In case of idle state Non-Access Stratum (NAS) information has to be provided else an Access Stratum (AS) ID (the C-RNTI) is used. The contention resolution is performed, that is, the NB addresses the UE by its ID.

The non-contention based access is collision-free. For example, during handover a reserved preamble will be issued by the target eNB and signalled to the UE by the source eNB. It is (temporarily) dedicated to this UE. No contention resolution is needed as the preamble shall not be used by other UEs.

Normal DL/UL transmission can take place after the random access procedure is completed. In total there are 64 preamble sequences per cell. Preambles are grouped to indicate the length of the needed resource. Some preambles are reserved for contention-free access.

5.5.2 PRACH Configuration

The preamble is sent on the RACH mapped to a physical channel, the so-called PRACH (Physical Random Access Channel). A single PRACH resource in time and frequency grid is given by one, two or three subframes and a bandwidth (BW) of six resource blocks (RB). The BW is kept constant, the exact position of the six RB within the cells BW is given by *prach-FreqOffset*. The different options for the time duration are given by the preamble formats as defined by (3GPP TS36.211, 2011). Table 5.8 is a summary of FDD preambles format and resulting max cell radius.

If Preamble Formats 2 and 3 are in use, the 800 μs sequence is sent twice (hence the duration is doubled), in this way the received energy can be increased by 3 dB.

The preamble format together with the number of PRACH resources ('access slots') per time including the exact time occurrence is determined by the PRACH Configuration Index:

Table 5.8 Preamble formats for FDD and resulting max cell radius

Preamble Format	Duration Total	CP/µs	Preamble Sequence Duration	Guard Time	Max Cell Radius
	ms (sub-frames)	µs	µs	µs	km
0	1	103.1	800.0	96.9	14.5
1	2	684.4	800.0	515.6	77.3
2	2	203.1	1600.0	196.9	29.5
3	3	684.4	1600.0	715.6	101.9

prach-ConfigIndex. The frequency of access slots and consequently the overall capacity can be configured within a large range, minimum capacity being ½ slot per frame (one slot every second frame, 20 ms). The maximum capacity is given by 10 access slots per frame for Preamble Format 0. Preamble Formats 1, 2 and 3 are limited by 5, 5, 3 occurrences per frame, respectively. Note, in FDD mode there is maximum one PRACH slot at a given time instance. Preamble Format 4 is used in TDD mode. In TDD more than one slot may exist at a given time instance.

In total 64 different preamble sequences per cell are available. The preamble sequences are based on Zadoff-Chu root sequences and their cyclic shifts. The overall number of defined root sequences is 838. The number of preambles which can be generated per root sequence depends on the required zero correlation zone. This is due to time uncertainties, mainly determined by cell size and additional delay spread. In the normal speed case the unrestricted set may be used. For a cell radius up to about maximum 760 m all preamble sequences can be generated from a single root sequence by deploying 64 cyclic shifts. The other extreme is given by a cell radius >58 830 m; only one cyclic shift per root sequence is possible and so each preamble sequence uses a different root sequence.

As detection is sensitive to Doppler shifts special rules are defined for cells where high UE speed and fading is present. Therefore high speed cases impose a limitation of the usable cyclic shifts and are enforced by utilisation of the restricted set, signalled via a parameter *high-SpeedFlag* (true, false). Which root sequence(s) is (are) used within a cell is signalled to the UE via *rootSequenceIndex* and the *zeroCorrelationZoneConfig*. The value rootSequenceIndex gives the first (logical) root sequence index. Hence these three configuration parameters are sufficient to provide the UE with full knowledge about the available 64 preambles in a cell.

A PRACH resource is defined by timing, frequency, root sequence and cyclic shift. In the assignment process of PRACH resources to a cell the neighbourhood needs to be taken into account: A sufficient re-use distance helps to avoid/minimise cases where random access attempts are erroneously detected, that is, in an unintended cell.

5.5.3 RACH Configuration

As mentioned above, 64 different preamble sequences are used per cell. These 64 sequences are subdivided in three sets:

- Contention based (= non-dedicated preambles) Group A;
- Contention based (= non-dedicated preambles) Group B;
- Contention free (= dedicated preambles).

The sizes of the different sets are signalled as follows: the number of contention based preambles (sum of Group A and B) is given by *numberOfRA-Preambles* (4, 8, 12, 16, 20, 24, 28, 32, 36, 40, 44, 48, 52, 56, 60, 64), the rest is used for dedicated preambles. The number of the Group A within the contention based set is *sizeOfRA-PreamblesGroupA* (4, 8, 12, 16, 20, 24, 28, 32, 36, 40, 44, 48, 52, 56, 60, 64), or, if not signalled, it equals the full contention based set.

For the contention based procedure a Group B preamble is used:

- if Group B is not empty and
- in case a large Message 3 (larger than messageSizeGroupA/bit) has to be delivered and
- the pathloss (PL) is less than a pathloss threshold (PL_THR).

The pathloss threshold is given by:

$$PL\_THR = P_{CMAX} - P_{O\_PRE} - \Delta_{PRE\_Msg3} - P_{offsetB}$$
P_{CMAX}: UE max power/dBm
P_{O_PRE}: *preambleInitialReceivedTargetPower*/dBm
Δ_{PRE_Msg3}: *deltaPreambleMsg3*/dB
$P_{offsetB}$: *message PowerOffsetGroupB*/dB

The preamble transmission may require several attempts if no response could be detected by the UE. For successive preamble transmission an increased transmission power may be used ('power ramping'), see Figure 5.75. The initial power of the first attempt is determined by open loop power control (3GPP TS36.213, 2011):

$$P_{PRACH} = min(P_{CMAX}, \text{preambleInitialReceivedTargetPower} + PL).$$
The power ramping step size is given by *powerRampingStep*/dB.

The setting *preambleTransMax* defines the maximum number of attempts. In order to observe the response from the eNB the UE monitors the PDCCH during a given time interval of length *ra-ResponseWindowSize*/subframes. This interval starts three subframes after the subframe where the preamble ends. Note, the eNB response may contain a backoff indicator (0.960 ms) in order to postpone and randomly distribute further attempts.

A *mac-ContentionResolutionTimer*/subframes is used to supervise the last step of the RA procedure (3GPP TS36.321, 2011). Message 3 may deploy Hybrid ARQ (HARQ), the maximum number of HARQ transmissions given by *maxHARQ-Msg3Tx*.

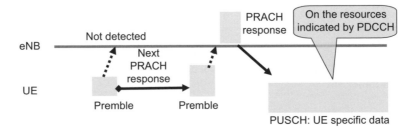

Figure 5.75 RA procedure. Physical layer view.

5.5.4 RACH/PRACH Configuration Example

The RACH/PRACH configuration is Sys Info, broadcast within SIB2 (3GPP TS36.331, 2011).
Example of involved parameter settings:

numberOfRA-Preambles n40
sizeOfRA-PreamblesGroupA n28,
messageSizeGroupA b144,
messagePowerOffsetGroupB dB10
powerRampingStep dB2,
preambleInitialReceivedTargetPower dBm-90
preambleTransMax n5,
ra-ResponseWindowSize sf10,
mac-ContentionResolutionTimer sf64,
maxHARQ-Msg3Tx 3,
rootSequenceIndex 12,
prach-ConfigIndex 4,
highSpeedFlag FALSE,
zeroCorrelationZoneConfig 12,
prach-FreqOffset 3,
referenceSignalPower d Bm 20,
deltaPreambleMsg3 dB 1.

In this example there are 40 preambles for the contention-based RA. For small message sizes 28 preambles are in use (Group A), Group B consists of $40 - 28 - 12$ preambles. Group B preamble can be used if more than 144 bits need to be sent and the pathloss is smaller than PL_THR $= 23\,\mathrm{dBm} - (-90\,\mathrm{dBm}) - 1\,\mathrm{dB} - 10\,\mathrm{dB} = 102\,\mathrm{dB}$ (assuming a UE max power of 23 dBm).

The initial preamble power is given by $-90\,\mathrm{dBm}$ plus measured and higher layer filtered pathloss [dB]. Each further transmission step increases power by 2 dB. This naturally only applies in cases where the UE is not limited by its max power.

In the discussed example, a maximum of five preamble transmissions are allowed, if there is no response after the fifth attempt the RA procedure is aborted and failure is detected.

Following each preamble transmission the UE observes during 10 subframes the response on PDCCH (starting three subframes after preamble end). After Message 3 transmission the contention resolution should be done within 64 subframes. A maximum of three HARQ transmissions are configured. The setting *prach-ConfigIndex 4* determines Preamble Format 0 (1 ms length) as well as the time instance: each sub-frame four provides an access slot, that is, there is one slot every 10 ms. The RB number to use is 3..8 (*prach-FreqOffset 3*), the lower 0..2 are most probably used for PUCCH. From the parameter *zeroCorrelationZoneConfig 12* follows a cyclic shift separation of 119 (see the table in 3GPP TS36.211, 2011) which yields seven preambles per root sequence. As 64 are to be defined, 10 root sequences are consumed by this cell setting: logical index 12..21, the start given by the *rootSequenceIndex 12*. Parameter *referenceSignalPower dBm 20* is needed for PL estimation.

5.5.5 RA Performance

PRACH receiver performance requirements have been defined by 3GPP TS36.104 (2011) and 3GPP TS36.141 (2010) in terms of preamble false alarm probability to be less than or equal to 0.1% and furthermore a preamble detection probability of 99% or higher. Hence, the missed detection probability being less than or equal 1%. These requirements are valid for given conditions (SINR etc.).

In real deployment scenarios the conditions vary as channel conditions, UE speed, SINR are dependent on transmitted power, load, geometry, and so on. Therefore a tuning of parameters might be advantageous. In particular the power settings can have a large impact on this aspect of the RA performance. There is a trade off between the proportions of failed RA procedures, the duration of the RA procedure (depending on the number of preamble attempts) and the introduced interference by PRACH transmissions. Note, the latter is less critical than for example, in CDMA based technologies as SC-FDMA is used in LTE UL and orthogonality is maintained. However, low preamble power might be beneficial to lower the inter-cell interference. Related parameters are: *powerRampingStep*/dB *preambleInitialReceivedTargetPower*/dBm, *preambleTransMax*. UE reports upon the number of transmitted preambles may be requested by the eNB and can be utilised as input in a RACH-optimising SON algorithm by the eNB (see the following).

Another possible cause of performance degradation is high RACH load. The contention based RA procedure follows a slotted-aloha principle with an inherent collision risk. A collision in this sense is the coincidence of the same preamble sequence in a given cell. As a random selection is performed by the UEs there is a certain probability that the same preamble is transmitted by two or more UEs. Maximum one UE might be successful in such cases of collision. The chance of collision increases with the load and determines the throughput in slotted-aloha based access systems:

$$Y = A \, e^{-A/N}.$$
Y: traffic carried/1.
A: traffic offered/1.
N: number of parallel channels/1 (i.e. the number of available sequences)

Note, A and N are given relatively to the maximum channel capacity: $= 1$. For the unloaded system $(A \rightarrow 0)$ the collision probability $\rightarrow 0$, hence $Y = A$. In an overloaded system $A \rightarrow \inf$. I.e. there are no collision free attempts and $Y \rightarrow 0$ (assuming that none of the colliding attempts can be detected).

It should be noted that a collision not necessarily means a failed RA procedure as *preambleTransMax* chances exist for each UE to get the preamble transmitted. Furthermore, there is a chance that one of the involved UEs can be decoded. A clear disadvantage is, however, obvious: The more repetitions are needed the higher the introduced delay and additionally also the interference may increase. To counteract these effects an extension of the number of PRACH resources might be useful. The trade off is the possible loss of PUSCH capacity. Note, the PRACH time-frequency resources could be used by PUSCH scheduling, the disadvantage being higher interference in the considered RBs.

In order to measure the performance with regards to collision events the eNB could simply measure Y, the traffic carried on PRACH, and use the calculation above to estimate collision probabilities. Additionally UE reports on MAC detected collisions can be requested and utilised.

The described performance aspects so far can be evaluated and potentially tuned by the eNBs itself, that is, in a local way. The situation is different if the optimisation of the PRACH resource assignment is considered. As mentioned earlier, PRACH resources of neighbouring cells should be separated. This can be achieved either in terms of timing, frequency or by deploying different root sequences. A deterministic time separation is only possible in case of synchronisation of the involved cells, for example, different sectors of the same eNB but also in TDD mode. Separation based on frequency or different root sequences could be utilised in any case.

It becomes obvious that in this optimisation context a local perspective is not sufficient, but a rather network-orientated approach is required. In order to support decentralised SON an inter-eNB exchange of the related parameters is supported (see the following).

5.5.6 Self-Optimisation Framework

To enable automatic management of RACH that would react to changing network situation and thus would continuously optimise configuration of RACH, a SON RACH optimisation algorithm uses two separate mechanisms.

1. UE reporting.
2. Inter-eNB information exchange.

RACH optimisation, as a SON use case, assumes that an eNB continuously adjusts the RACH configuration in its cells to changing network situation. In order to close the control loop and to receive feedback, the eNB may request RACH performance information from any UE that has just set up a connection in one of its cells. Then, in order to coordinate RACH configuration with its neighbours, the eNB may provide RACH configuration of its cells at X2 setup and later, whenever the configuration changes. This mechanism is presented in Figure 5.76.

5.5.7 UE Reporting

The UE reporting is a part of RRC specification. The mechanism mandates the UE to report the number of preambles that its Medium Access Control (MAC) layer issued before the RACH

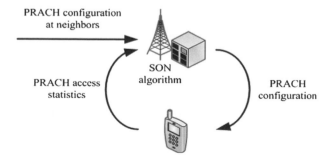

Figure 5.76 RACH optimisation cycle.

access completed successfully. Additionally, if MAC detected collision at any of the preambles, it shall be indicated in the report (a contention flag is set to 'true'). In order to avoid reporting to eNBs that do not support SON or just RACH optimisation, the reporting has been defined as a separate procedure. If the eNB that the connection has just been set up to is interested in RACH statistics, it may request the report, which the UE is obliged to provide.

If UE reporting is provided, it is possible to analytically measure the Access Probability (AP), which can be used as an optimisation criterion since it has a direct impact on QoS (e.g. delay), based on collected statistics in a given time interval. Access probabilities for contention-based RA (CBRA) and non-contention-based RA (NCBRA) at the attempt m are given in Equations 5.1 and 5.2 respectively using input performance metrics Detection Miss Probability (DMP) *DMP*, Collision Probability (CP) *CP* for CBRA and Blocking Probability (BP) *BP* for NCBRA (Amirijoo *et al.*, 2009).

$$AP_{CBRA,m} = 1 - \prod_{i=1}^{m}(DMP_{CBRA,i} + (1 - DMP_{CBRA,i}) \cdot CP) \tag{5.1}$$

$$AP_{NCBRA,m} = 1 - \prod_{i=1}^{m}(DMP_{NCBRA,i} + (1 - DMP_{NCBRA,i}) \cdot BP) \tag{5.2}$$

In case DMP and AP performance indicators are monitored by the local SON module at eNB, transmission power parameters (e.g. initial power, and power ramping step) might be self-tuned (Amirijoo *et al.*, 2009) as well as RACH preamble allocation period and preamble split parameters can be optimised (Yilmaz *et al.*, 2011a). Furthermore, if RACH performance is able to be kept above the given operator's targets as shown in Figure 5.77, PUSCH capacity can be extended too as given in Figure 5.78 where virtual load indicates the normalised capacity requirement for the complete service satisfaction.

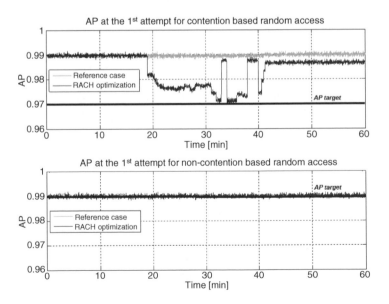

Figure 5.77 Access probability at the first attempt (Yilmaz *et al.*, 2011a). Reproduced with permission of © 2011 IEEE.

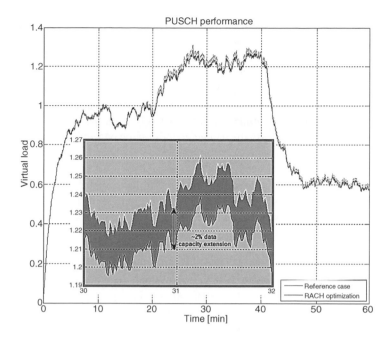

Figure 5.78 Optimisation of PUSCH performance (Yilmaz *et al.*, 2011a). Reproduced with permission of © 2011 IEEE.

5.5.8 Inter-eNB Communication

The reporting mechanism presented above enables eNB to monitor performance of RACH access. However, conflicting RACH configuration in other cells may affect the performance, too. Furthermore, static coordination based on network planning is not sufficient in case local configuration may change dynamically. Therefore, in order to provide the SON algorithm with information about current setting at neighbouring eNBs, the RACH configuration can be informed to a neighbour at X2 interface setup procedure or whenever the configuration of an eNB changes. In each of these cases, the X2 interface signalling (3GPP TS36.423, 2011) allows to include following information:

- Root sequence index.
- Zero correlation zone configuration.
- High speed flag.
- PRACH frequency offset.
- PRACH configuration index (TDD mode only).

These parameters are defined in 3GPP, TS36.211 (2011); in the following their meaning and possible usage for RACH optimisation is briefly described.

Root sequence index
PRACH uses Zadoff-Chu sequences to generate the preambles. The main reason is that ZC sequences feature relatively low peak-to-average-power ratio, thus facilitating implementation of the amplifier at the terminal. The sequences are generated from a root sequence.

In LTE the root sequences are grouped into 838 logical root sequences and the index of the one that is to be used is broadcast in every cell. Neighbouring eNBs can utilise this information to avoid scheduling overlapping sequences.

Zero correlation zone configuration

Based on the selected root sequence index, the preambles are generated using a given zero correlation zone and a cyclic shift. The zero correlation zone is used to guarantee orthogonality of the generated sequences irrespectively from the UE time uncertainty and the transmission delay. Therefore, the value depends on the particular conditions in the cell, for example, cell size and synchronisation source, and is broadcast in every cell.

High speed flag

Correlation between cycles may be affected also by terminal speed. Therefore, in cells where UEs may be carried with high speed, the length of cyclic shift selected with given zero correlation zone is further restricted. Whether in given cell the UEs should use unrestricted or restricted lengths depends on the broadcast flag.

PRACH frequency offset

The frequency the cell uses is divided into PRBs, as described in Chapter 2. The cell must make it known to the terminals which PRB is available for RACH access. For this purpose the frequency offset is used in FDD mode and the value is broadcast in every cell. Neighbouring eNBs can utilise this information to avoid scheduling the same PRBs for RACH access.

PRACH configuration index (TDD mode only)

In TDD mode, the frequency offset cannot be derived only based on the PRB number, but also PRACH configuration index must be given. The index of the PRACH format to be used is always broadcast in every cell, but in order to derive the PRBs that are occupied by PRACH its value is needed only in cells working in TDD mode.

5.6 RRM and SON (Interference Coordination, P0 Optimisation)

The difference between Radio Resource Management and SON is not black and white. Here, whether something belongs to RRM or SON depends mainly on the time scale of the dynamics. If something is changed fast, like on ms basis, it belongs to RRM, while if the actions are done every couple of seconds or longer, one can refer to it as SON.

However, many RRM algorithms contain parameters, which can be updated at a slower rate, which can be done by SON algorithm. In this section a couple of examples are given for this kind of cases. First in Section 5.6.1 interference coordination is described, while in Section 5.6.2 optimisation of the P0 parameter, which is used in uplink power control is described.

5.6.1 Interference Coordination

The basis of Interference Management (ICIC, Inter Cell Interference Coordination) is the coordination of the radio resources used by neighbour cells in a geographical area. The objective of these techniques is to control inter-cell interference and improve the SNR at the cell edge. Typical strategies utilise ordering or grouping users on the resources in a smart way. Those can be further enhanced by reserving PRBs for certain groups of users or using different power coordination techniques. Figure 5.79 shows an example of such an ICIC coordination method: the inter-cell interference is kept low in the part of the spectrum (dark area) where the

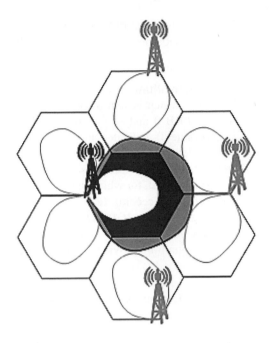

Figure 5.79 Example of ICIC coordination.

cell edge users are likely to be scheduled and neighbour cells do not schedule users in the light grey area on the same frequency resources allocated to users in the dark area.

ICIC can be done in both uplink and downlink for the data channel (PDSCH and PUSCH) or uplink control channel (PUCCH). ICIC is not possible on the downlink control channel (PDCCH) since the control channel elements are spread over the full band and several OFDM symbols.

There exist dynamic, static and semi-static ICIC coordination schemes. Dynamic schemes rely on frequent adjustment of parameters with the use of signalling between cells. For cells at different sites this requires X2 signalling (Figure 5.80). Static ICIC refers to schemes where the parameters are not changing and semi-dynamic refer to schemes with slow adaptation of the parameters based on signalling between the cells.

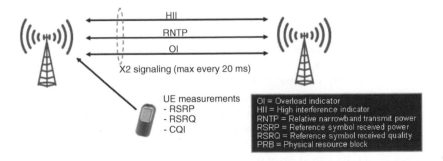

Figure 5.80 ICIC X2 signalling.

To support dynamic ICIC schemes between different sites and parameter adjustment the following X2 signalling have been standardised in Release 8 (3GPP TS36.423, 2011). Two kinds of parameters exist:

1. **'Scheduling announcements' to facilitate proactive coordination**
   HII: High Interference Indicator, which is sent as bitmap. Each position in the bitmap represents a PRB (first bit = PRB 0 and so on), for which value '1' indicates 'high interference sensitivity' and value '0' indicates 'low interference sensitivity'. See 3GPP TS36.423 (2011) for further details.
   RNTP: relative narrowband transmit power. Each position in the bitmap represents a n_{PRB} value (i.e. first bit = PRB 0 and so on), for which the bit value represents $RNTP (n_{PRB})$. Value 0 indicates 'Tx power not exceeding RNTP threshold'. Value 1 indicates 'no promise on the Tx power is given'. See 3GPP TS36.423 (2011) for further details.
2. **Indicators to facilitate reactive coordination**
   OI: Overload Indicator. This signalling provides, per PRB, a report on interference overload.

Note that X2 signalling suffers from certain delay, so instantaneous adjustments are not possible. Additionally, this signalling costs X2 bandwidth (it depends on the frequency of the signalling).

To illustrate the potential of ICIC here some uplink results are shown. Focus is on the uplink since there the potential is largest, since in the uplink UEs in power limited areas, where the interference is largest, cannot expand to using many PRB due to limited transmission power.

In the uplink, the UEs transmit power is determined by open loop power control as specified in the 3GPP specifications, characterised by the following equation (in dB scale):

$$P_{PUSCH} = \min \{P_{MAX}, 10.\log_{10} (M) + P_0 + \alpha.PL + \Delta_{MCS} + f(\Delta_{PUSCH})\}$$

where P_{MAX} is the maximum UE transmission power, M is the number of assigned Resource Blocks (RB), P_0 has both a UE specific component and a broadcast one, α is the cell-specific pathloss compensation factor, PL is the downlink pathloss estimated by the UE based on the transmit power of the reference symbols, Δ_{MCS} is a UE-specific parameter depending on the chosen Modulation and Coding Scheme (MCS), $f(\Delta_{PUSCH})$ is a UE-specific correction value also referred as TPC, and $\Delta_{PUSCH} = [-1;0;1;3]$ dB in the case of cumulative closed loop power control commands.

Interference coordination in the uplink can be done through:

- Hard splitting the frequencies and use a frequency reuse like in GSM. This lowers the interference but has a negative effect on the cell and user performance, due to very negative effect on the number of available PRBs.
- Using a soft frequency reuse pattern, where the part of the band, referred to as preferred band, where the cell edge users, who are causing most interference, are coordinated between the cells such that a frequency reuse for those users is used.
- The PRBs which can be used for the cell edge users can be updated based on X2 signalling. The key question is how to identify the users to be placed in the preferred band. Some key strategies are shown in Figure 5.81.

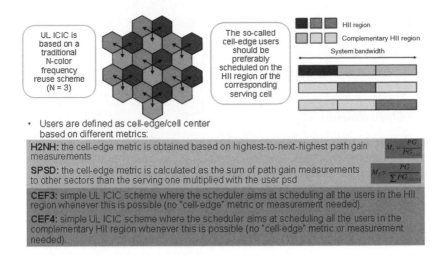

Figure 5.81 UL ICIC.

Figure 5.82 shows the normalised effective spectral efficiency for the 95% coverage (i.e. 5% outage) of the above described algorithm examples. Given these results, it can be observed an increasing gain from UL ICIC as the fractional load is decreased. As an example, no significant gain is observed at high load, while a 22% gain is found at 25% fractional load (25% of the available PRBs are used). It is furthermore observed that the best UL ICIC performance benefit is obtained by simply aiming at scheduling all the users in the HII (High Interference Indicator) bandwidth region without relying on any UE measurements for cell-edge identification.

Note that the reference in these results is a channel aware proportional fair scheduling. Also within the preferred band and the non-preferred band proportional fair scheduling is used.

The previous text shows that there are small gains achievable by interference coordination. It should be noted that the gains of interference coordination may be larger in case of heterogeneous networks. The SON component in the interference coordination comes from coordination of the preferred frequency bands in the neighbouring cells. This can be

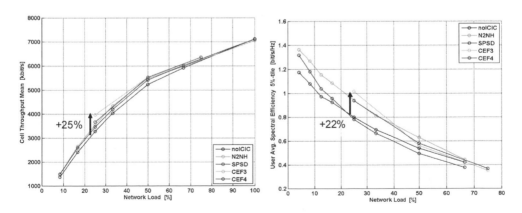

Figure 5.82 Spectral efficiency with different ICIC techniques.

coordinated through distributed coordination by the use of X2 based signalling or by a centralised solution where the preferred bands are planned in a central unit and pushed out to the different eNBs. The centralised solution can make use of input from planning tools, knowing the interference coupling between cells.

5.6.2 P0 Optimisation

The power control in LTE UL has an open loop and a closed loop component. The open loop component is meant to compensate the slow variations of the received signal due to the pathloss including shadowing, while the closed loop component is meant to further adjust the user's transmission power to compensate for errors and rapid variations as well as potentially optimise the system performance. The way to do such corrections and optimisations is left up to the equipment manufacturer. The setting of the UE Transmit power P_{PUSCH} for the PUSCH transmission in a given subframe is defined in the equation given in the previous section.

$$P_{PUSCH} = \min\{P_{MAX}, 10.\log_{10}(M) + P_0 + \alpha.PL + \Delta_{MCS} + f(\Delta_{PUSCH})\}$$

The Fractional Power Control algorithm aims at partially compensating the pathloss of users in such a way that it will impact users located at the cell edge more than those located close to the cell centre. The coverage (defined as the fifth percentile user throughput averaged over the session time) versus capacity (defined as the average cell throughput) mapping is shown in Figure 5.83 for different values of α. The dots on each curve are obtained by increasing the

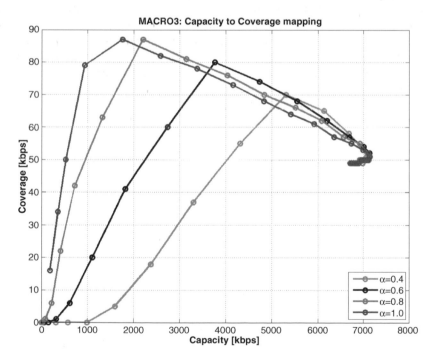

Figure 5.83 Coverage versus capacity for FPC. Macro 3 Scenario 6.

value of P0 with steps of 2 dB only. The coverage performance tops before that of the capacity and are obtained from different values of P0 ([−108, −84, −64, −36] dBm) depending on α ([1,0.8, 0.6, 0.4]). These results are obtained in Macro3 scenario using a fixed transmission bandwidth of 6 RBs with 10 UEs per cell.

This gives the reference performance case and illustrates that the throughput performance is highly dependent on the value of P0. Small variations of the settings have a significant impact on the throughput performance.

The optimal value of P0 is depending on the load of the system as illustrated in (Boussif *et al.*, 2010), but also to the environment. Small cells need a different P0 value as large cells. For setting the P0 in each cell in a network there are several possibilities:

- **Setting a fixed value P0 for all cells:** This is a simple method, which requires no network planning. It, however, is unlikely to produce the optimal results, since cells in the network are typically different.
- **Using the path loss information:** Path loss information per cell can be achieved through RSRP measurements made by the UEs. Then the P0 can be set according to the following equation:

$$\text{P0} = \text{Pmax} - \text{MdB} - \alpha \text{L@95\%}$$

Where Pmax is the maximum UE power, MdB is the number of PRBs used by the UE on average and L@95% is the 5% worst path loss estimated from RSRP measurements or achieved from a network planning tool. The thinking behind this algorithm is that only 5% of the users run into power saturation in each cell using this algorithm.
- **Optimising the SINR:** This can for instance be done by looking at the worst 5% SINR of every cell and associate a cost to that as shown in Figure 5.84. The information about the SINR can be achieved through planning tools or through measurements from the UEs (RSRQ) which give an indication. Then a heuristic search algorithm can try to optimise the cost function by tuning the P0 of the cells. Note that increasing P0 in one cell may improve the SINR in that cell but at the same time cause more interference in other cells. In the example cost function shown in Figure 5.84, there is a certain 5% lower bound SINR and an

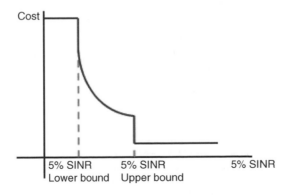

Figure 5.84 Cost function with regards to SINR.

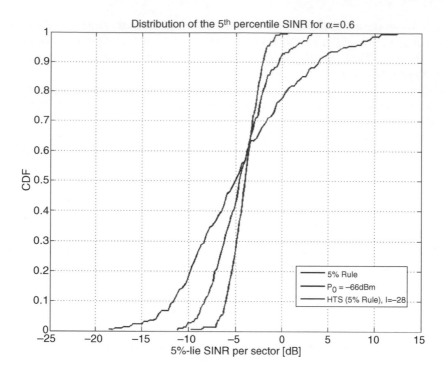

Figure 5.85 P0 optimisation methods comparison.

upper bound. Between those two values, cost function is decreasing. The effect of this is that the algorithm will spend most of the effort on the improving the low 5% SINR values, that is, on the cells with the lowest cells edge SINR, since there the gain is largest.

The three methods mentioned were evaluated offline in simulators based on pathloss information of a 3G network of major European city (Figure 5.85). The fixed P0 value of −66 dBm has been chosen since it gives the best result of all fixed P0 values for this network. It can be seen that 5% rule algorithm does not improve the worst cells compared to the fixed value, but leads to an improvement of the cell edge SINR of the best cells. The heuristic search tree (HTS) algorithm is best in improving the worst cells.

With this it is shown that even though power control clearly can be seen as RRM, the parameters in the different algorithm can be updated by SON based algorithms.

References

3GPP TR32.826 (2010) Technical Specification Group Services and System Aspects Telecommunication Management, *Study on Energy Savings Management (ESM)*, ver.10.0.0., Release 10, 30 March 2010. Available from http://www.3gpp.org/ftp/Specs/latest/Rel-10/32_series/32826-a00.zip [accessed 30 June 2011].

3GPP TR36.814 (2010) Technical Specification Group Radio Access Network (E-UTRA), *Further Advancements for E-UTRA Physical Layer Aspects*, ver.9.0.0., Release 9, 30 March 2010. Available from http://www.3gpp.org/ftp/Specs/latest/Rel-10/36_series/36814-900.zip [accessed 30 June 2011].

3GPP TS25.331 (2011) Technical Specification Group Radio Access Network, *Radio Resource Control (RRC) - Protocol Specification*, ver.10.4.0., Release 10, 1 July 2011. Available from http://www.3gpp.org/ftp/Specs/latest/Rel-10/25_series/25331-a40.zip [accessed 30 June 2011].

3GPP TS32.522 (2011) Technical Specification Group Services and System Aspects Telecommunication Management, *Self-Organizing Networks (SON) Policy Network Resource Model (NRM) Integration Reference Point (IRP); Information Service (IS)*, ver.10.2.0., Release 10, 17 June 2011. Available from http://www.3gpp.org/ftp/Specs/latest/Rel-10/32_series/32522-a20.zip [accessed 30 June 2011].

3GPP TS36.133 (2011) Technical Specification Group Radio Access Network (E-UTRA), *Evolved Universal Terrestrial Radio Access (E-UTRA); Requirements for Support of Radio Resource Management*, ver.10.3.0, Release 10, 21 June 2011. Available from http://www.3gpp.org/ftp/Specs/latest/Rel-10/36_series/36133-a30.zip [accessed 30 June 2011].

3GPP TS35.304 (2011) Technical Specification Group Radio Access Network, *User Equipment (UE) Procedures in Idle Mode and Procedures for Cell Reselection in Connected Mode*, ver.10.1.0., Release 10, 24 June 2011. Available from http://www.3gpp.org/ftp/Specs/latest/Rel-10/25_series/25304-a10.zip [accessed 30 June 2011].

3GPP TS36.300 (2011) Technical Specification Group Radio Access Network, *E-UTRA and E-UTRAN Overall description, Stage 2*, ver.10.3.0., Release 10, 22 June 2011. Available from http://www.3gpp.org/ftp/Specs/latest/Rel-10/36_series/36300-a40.zip (accessed 30 June 2011).

3GPP TS36.211 (2011) Technical Specification Group Radio Access Network, *E-UTRA Physical Channels and Modulation*, ver.10.2.0., Release 10, 22 June 2011. Available from http://www.3gpp.org/ftp/Specs/latest/Rel-10/36_series/36211-a20.zip (accessed 30 June 2011).

3GPP TS36.213 (2011) Technical Specification Group Radio Access Network, *E-UTRA Physical Layer Procedures*, ver.10.2.0., Release 10, 22 June 2011. Available from http://www.3gpp.org/ftp/Specs/latest/Rel-10/36_series/36213-a20.zip [accessed 30 June 2011].

3GPP TS36.104 (2011) Technical Specification Group Radio Access Network, *E-UTRA, Base Station (BS) Radio Transmission And Reception*, ver.10.3.0., Release 10, 20 June 2011. Available from http://www.3gpp.org/ftp/Specs/latest/Rel-10/36_series/36104-a30.zip [accessed 30 June 2011].

3GPP TS36.141 (2011) Technical Specification Group Radio Access Network, *E-UTRA Base Station (BS) Conformance Testing*, ver.10.3.0., Release 10, 20 June 2011. Available from http://www.3gpp.org/ftp/Specs/latest/Rel-10/36_series/36141-a30.zip [accessed 30 June 2011].

3GPP TS36.304 (2011) Technical Specification Group Radio Access Network, *Evolved Universal Terrestrial Radio Access (E-UTRA); User Equipment (UE) procedures in idle mode*, ver.10.2.0, Release 10, 21 June 2011. Available from http://www.3gpp.org/ftp/Specs/latest/Rel-10/36_series/36304-a20.zip [accessed 30 June 2011]

3GPP TS36.321 (2011) Technical Specification Group Radio Access Network, *Evolved Universal Terrestrial Radio Access (E-UTRA); Medium Access Control (MAC)*, ver.10.2.0, Release 10, 21 June 2011, Available from http://www.3gpp.org/ftp/Specs/latest/Rel-10/36_series/36321-a20.zip [accessed 30 June 2011].

3GPP TS36.331 (2011) Technical Specification Group Radio Access Network, *Evolved Universal Terrestrial Radio Access (E-UTRA); Radio Resource Control (RRC) - Protocol specification*, ver.10.2.0., Release 10, 24 June 2011. Available from http://www.3gpp.org/ftp/Specs/latest/Rel-10/36_series/36331-a20.zip [accessed 30 June 2011].

3GPP TS36.423 (2011) Technical Specification Group Radio Access Network, *E-UTRAN X2 Application Protocol (X2AP)*, ver.9.6.0., Release 10, 24 June 2011. Available from http://www.3gpp.org/ftp/Specs/latest/Rel-10/36_series/36423-a20.zip [accessed 30 June 2011].

3GPP TR36.927 (2011) Technical Specification Group Radio Access Network, *Potential Solutions for Energy Saving for E-UTRAN*, ver.10.0.0., Release 10, 24 June 2011. Available from http://www.3gpp.org/ftp/Specs/latest/Rel-10/36_series/36927-a00.zip [accessed 30 June 2011].

3GPP TR36.942 (2010) Technical Specification Group Radio Access Network, *Evolved Universal Terrestrial Radio Access (E-UTRA); Radio Frequency (RF) System Scenarios*, ver.10.2.0., Release 10, 6 January 2011. Available from http://www.3gpp.org/ftp/Specs/latest/Rel-10/36_series/36942-a20.zip [accessed 30 June 2011].

Aha, D.W. (1991) Case-based learning algorithms. In Proceedings of Case-Based Reasoning Workshop, pp. 147–158.

Alpaydin, E. (2010) *Introduction to Machine Learning*, 2nd edn, The MIT Press, Massachusetts, pp. 486–489.

Amirijoo, M., Frenger, P., Gunnarsson, F. *et al.* (2009) Towards random access channel self-tuning in LTE. in Proceedings of IEEE Vehicular Technology Conference (VTC), Spring 2009.

Boussif, M., Rosa, C., Wigard, J. and Mullner, R. (2010) Load adaptive power control in LTE Uplink. Proc of IEEE European Wireless Conference, pp. 288–293.

Dottling, M. and Viering, I. (2009) Challenges in mobile network operation: Towards self-optimizing networks. In Proceedings of IEEE International Conference on Acoustics, Speech and Signal Processing, Taipei, Taiwan, April.

EARTH INFSO-ICT-247733 (2010) Deliverable D2.3, December 31, 2010. Available from https://bscw.ict-earth.eu/pub/bscw.cgi/d31515/EARTH_WP2_D2.3.pdf, [accessed 15 March 2011].

ETSI TS 102 706 (2007) *A WRI/WBCSD GHG Protocol Initiative Calculation Tool*, www.ghgprotocol.org [accessed 15 March 2011].

Holma, H. and Toskala, A. (eds) (2011) *LTE for UMTS: Evolution to LTE-Advanced*, John Wiley & Sons, Ltd., Chichester.

Islam, M.N.ul, Abou-Jaoude, R., Hartmann, C. and Mitschele-Thiel, A., (2010) Self-optimisation of Antenna Tilt and Pilot Power for dedicated channels, Modeling and Optimisation in Mobile, Ad Hoc and Wireless Networks (WiOpt). Proceedings of the 8th International Symposium, June.

Jørgensen, N.T.K., Laselva, D. and Wigard, J. (2011) On the potentials of traffic steering techniques between HSDPA and LTE. In proc. of VTC Spring 2011, Budapest.

Laiho, J., Wacker, A. and Novasad, T. (2006) *Radio Network Planning and Optimisation for UMTS*, 2nd edn, John Wiley & Sons, Inc., New York.

Langley, P. and Iba, W. (1993) Average-case analysis of a nearest neighbor algorithm. In Proceedings of the 13th IJCAI, pp. 889–894.

Lee, W.C.Y. (1998) *Mobile Communications Engineering*, McGraw-Hill, New York.

Mogensen, P., Wei Na, Kovacs, I.Z., Frederiksen, F., Pokhariyal, A., Pedersen, K.I., Kolding, T., Hugl, K., Kuusela, M. (2007) LTE Capacity compared to the Shannon Bound. In Proceedings of IEEE Vehicular Technology Conference (VTC), Dublin, May.

Nihtilä, T., Turkka, J. and Viering, I.(May 2011) Performance of LTE self-optimizing networks uplink load balancing. Accepted to IEEE Vehicular Technology Conference Spring 2011, Budapest.

Schwartz, M. (1988) *Telecommunication Networks: Protocols, Modeling and Analysis*, Addison-Wesley, Reading, Massachusetts.

Siomina, I., Värbrand, P. and Yuan, D. (2006) *Automated Optimisation of Service Coverage and Base Station Antenna Configuration in UMTS Networks*, Linköping University, Linköping.

Turkka, J. and Lobinger, A., (2010) Non-regular layout for cellular network system simulations. In Proceedings of the International Symposium on Personal, Indoor and Mobile Radio Communications (PIMRC '10, September), Istanbul.

Uehara, K. (1998) Random case analysis of inductive learning algorithms. In Proceedings of the First International Conference on Discovery Science.

Wang, Y., Yang, X., Ma, A. and Cuthbert, L. (2009) Intelligent resource optimisation using semi-smart antennas in LTE OFDMA systems. in Proceedings of IEEE International Conference on Communications Technology and Applications (ICCTA, October).

Yilmaz, O.N.C., Hämäläinen, S. and Hämäläinen, J. (2009a) Analysis of antenna parameter optimisation space for 3GPP LTE. In Proceedings of IEEE Vehicular Technology Conference (VTC), Anchorage, Alaska, September.

Yilmaz, O.N.C., Hämäläinen, S. and Hämäläinen, J. (2009b) Comparison of remote electrical and mechanical antenna downtilt performance for 3GPP LTE. in Proceedings of IEEE Vehicular Technology Conference (VTC), Anchorage, Alaska, September.

Yilmaz, O.N.C., Hämäläinen, S. and Hämäläinen, J. (2010) Self-optimisation of Remote Electrical Tilt. In Proceedings of IEEE Personal, Indoor and Mobile Radio Conference (PIMRC), Istanbul, September.

Yilmaz, O.N.C., Hämäläinen, J. and Hämäläinen, S. (2011a) Self-optimisation of Random Access Channel in 3GPP LTE. International Wireless Communications and Mobile Computing Conference (IWCMC), Istanbul.

Yilmaz, O.N.C., Hämäläinen, S. and Hämäläinen, J. (2011b) *Optimization of Adaptive Antenna System Parameters in Self-organizing LTE Networks*, to be submitted to *Wireless Networks*, Springer, Heidelberg.

Zheng, Y.R. and Xiao, C. (2002) Improved models for the generation of multiple uncorrelated Rayleigh fading waveforms. *IEEE Communications Letters*, **6**(6), 256–258.

6

Self-Healing

Szabolcs Nováczki, Volker Wille, Osman Yilmaz, Seppo Hämäläinen
and Henning Sanneck

Cellular networks are very large and extremely complicated systems. For systems of this size and complexity it is not uncommon for faults to occur. Faults can appear at several functional areas of a complex cellular network, however, the most critical domain from a fault management viewpoint is the Radio Access Network (RAN). Every base station is responsible to serve a dedicated space of the coverage area with little, if any, redundancy. If one of these network elements is not capable to fulfil its responsibilities due to the presence of a fault, there will be no other entity to offer service until the fault is rectified. During the resulting period of *degraded* performance, users are not experiencing services with acceptable availability, reliability or quality-of-service, which may cause serious revenue loss for the operator.

The problem for the network operator is, first of all, that there is huge number of network elements (i.e. base stations) each of which can go into a state of degradation. Such degradations manifest themselves in variations of several KPIs and raised alarms which are not easily mapped to a specific cause. These facts imply that considerable *manual workload* is required to manage this part of the system, because troubleshooting engineers need to permanently analyse performance data. 'Manual work' often also means that the degradation time period mentioned above can be significant.

In Section 6.1 the self-healing use cases and the process for self-healing are introduced using the 3GPP framework. Sections 6.2 and 6.3 cover the main aspects of degradation detection and diagnosis respectively (the advanced topic of fault prediction is briefly introduced as well). In Section 6.4 the compensation of outages is discussed.

LTE Self-Organising Networks (SON): Network Management Automation for Operational Efficiency, First Edition.
Edited by Seppo Hämäläinen, Henning Sanneck and Cinzia Sartori.
© 2012 John Wiley & Sons, Ltd. Published 2012 by John Wiley & Sons, Ltd.

6.1 Introduction

Within the telecommunications industry, the operation of the network and in particular troubleshooting, are seen as major cost factors. Therefore attempts are being made to provide functionality to reduce the troubleshooting effort. At the time of writing there is one main 3GPP technical specification (TS) (3GPP TS32.541, 2010) that describes the concepts and requirements related to self-healing. Besides this there are four documents specifying the Information Service (IS) (3GPP TS32.522, 2011; 3GPP TS32.762, 2010) and Solution Sets (SS) (3GPP TS32.526, 2011; 3GPP TS32.766, 2010) for the Self-Organising Networks (SON) Policy Network Resource Model (NRM) Integration Reference Point (IRP). The main TS defines the high-level concepts, requirements, the associated logical architecture and reference model as well as the use cases related to self-healing functions for 3GPP networks.

6.1.1 3GPP Use Cases

6.1.1.1 Self-Recovery of NE Software

In many cases faults can be healed if an earlier software version or configuration is loaded in a NE. Thus it is often desired to recover the NE software to the initial or the previous software level, to ensure the NE runs normally. If the NE Software Self-Recovery is triggered to heal the fault, the version of software is verified. The faulty software is removed and the backup is loaded. The configuration data is also checked and if it is found to be incorrect, the configuration data is restored or reconfigured. If necessary the process is restarted. Finally the results are verified and the IRPManager is informed if the fault could be healed or not.

6.1.1.2 Self-Healing of Board Faults

The other usual fault type is the breakdown of a hardware element in the NE. This use case aims to mitigate or solve these faults automatically to avoid user impact (e.g. system switches to a standby board automatically when the active board malfunctions). First, the redundancy information is collected about the faulty board. If there is a backup board that can take over the tasks of the faulty one, a switchover is initiated. If it is not possible to delegate the tasks of the faulty element, the IPR Manager is notified.

At the time of writing only one cell outage scenario is considered in 3GPP TS32.541 (2010): in this scenario, there is a *loss of total radio services* in the outage cell, UEs cannot establish or maintain any of the Radio Bearers via that particular cell. Here, the use cases currently associated with *cell outage management* are listed:

6.1.1.3 Cell Outage Detection

The system detects a cell outage, for example, 'sleeping cell', cell out-of-service, and so on, automatically (a 'sleeping cell' is a cell which carries no traffic (despite users being present requesting service) but not raising any alarm). Several system variables, performance indicators, alarms are continuously monitored and compared against thresholds and profiles. If an alarm appears or one of the observed variables violates the associated threshold, the proper corrective action is triggered and/or the operator is informed about the detection.

6.1.1.4   Cell Outage Recovery

The system recovers a cell outage automatically. Based on detection and diagnosis result the best available recovery action (e.g. a cell reset) is performed and the operator is notified about the results.

6.1.1.5   Cell Outage Compensation

This use case defines the system's ability to compensate a cell outage automatically to maintain as much as possible normal services to subscribers. First the actual situation is studied by collection of the available configuration information. Then the associated cells are reconfigured to improve service quality in the coverage area of the cell in outage (cf. Section 6.4).

6.1.1.6   Return from Cell Outage Compensation

The system returns to normal operation after a cell outage compensation. If the fault that caused cell outage and triggered outage compensation is solved it is required to automatically reconfigure cells that take part in the compensation.

6.1.2   3GPP Self-Healing Process and its Management

The self-healing functionality is provided by several logical function blocks shown at the left hand side of Figure 6.1. The function blocks are providing either monitoring functions or the

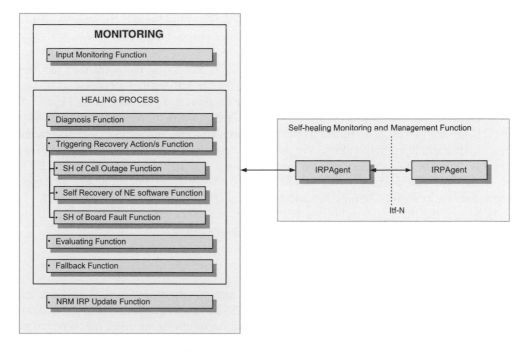

Figure 6.1   Self-healing process.

healing process itself. They are located at either the Domain Manager (DM, cf. Section 2.1.9), the NE level or both levels.

The general self-healing process includes the following steps. Note that the details and the order of steps within the healing process might be different depending on the individual self-healing use cases.

The 'Input Monitoring' Function continuously checks if any of the pre-defined conditions appear in the network that should trigger the healing process. In case a trigger condition is met the corresponding self-healing process is triggered.

Depending on the trigger the 'self-healing process' gathers more necessary information about the fault (e.g. performance indicators, radio measurements, system variables, test results, etc.). These building blocks cover the *detection* functionality introduced in Section 6.2.

Based on the trigger condition and the additional information, analysis and diagnosis of the fault are performed to find the root cause of the detected fault ('*Diagnosis* Function', Section 6.3).

The diagnosed root cause can be solved either automatically or manually. In case there is an automatic recovery action available it will be executed first. There might also be a need to backup configuration data prior to the execution of the recovery action so that it is possible to fallback to this configuration.

After the corrective actions have finished there is a need to evaluate ('Evaluating Function') the result, that is, to check if the fault disappeared and no other problem is introduced. If the fault is not solved and the stop conditions are not met, the self-healing process performs a new healing iteration. If the fault is solved or some stop condition prevents new iterations, a notification is emitted to report the result of the self-healing process. In some cases it is necessary to fall back to the backup configuration ('Fallback Function').

Logs might be created during the whole procedure documenting the performed recovery actions and the occurrence of important events during the self-healing process.

This *Self-Healing Monitoring and Management Function* (cf. right hand side of Figure 6.1) supervises the self-healing process and provides the operator with the necessary information of the self-healing process. This function shall be able to get information about all other functional blocks. In addition to this, it allows the operator to control the execution of the self-healing process. This function consists of two logical entities: The IRPManager (cf. Section 2.1.9) represents the NM portion of the Monitoring and Management Function. The IRPAgent represents the portion of Monitoring and Management Function operating below Itf-N.

In general the IRPAgent enables the IRPManager to acquire various information about the self-healing functions as well as control them as shown in Figure 6.2.

6.1.3   Cell Degradation Management

Cell degradation management use cases are the main focus of self-healing research and standardisation (Van den Berg *et al.*, 2008) (3GPP TS32.541, 2010). While 3GPP use cases are focused on 'cell outage' (cf. Section 6.1.1 earlier), in the following we adopt the more general concept of 'cell degradation' (which includes the 'outage' case, cf. the definition below). Note that for treating 'degradations' the use case 'diagnosis' (which is in fact not defined in 3GPP TS32.541, 2010) to find out a detailed root cause is much more relevant than if just 'outages' are considered.

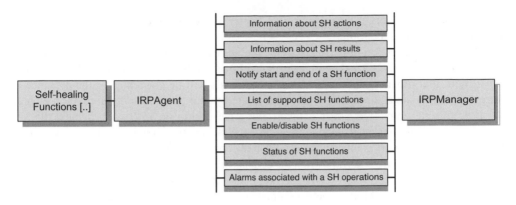

Figure 6.2 Main requirements of the Monitoring and Management Function.

The traditional cell degradation management process consists of four tasks (Figure 6.3). First (in addition to analysis of the alarm list), performance data must be measured. This data is usually collected and stored in performance databases. The absence of a centralised RAN network element in LTE makes it somewhat more difficult to collect the data or to shift some part of the cell degradation management functionality to the NEs. Performance information is analysed by engineers to spot cells, so that the root cause of the degradation can be identified. It is common practice that a detected fault is not diagnosed at all. In these cases a cell reset may be issued 'blindly' as it solves most of the simple, software-related issues. Finally corrective actions must be deployed to repair or compensate the fault. As mentioned above for conventional macro network scenarios (cf. Section 2.2), usually there is no redundancy available which can be taken into account for the solution deployment. This is changing for

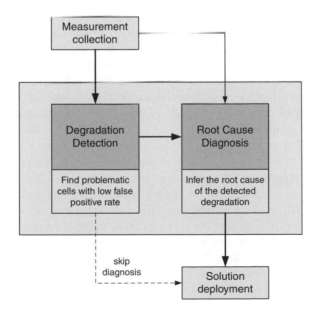

Figure 6.3 Block diagram of cell degradation management process.

'Heterogeneous Networks' scenarios (cf., Chapter 10) where multiple resource layers are available at the same location.

Without SON, detection of performance degradations in cellular networks is an activity that is only partially automated and involves personnel from different departments. The operations department that monitors the network on a 24/7 basis will be checking for the presence of alarms that indicate equipment issues either related to hardware or software. Once such an alarm is present, corrective action is taken. For a new alarm on a network element the first corrective action is often a cell reset. If this action has cleared the fault then no further action has to be taken and no effort is invested to establish the causes of the fault. Commonly the system that has reset a cell also tracks how often the cell has been reset. It will not attempt another reset if a given number of resets had already been applied within a given time period and the fault still has not been rectified. Under such circumstances the cell is brought to the attention of operations personnel to manually investigate the case in detail. The analysis is first attempted remotely as this is a cost effective approach. But in case the fault can not be fixed in this way, engineers will be sent to the site to investigate the case. It is not uncommon that several site visits are required to fix a complicated issue. Such cases often require the replacement of several hardware units until the faulty component has been identified.

In the network optimisation and performance department, performance degradation detection is not driven by the presence of alarms, but by degradations in Key Performance Indicators (KPIs). Engineers often focus their efforts on cells that have the worst performance in the network during a given time period (e.g. the top 10 cells having the highest number of dropped calls during a given week); or cells that experience the largest change in performance from one time period to the next (e.g. the top 10 cells that have the largest change in the number of dropped calls from one day to the next).

Once detailed investigations into a particular cell are launched, there is a need to consider both alarms and KPIs because neither category alone provides sufficient information regarding all possible causes of poor performance. Often alarms are useful to indicate HW related issues while degradation in KPIs without the presence of alarms indicates other influences such as interference. These influences are typically not related to HW, but possibly linked to poor planning/configurations settings or SW malfunctions. It is also conceivable that a HW fault might not directly lead to significant performance degradation due to node-internal fault handling. Similarly not all poor performance is due to HW faults. So both indicators provide complementary information for identifying the cause of poor network performance.

The methods in use by operators today have evolved slowly from the early days of GSM and indicate a migration of approaches applied in 3G also to LTE rather than a radical change in methodology. It is also worth stressing that each operator has its own way to identify poor performance. The general idea mentioned above (poorest performance and/or largest step changes) is usually applied, but differences exist regarding the considered time period and the incorporation of past rankings into the most recent ranking. For example, it is possible that only data from the past two weeks are combined when creating a ranking, or that each week is considered independently and then the rankings from the two weeks are combined. Each method leads to slightly different results so that different top-x list are created by these different ways of applying 'the same' method. Note that these only semi-automated methods need to be made aware of 'special events' which can occur during a limited time periods and cause performance degradation. For automated methods such information is also required as context for the self-healing process.

Often the deployed methods try to trade off speed of detection with detection accuracy. The longer a KPI is degraded the higher the probability that the degradation is due to an actual fault and is not only driven by a temporary issue such as a traffic peak, for example, caused by a traffic accident or road closure. In this particular context it is desirable to consider a long time period before taking any corrective action involving manpower, because this will filter out short-term issues caused by temporary events and thus improve detection accuracy. However, the longer the time period considered in the decision making, the longer subscribers have to cope with unacceptable performance if a real issue is present, which will inevitably lead to an increase of customer complaints. Hence in this particular context quick action is desirable. Therefore operators have to adapt their methods for fault detection to address the 'speed versus accuracy' problem.

An efficient solution approach, which builds directly on the fault detection, is the automated reset of HW based on the presence of alarms and/or KPI degradation. This approach is typically more suitable for dealing with equipment issues that are indicated by alarms, as for these the chance that the reset of the network element corrects the issue is relatively high. Performance degradation due to external issues such as poor planning/configuration can seldom be fixed by resets, because the reset does not address the cause of the problem. Nevertheless as the reset can be automated and scheduled to occur during night time, this is a course of action which can be tried even if the success probability is relatively low, because the cost associated with the action is relatively low (it is not negligible because there is low probability risk that the state after the reset is even worse than before). It will also prove to the troubleshooting engineer that the performance degradation is not likely to be caused by a temporal equipment issue that can be solved by a reset. Note that cell resets are not necessarily triggered right after the fault has been detected. In most cases cell resets must be carefully planned and scheduled at preferably no or low traffic periods.

In contrast to the fault detection in which certain tasks are automated already, root cause diagnosis is practically a manual activity. The speed of establishing the cause of the problem is directly linked to the experience of the engineer and the quality of information that the deployed tools provide. The skills gained by an engineer are solely derived from dealing with many such cases as there is no other way of obtaining these skills on a theoretical basis. It is useful to bear in mind that the observations taken for the same fault can vary widely between equipment vendors and sometime even between software versions of the same vendor.

It is important to note that despite the steps described above to characterise the troubleshooting activities, the actual realisation of this process can vary highly from operator to operator.

Before getting into the details of degradation detection and diagnosis the most important terms that appear during the further discussion are summarised in the following:

- *Symptoms* are anomalies in performance indicators and presence of alarms usually caused by faults but sometimes by unusual (but not faulty) network/user behaviour. Thus, anomalies may appear in performance data during normal (i.e. healthy) operation so that these situations can be confused with fault conditions.
- *Degradation* refers to a significant difference between the expected and actual performance of a cell, that is, the cell is not in a 'healthy' state where performance can be improved with self-optimisation, but it is clearly below the expected minimum performance (e.g. users in a cell experiencing call drops due to a misconfiguration versus tolerable QoS impairments for a cell with high load level). Degradations cannot be measured directly but may be displayed in

one or several symptoms depending on the underlying fault. Degradations can be further classified based on their severity: degradations can range from performance problems to *outage* scenarios where the faulty cell is not capable to deliver any traffic at all. Obviously, a complete cell outage is easier to detect than a 'sleeping cell' outage, which in turn is easier to detect than a partial performance degradation. On the other hand, outages usually require on-site repair actions or complex compensation with other resources (cf. Section 6.2) whereas less severe degradations can often be solved by reconfiguring or resetting a cell.

- *Fault* may refer to any hardware, software, configuration and planning error that causes degradations. The severity of a degradation is determined by the severity of the fault causing it. Some faults cause just performance reduction while others result in complete outage of the network element. Moreover faults can evolve over time first causing relatively minor problems and then slowly escalating into serious outages.

The problems that may appear in cellular networks are very diverse. The most common categories of faults are:

- *Hardware faults* can appear at physical elements of the network entities and usually manifests itself in an explicit (although low level) alarm. Problems are often related to the antenna and its connection to the base station, but also components of the base station or transmission line.
- *Software faults* appear in program execution at network elements and can also affect the service performance in several ways.
- Networks planning and configuration is done by humans which is a considerable source of *planning and configuration faults*.
- *Environmental changes* can significantly disturb the radio environment. Construction and demolishion of buildings can cause unforeseen changes in coverage and capacity and thus lead to performance degradation.

6.2 Cell Degradation Detection

The first step in the cell degradation management process is the detection of cells with degraded performance. This is often referred to as 'fault detection': a term that suggests that these algorithms are capable to detect and differentiate particular faults. This is not the case or the goal of these algorithms. The goal of this class of algorithms is to detect that there is degradation. The identification of a particular fault causing that degradation is the task of the diagnosis process. In other words, detection algorithms identify symptoms and correlate them to come to a decision on the presence of a degradation. Diagnosis combines degradation detection results to identify the actual root cause (i.e. the particular fault) that manifested itself in the detected symptoms.

As mentioned earlier, the detection process, no matter if it is manual or automated, operates on performance indicators. The main classes of performance indicators widely used during fault management processes are:

- **Counters:** A counter represents a single value that is maintained by the firmware or software of a network element (i.e. an eNodeB). An example is an event counter (fault counter) that incrementally counts specific events.

- **Key Performance Indicators (KPI) and Key Quality Indicators (KQI):** A KPI uses one or more counters/measurements as input and calculates a value according to a well-defined (often standardised; 3GPP TS32.410, 2011) formula. KQIs usually take KPI aggregates to provide a broader view of the network performance and are rarely useful to assess cell level performance.
- **Alarms:** An alarm is emitted by a network element towards the OAM system, triggered by predefined events. HW and SW faults are usually covered by alarms but also certain performance indicator-based events (e.g. violation of a threshold by a KPI) can trigger alarms.

Performance indicators to assess the performance of a cell can be acquired from different sources:

- The base station controlling the cell.
- User Equipments (UEs) served by the cell or adjacent cells (3GPP TR36.805, 2010).
- Adjacent or other cells (being located on the same or another base station).
- Core network elements, gateways, and so on.

The main goal of anomaly detection is to decide if the actual values of indicators fit into what is expected or not. Note that the definition approaches the problem as detection of performance indicators leaving the normal, *'healthy' operation domain* and not as detection of performance indicators entering into a domain that is defined to be faulty. At first this seems to be a subtle difference in how the problem is approached, but in fact this is a key design decision when automation comes into the picture. In the first case the normal operation domain must be defined while in the second case the domain of values of performance indicators in case of particular fault cases have to be stated. Cellular networks are highly complex systems and performance indicators derived from them are stochastic in nature thus in most cases (see the explanation that follows) statistical methods are needed to describe either the normal or the faulty operation domains. The second option is problematic as faults are divergent and rare events during the lifecycle of cellular networks thus it is not reasonable to use statistical calculations due to the low number of available samples referring to fault scenarios. In contrast normal operation can be safely approached by statistical methods thanks to the vast number of available performance data.

Traditionally the term *profile* refers to the data structure that reflects the temporal fluctuation of a performance indicator during a time period (e.g. a day, with one hour granularity). In this book, profiles are used with an extended scope as *any method that defines the normal operation domain of a performance indicator*. Profiles play an essential role in automated anomaly detection.

The type of profile to use for a particular performance indicator depends on its temporal structure during healthy operation (Figure 6.4):

- **Absolute threshold profile:** A wide category of indicators represent performance information that should not exceed or fall below a certain threshold (Figure 6.4a). Typical examples are fault frequencies (i.e. number of faulty cases/number of all cases) and fault rates (i.e. number of faults during a time period). The profile for these indicators is a simple absolute threshold value (e.g. the number of dropped calls should not exceed 1% of all calls).

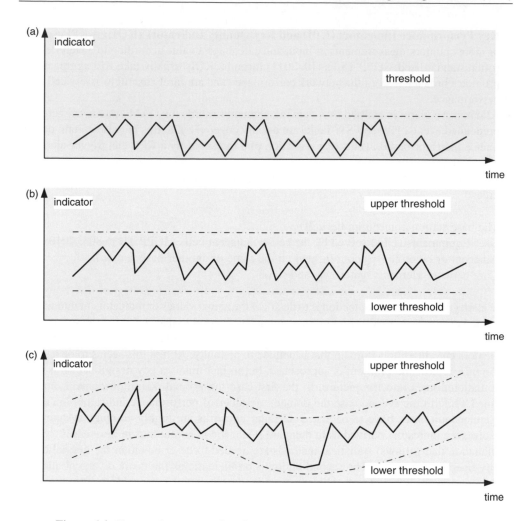

Figure 6.4 Temporal structure of performance indicators and corresponding profiles.

- **Statistical profile:** Another category includes indicators that oscillate around an expected (mean, average) value (Figure 6.4b). A trivial profile here can be a threshold pair that defines the maximum acceptable deviation from the mean value. The upper and lower thresholds can be symmetric or asymmetric; moreover, one of them can disappear in extreme cases. The maximum deviation can be defined in several ways. One example is to set the thresholds to some low multiple (e.g. 1, 2 or 3) of the standard deviation.
- **Time-dependent profile:** There are indicators that show temporal periodicity, usually driven by user behaviour (Figure 6.4c). This category thus carries the intuitive meaning of 'profile'. These profiles are data structures that represent the expected value of the corresponding indicator with a certain temporal granularity (e.g. how the expected load of a cell changes during a working day with 15 min granularity). Such indicators can be associated with more than one profile (e.g. one for working days and another weekend).

It is important to note here that not every indicator can be represented with one of the introduced profile types. There are indicators, especially at cell level, which are driven by actual user demands but which, however, do not exhibit strong periodicity. As a result the normal operation domain of these indicators does not show any reliable temporal structure. An example is throughput which usually shows strong temporal periodicity at traffic aggregation nodes (e.g. RNCs in 3G networks) but much weaker periodicity at individual cell level (typically only a coarse difference between night time and the rest of the day is visible). Therefore troubleshooting experts cannot assess these indicators independently, but need additional information on the overall status of the cell.

In summary anomaly detection is achieved in three main steps:

1. Performance indicators must be defined that are of interest.
2. The normal, expected operation domain must be defined for each indicator.
3. During operation actual values of indicators must be evaluated against the corresponding profiles.

In current troubleshooting most of these tasks are aided by a certain level of automation: collection of performance indicators, profile construction, evaluation of actual performance information against profiles are all done automatically. There is still need to improve the current situation as the output of these detection procedures typically produces too many 'false positive' alarms. The problem is that performance indicators often show unusual behaviour even if there is no fault and detection algorithms are not capable to distinguish between real faults and unusual network behaviour. Thus the main goal of degradation detection research is to improve this classification capability.

The cornerstone of improved automated detection methods is that deeper statistical knowledge must be gained about performance indicators by applying more elaborate statistical models to define the normal operation domain (i.e. profile) of indicators which better separates faulty situations from unusual but healthy cases.

Automated anomaly detection method can be classified into two major classes each comprising several methods published in literature (Chandola et al., 2009):

Univariate techniques take a specific indicator into account and make a decision if it behaves as expected or not. These algorithms operate on the past observation of the corresponding indicator and build a statistical model (e.g. only mean and standard deviation in the simplest case) representing the normal operation domain. The main property of these methods is that they are not capable to differentiate anomalous situations caused by real faults and unusual (but still normal) cell behaviour. The detection accuracy is more dependent on the underlying statistical model (i.e. the profile) than the actual detection decision meaning that the challenging task is to construct/learn a good profile. If once it is in place, the decision whether the actual performance is according to the profile or not, is a relatively simple task.

Known univariate techniques are:

'Competitive neural algorithms' are used in (Barreto et al., 2004) for fault detection and diagnosis in third-generation (3G) cellular networks.

The class of 'Operational Fault Detection' (OFD) algorithms are introduced in (Cheung et al., 2005). OFD algorithms analyse performance indicators detect fault signatures without the need for operators to manually set thresholds. Instead a generalised likelihood test method is used to adaptively determine thresholds. This is used to build the profile of the system under analysis by either looking at its earlier behaviour or comparing it to similar systems. The assumption is that during normal conditions systems tend to behave as earlier and similarly to other systems.

The same OFD approach is followed in (Rao, 2006) where a statistical hypothesis test framework is defined for determining faults. The thresholds are determined by learning expected deviations during a training phase. Fault thresholds can adapt to spatial and temporal mobile traffic patterns.

In (Zanier et al., 2006) a method to detect coverage and dominance problems and identify interferers in WCDMA networks is introduced. Instead of relying on drive tests or simulations, certain metrics are calculated for every cell and cell pair by processing signalling messages exchanged through the radio interface during normal network operation reflecting real traffic distributions and geographical user locations.

(Mueller et al., 2008) presents and evaluates a cell outage detection algorithm, which is based on the neighbour cell list reporting of mobile terminals. The outage cells considered are the ones that do not carry any traffic and are inoperable. The idea behind the presented outage detection algorithm is to use Neighbour Cell List (NCL) reports to create a graph of visibility relations, where vertices of the graph represent cells or sectors. Edges of the graph are generated according to the received NCL reports, where an edge weight determines the number of mobile terminals that have reported a certain neighbour relation. The outage detection problem thus translates into a classification problem on change patterns of the visibility graph. The algorithm is presented as having overall good detection sensitivity, however, with a tendency towards the generation of false alarms which is too high for practical applications.

Multivariate anomaly detection methods that are capable to assess the performance of a whole cell as multiple indicators are processed and evaluated against each other. There are lot of multivariate anomaly detection methods mainly applied in other engineering problems than anomaly detection in cellular networks, however, some recent publication address cellular networks with similar approach:

In (Turkka et al., 2011) the authors introduce a novel concept of using diffusion maps for dimensionality reduction when tracing problems in 3G radio networks. The goal is to identify base stations with anomalous behaviour from a large set of data as well as find out reasons why the identified base stations behave differently than others.

A similar approach is followed in (Chernogorov et al., 2011) to detect the presence of sleeping cells (i.e. cells that does not provide any service though seem to be healthy) in the network and to define its location.

Self-Organising Maps (SOM) are neural network-based algorithms that can effectively visualise multivariate data. In Laiho et al. (2005) and Wietgrefe et al. (1997) the authors showed that the SOM approach together with a conventional clustering method can effectively be used to simplify and focus network analysis. These algorithms help in visualising and grouping similarly behaving cells. Thus, it is easier for a human expert to discern different states of the network. This makes it possible to perform faster and more efficient trouble shooting and optimisation of the parameters of the cells. The presented methods are applicable for different radio access network technologies.

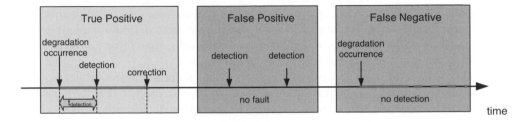

Figure 6.5 Degradation detection accuracy and delay.

Regardless of the detection method there is a need to define the actual detection *performance metrics* that can be used to validate and compared different techniques. The two most important metrics are *detection accuracy* and *detection delay* (Figure 6.5):

Detection accuracy is composed of two main aspects. On the one hand a detection method is reliable if it signifies degradation when there is an actual anomaly in the system (i.e. true positive). A detection event that notifies the operator about a degradation while there is no existing anomaly (e.g. due to some unusual but normal behaviour) is defined as a false positive detection. On the other hand it is essential to have a detection method that reveals actual anomalies and does not let them stay undetected. This case when an anomaly appears in the system but the detection method fails to notice it is called a false negative case. One of the main challenges of detection systems is to find a good trade-off between these two aspects of detection accuracy (i.e. the sensitivity trade-off).

Detection delay is defined as the time interval from the actual anomaly occurrence until it is detected. Detection delay is computed only for true detections (true positives), where an anomaly is detected within the anomaly duration. It may occur that an anomaly is detected outside the anomaly duration period. This may be due to a delay inside the detector, that is, it may take some time for the detector to output a result, or it may be due to a false positive. As it is not straightforward to distinguish between these two cases, it is assumed that detections outside the anomaly duration interval are false detections and only consider detection delay for true detections, that is, detections within the anomaly duration interval.

Besides the main characteristic of a detection method there are some other characteristics of a degradation detection solution that need to be considered with regards to practical deployment:

Severity indication accuracy: One anticipated output of the detector is an indication of the severity of the degradations observed. This information is one of the key inputs when scheduling corrective actions thus reliable severity indication is essential.

Signalling Overhead introduced by fault management functions can be divided into two categories; namely, one regarding the transport network and one dealing with overhead caused over the radio interface. The transport network signalling overhead should capture all data that is transmitted over the transport network, for example, between eNodeBs (X2), between eNodeB and MME (S1), and eNodeB and OSS (Itf-S). Measurements depend on the architecture, that is, whether the solutions are centralised or distributed (or hybrid), cf. Section 3.4. In any case, one approach is to measure the estimated overhead incurred over all interfaces. Signalling overhead also includes additional communication needed between a UE and an eNodeB in order to facilitate degradation detection. This can include, for example, setup and delivery of needed UE measurements.

Processing Overhead refers to the amount of processing needed to detect the degradation. The execution time of a particular algorithm typically depends on the size of the input data, for example, number of measurements, and can be asymptotically logarithmic, polynomial, or even exponential. An analysis of asymptotic execution time gives an insight in the processing demand of an algorithm. A theoretical analysis of the algorithms may be performed.

6.3 Cell Degradation Diagnosis and Prediction

In the previous subsection it has been shown, that degradation detection is already aided by certain level of automation in commercial cellular network management. In contrast, cause diagnosis is typically still a manual activity and it is based on the hard earned domain knowledge of troubleshooting experts. This knowledge is gained by years of troubleshooting experience, it is often rather subjective, operator-specific and rarely explicitly formulated nor properly documented.

The traditional manual workflow of fault diagnosis, that represents the way of human thinking when facing the problem of diagnosis, can be outlined as shown (Figure 6.6).

The output of the degradation detection process informs the fault management system about a possible problem in the network, and triggers the diagnosis process. The trigger might be a

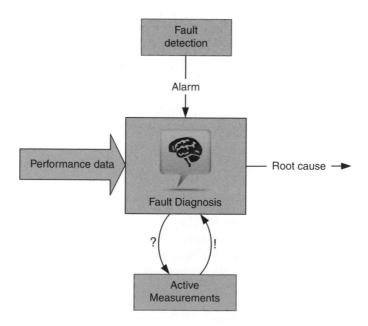

Figure 6.6 Block diagram of cell degradation diagnosis process.

SW/HW alarm or a KPI violating a threshold. Troubleshooting experts analyse available performance data, looking for symptoms that characterise a certain fault. Based on the kinds of symptoms observed, the most possible root cause is mapped to the observations. Usually the amount or type of performance data that triggered the diagnosis process is not enough to identify a certain root cause with sufficient confidence. Experts use available information to narrow down the group of possible faults but active measurements are needed to get more information about the situation to identify the exact root cause. Moreover, there are cases when it is not possible to remotely acquire additional performance information and a drive test must be scheduled.

The level of expertise of the troubleshooter as well as the granularity of information available for analysis (i.e. performance information aggregated at an RNC is not always useful to spot a specific cell experiencing degradation) highly influences the time required for and actual quality of diagnosis. Finally the tools used to visualise the relevant information is another fundamental aspect. It is a highly challenging task that requires complex under-standing of the system itself, the potential faults as well as the tools used to analyse performance metrics. This expertise is an important asset one can only earn during years of troubleshooting work.

The increasing size and complexity of cellular networks calls for efficient automated troubleshooting tools and methods realised within the SON framework. A high level of automation in troubleshooting reduces the time required to find the fault that causing a certain problem. An immediate consequence of automation is enhanced network performance as degradation time will be limited significantly. Another effect of automation is that it reduces cost associated with troubleshooting as fewer personnel are able to do the same work. A lower entry skill-level is required for troubleshooting personnel as expert knowledge is stored in automated systems that can aid less qualified engineers to find faults and root causes. However, the tough cases, which are hard to diagnose and require a deep understanding, probably have to be analysed still by highly skilled troubleshooting experts.

The complete automation of the diagnosis process is of course the envisaged long term result of this evolutionary process. The first step can be described as learning/customisation phase where knowledge of experts is used to train the automated system. In the learning phase the system is not capable of aiding the troubleshooting process. When the system has gained a sufficient knowledge base it can be launched to the second phase, called the assistance phase. During the assistance phase the expert system is capable to give advice to troubleshooting experts about the most probable faults but the final decision is still in the hands of human engineers who can give feedback about the advice given by the expert system. This feedback can be used to further refine and fine tune the stored knowledge. Finally as the system and the confidence about its decisions evolve, it can take over tasks from human operators and trigger corrective action automatically. A long-term vision is that human operators need to supervise the automated diagnosis system by defining high level policies guiding the automated troubleshooting process while leaving the details to the expert diagnosis system (cf. Section 11.3.3).

The most critical step in this process from an automation point of view is the association between observations and root causes that is commonly referred to as *expert knowledge*. It has been shown that during the diagnosis process engineers collect all available observa-tions and *map* them to the most probable root cause according to their troubleshooting experience. This is a highly intuitive process and the expert knowledge is gained by years of troubleshooting. Consequently most automated methods focus on the collection and integration of expert knowledge into a framework that is capable to perform the same reasoning which is basically a machine learning problem. Another important general aspect of expert systems is the capability to adapt to different environments (i.e. operators, networks, technologies). The expert system must be able to specialise itself to learn the particularities of the network it diagnoses.

In the following, some approaches addressing automated troubleshooting in cellular networks are introduced. All these systems aim to extract, store and utilise the knowledge of troubleshooting experts with different mathematical background behind them.

6.3.1 Rule Based Systems

The first systems that attempted to store expert knowledge and utilise this to diagnose faults in complex environments were rule based systems. These systems are based on learning IF(A)-THEN(B) rules, where A is an assertion on the set of observations and B is the root cause or an action. During operation, if a stored assertion appears to be fulfilled, the expert system can infer the possible root cause or suggest checking another assertion. Troubleshooting examples can be: 'IF a certain hardware alarm is received THEN check the hardware component annotated by the alarm' or 'IF there are too many failed handovers THEN check if handovers occur according to the planned adjacencies'.

Rule based experts systems are usually easy to train as in most cases operators use troubleshooting workflows that are based on rules similar to the ones mentioned above. This is a great advantage as these systems provide a logical way of reasoning aligned with human reasoning which makes inference models easy to build.

While rule based systems are natural way to mimic human troubleshooting reasoning process and have proven their effectiveness in small deterministic environments they have several drawbacks when applied in more complex systems like cellular networks. On one hand too many rules are required to catch every possible problem which makes rule sets unmanageable and inference being slow. On the other hand, troubleshooting experts are usually uncertain in their reasoning. Conclusions drawn from observed state of the system are not completely certain. A set of symptoms can indicate several problems each with some probability thus rules affecting the reasoning of expert looks like: 'IF there are hardware alarms THEN it is very probable that there are some broken hardware elements'. Certainly experts will first look for hardware faults but it can turn out that the root cause of the observed phenomena is something else. The described uncertainty cannot be expressed by rule based systems. There are ways to extend rule based systems to be able to express uncertainty (e.g. using uncertainty factors of fuzzy logic) but this results even more complicated rules that can come to wrong decisions.

An automated root cause diagnosis framework is presented in (EMC, 2009) that builds upon and extends the rule based system approach, called codebook based reasoning. The framework utilises the fact that every fault in a complex system has its unique signature. The framework stores the signatures of known faults and matches actual system information (i.e. the actual signature) to the stored information. The diagnosis looks for the closest matching signature thus it does not require an exact match. This enables the approach to work fast and be effective even with incomplete information.

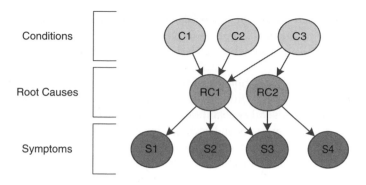

Figure 6.7 The general structure of Bayesian Networks.

6.3.2 Bayesian Networks

In contrast to rule based systems, Bayesian Networks (BN) are capable of expressing uncertainty as they find their roots in classical probability calculus. The name comes from the fundamental probability principle called Bayes' theorem, which expresses the conditional probability of a random variable given the value of another variable. This can be extended to represent conditional probability dependencies between several random variables which can be essentially visualised by a directed, acyclic graph shown in Figure 6.7.

A BN consists of a set of variables or nodes (which have different discrete states, but continuous models are also possible [18]) and a set of directed edges between these variables. Directed edges normally reflect cause-effect relations within the domain. The strength of an effect is modelled as a conditional probability. When applied to troubleshooting, the nodes can represent certain conditions, the faults (i.e. root causes) and their symptoms. Conditions are used to represent the global properties of the environment that update the context in which the association between faults and symptoms are studied (e.g. the operational state of a cell, locked/unlocked). The structure of the network is built by acquiring expert knowledge: defining the performance indicators and root causes of interest and capturing their relations by defining the connections and conditional probability tables. As an example a hardware fault might be the root problem causing the identified performance indicators being degraded (thus resulting in 'symptoms'). The strength of the effect of that certain fault is expressed by the corresponding conditional distribution table entry of each symptom variable. Note that a certain problem can affect several symptoms and one symptom can be affected by several root causes. The conditional probability tables are used to calculate the probability of the actual state (i.e. the probabilities of the state of the dependent nodes) given the evidences (i.e. proven states of parent nodes).

The essence of BN is that this inference can be performed in both directions. Thus, during diagnosis the expert system can tell the probability of root causes given the observation of symptoms (i.e. evidences). So once the experts have built up the BN from top-down and injected conditional probability knowledge, it is operated logically in the opposite direction to find the root causes.

Furthermore, BNs can be configured to evolve over time. The diagnosis accuracy improves with the number of cases observed. The system can be launched with a rough estimation of conditional associations which is fine tuned as more and more cases are diagnosed. This needs of course feedback from the operator about the observed cases. Yet, even the initial structure and conditional probability constellation can be learned automatically from observing example data (Neapolitan, 2004).

While BNs can cope with incomplete and uncertain knowledge, there is a clear drawback related to them. It is relatively difficult to build models (despite the available learning techniques), especially if several probabilities have to be specified. Thus, when creating a BN the trade-off between simplicity and accuracy have to be taken into account. Clearly, the model should accurately represent the underlying system with every possible condition, fault and symptom to be able to assure that the right cause of the problem is found in each case. This can lead to quite complex models where knowledge acquisition is an extremely challenging task, while the inference algorithms are getting complicated requiring considerable time and computational resources to complete. Therefore, the aim is to define a model as simple as possible, but being sufficiently accurate.

Furthermore, if a node has several ascendants defining or learning the conditional probability tables becomes a challenging task. The probability of all states of child nodes must be defined for every possible combination of states in parent nodes. This is problematic in two ways. On one hand, the size of the conditional probability table grows exponentially with the number of parents' states. On the other hand, experts cannot define probabilities with the required granularity. For example, even for an expert it is very difficult to define which is the probability of having a handover failure given that there is an interference problem and, at the same time, a coverage problem but not a hardware problem being present.

In order to simplify the model creation, two plain models are proposed: the *naïve model* and *noisy-or model*. Note that for the sake of simplicity, nodes representing conditions are excluded from the following models.

The *naïve model* (Figure 6.8) is the most common approach in diagnostic systems to simplify the model. This paradigm basically reduces the number of nodes by defining a single root cause node with each state corresponding to one possible root cause. This node is connected to each of the symptom nodes representing the conditional dependence amongst them. Note that the states of the RC node are mutually exclusive, that is, only one cause can be present at the same time. This reduces the expressive power of the model as it cannot handle multiple parallel fault situations.

The *noisy-or model* (Figure 6.9) or causal-independence models can be used instead of the naïve model to overcome the single fault restriction while keeping low model complexity. In this model paradigm every root cause is represented by a binary node (i.e. fault exists or not). As explained above, multiple root causes make it difficult to provide conditional probabilities for a

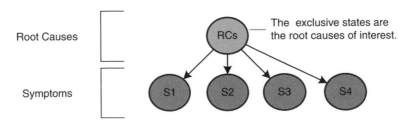

Figure 6.8 The naïve Bayes model.

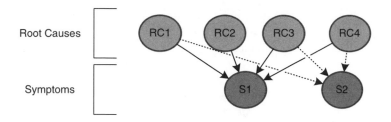

Figure 6.9 The noisy-or model.

symptom as each possible combination of the causes has to be expressed. The basic simplification assumption of the noisy-or model is that it captures the individual contribution of each cause to a certain symptom. Thus instead of reducing the number of nodes the noisy-or model directly reduces the size of the conditional probability tables.

There are several publications introducing BN for cellular network diagnosis:

In the work presented in (Barco *et al.*, 2005), a simple model based Bayesian classifier is presented for automated diagnosis in real networks. The model uses different types of causes, symptoms and conditions specifically for a GSM/GPRS network based on the probability of each cause given a set of symptoms and conditions.

The same approach has been applied for a diagnostic system for UMTS networks and presented in (Khanafer *et al.*, 2008).

For discrete models, there is a need to discretise continuous variables before adding them to a BN model. In (Khanafer *et al.*, 2008), two different discretisation methods for specifying thresholds have been applied to the Bayesian model: the percentile based discretisation and the entropy minimisation discretisation. Note that the method used affects the diagnosis accuracy and simplicity of the model. There is a clear trade-off between the requirements as fine granularity in discretisation improves the accuracy but introduces additional complexity (i.e. more states to be handled, thus increasing the size of the conditional probability tables). In (Barco *et al.*, 2008) the continuous and discrete models are compared and their respective pros and cons are summarised.

Inaccuracies that are introduced by imperfect discretisation highly affect the diagnosis accuracy of BNs. In (Barco *et al.*, 2007) the concept of 'Smooth Bayesian Networks' (SBNs) is presented to deal with inaccuracies of model parameters. SBNs are able to reduce the sensitivity of diagnosis accuracy. Results presented in (Barco *et al.*, 2007) shown that SBNs are successfully able to handle imperfect parameter settings and outperform traditional BNs.

6.3.3 Case Based Reasoning

A clear applicability barrier of BN based methods is their reliance on statistical knowledge about faulty scenarios: the conditional probability distribution of a performance indicator must be defined in case of every possible fault. As stated above faults are rather rare and very diverse events in the life of a cell compared to the healthy operation may occur. Statistical models based on this information can be very unstable and sensitive to minor changes in the input variables. In certain domains (e.g. medical diagnosis, troubleshooting of printers) the described approach

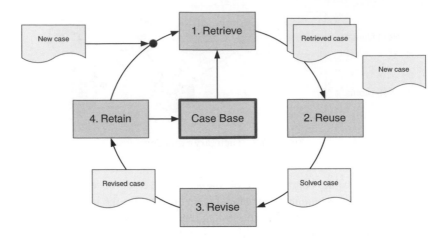

Figure 6.10 The CBR cycle (Aamodt and Plaza, 1994). Adapted with permission from IOS Press.

can be reliable due to the huge statistical information extracted from documented fault cases. This information is usually missing or insufficient to build a reliable diagnosis model in cellular networks. Thus other solutions operate on the following basic idea: use a degradation detection method that detect if performance indicators show some anomaly and maintain a database that maps anomaly patterns (i.e. a set of indicators showing anomaly) to fault cases. Note that this mapping also requires the knowledge of experts but in contrast to BN models there is no need to initially have reliable statistical knowledge on performance indicators. 'Case Based Reasoning' (CBR) methods are the most prominent examples in this category.

The CBR approach is a fundamentally different form of an expert system as those described so far. CBR systems do not rely on implicit knowledge that needs to be acquired from domain experts. Instead CBR systems store previously observed problematic situations (fault cases) and build an ever improving knowledge base. Upon solving a new problem, a similar prior case is retrieved and reused to give a solution to the new problem (Figure 6.10). Then the knowledge is updated with the new problem and the associated solution during operation (revise and retain), and similarly every new case is built in after it has been solved making every new solution attempt better than the previous one. From a diagnosis point of view, this is a classification of symptoms into groups associated to root causes. Every time a new fault appears it is compared to the already observed and diagnosed cases. The system retrieves the most similar prior case and the diagnosis result associated with it. Consequently the most probable root cause of the new fault will be the root cause of the most similar earlier case. Then the solution of the new case is revised to see if it is correct. Upon correct solution the new case is incorporated to the knowledge base. Otherwise manual troubleshooting is required to solve the case. The result of the manual process is also added to the knowledge base. Concrete implementations of CBR systems can be different in the way they represent and store the cases and the way they define the measure of similarity along which the most relevant prior case is retrieved. The selection of this approach depends very much on the characteristics of the problem domain and thus highly affects the diagnosis (i.e. classification) accuracy. Despite the successful application of CBR systems in other technology

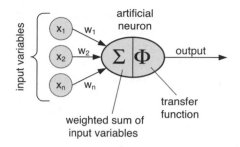

Figure 6.11 The general structure of an artificial neuron.

domains (e.g. Montani and Anglano, 2006), only recently CBR has been proposed to address fault diagnosis in cellular systems (Novaczki and Szilagyi, 2011).

6.3.4 Neural Networks

The structure and functional aspects of neurons in human brain inspired the mathematical model of neural networks (NN). NNs like BNs are usually represented by a set of inter-connected nodes, that is, the artificial neurons (Figure 6.11), organised in layers (e.g. input, output and hidden layers, cf. Figure 6.12). Every neuron is connected to all neurons in the next layer. Such connections between node pairs are associated with weights, where the weight defines the strengths of the association between the nodes. Each input variable is associated to one node in the input layer. During operation the actual values of input variables are fed into these nodes which calculate the weighted sum of the inputs and puts the results to the output of the node through a (usually non-linear) transfer function. The next layer receives the output of the input layer and data is propagated through the layers, until the final result appears at the output layer. Most importantly, a neural network can be trained to learn weights of connections. There are three main learning concepts for NNs: supervised, unsupervised and reinforcement learning. In case of supervised learning an input vector is provided at the input layer and it is associated with the expected response for each node at the output layer. The difference between the actual and expected output is used to fine-tune the weights. Unsupervised learning differs

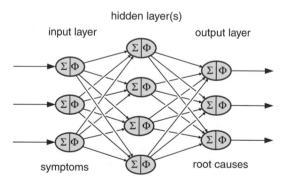

Figure 6.12 The general structure of artificial neural networks.

from the former paradigm in the fact that there is no external teacher who tells the correct output for a certain input. The network has to discover statistically important features of clusters of patterns presented at the input. Finally, in reinforcement learning the system has to learn what action to take to maximise a numerical utility function by trialling all possible actions.

The interpretation in the fault diagnosis domain is that a neural network's input variables are the symptoms observed in the system and the output variables are associated to the root causes, while the connections and their weight represent the expert knowledge. In general a diagnostic neural network is a classifier: different input patterns are classified into root causes. During operation a certain input constellation propagates through the network and sets the highest output value of the output node associated to the most probable root cause.

In (Barreto *et al.*, 2004) the authors present a fault detection and diagnosis system for 3G networks based on neural networks utilising competitive learning. Competitive learning means that the learning algorithm selects one node (i.e. the *winning neuron*) whose weight vector is the closest to the current input vector. The authors study and compare four competitive learning algorithms. The constructed neural model is trained with network performance data representing healthy operation of a CDMA2000 cellular system. The network is provided with so-called training vectors that contain several KPIs (e.g. number of users, downlink throughput, etc.). Then, a normality profile is constructed by using the sample distribution of the quantisation errors, which is the difference between the actual state vector and the weight vector of the winning neuron. The distribution of the quantisation errors is called the 'normality profile' of the system. The profile is associated to an empirical confidence interval which is used to assess the actual state vector. The actual vector is considered to healthy if it is within the confidence interval and faulty is it exceeds the upper or lower interval threshold. The quantisation errors can be used to generate diagnostic rules that help to find the root cause of the detected fault. In fact, the system presented in the paper does not indicate the root cause but selects one or more elements of the input state vector that are suspected to cause detecting the fault (i.e. deviation from the profile). This information might give a hint to the troubleshooting expert about the possible root cause of the observed fault.

An earlier work is presented in (Laiho *et al.*, 2005) where the authors demonstrate how Self-Organising Maps (SOM) can aid manual fault diagnosis by improving performance status visualisation. SOMs in principle are very effective in clustering and visualising high dimensional spatiotemporal data. The authors present how the vast amount of performance information collected at cellular network can be classified and visualised effectively to aid manual diagnosis.

(Wietgrefe *et al.*, 1997) presents an application of NNs for fault diagnosis (in the form of alarm correlation).

6.3.5 Active Measurements

As described above in the manual diagnosis process, experts first automatically receive notification about some main performance indicators showing anomalies. This initial information can be used to approximate a set of possible root causes (e.g. fault category). There is usually more information needed to single out the actual fault. This information is collected by running additional measurements and traces. This is an iterative process in which the experts narrow down the set of potential faults by acquiring specific information that helps to filter out unlikely faults. As new information arrives at the expert, potential faults that are less probable are dropped (by applying one of the methods introduced above) and additional measurements that will help to further narrow down the fault set are scheduled. The iterative process and the measurement taken at each stage can be considered to be a kind of expert knowledge which can be addressed by automatic methods as well:

In Varga and Moldovan (2007) the authors define a diagnosis framework for Ethernet networks where the iterative diagnosis process in applied using Petri-nets to define series of active measurements to process at each iteration to decide which is the most probable branch to go on with investigation. The clear advantage of the solution is that it requires only a small set of measurements running continuously to detect if there is a fault. Additional measurements are performed only on demand (i.e. when there is some problem) decreasing the baseline overhead introduced by the process.

In Heckerman *et al.* (1995) the authors extend BNs to achieve a decision theoretic-based troubleshooting. A diagnostic procedure is developed that not only identifies the most likely root cause given the actual symptoms but also generates a plan to refine this knowledge and take actions for repair.

6.3.6 Prediction

In contrast to degradation detection and diagnosis, prediction of degradations is not addressed in traditional manual troubleshooting processes in cellular networks at all. However, it is considered to be relevant within the SON self-healing framework and there are several research approaches for degradation prediction proposed. Therefore degradation prediction is envisioned to play a role in fault management for future 3GPP network generations.

Degradation prediction can be addressed in two fundamental ways. One major set of algorithms find their basis in traditional *reliability theory.* As these methods rarely consider the actual state of a dynamic system, they are not capable to reflect ever changing run-time situations and fault processes. Such methods have proven their incontestable necessity in long-term average reliability design and comparative analysis.

With the ever growing complexity of computer systems it has become essential to predict the behaviour (including fault processes) of the dynamic environment taking into account the actual state of the system. This goal is addressed by *online time failure prediction* methods. In this section online fault prediction methods are summarised, because these are expected to gain more importance in cell degradation prediction methods for 3GPP networks.

The authors in (Salfner *et al.*, 2010) have recently summarised online prediction methods and have built a valuable taxonomy of the field. In this section their approach is followed to introduce the reader to this research area.

Degradation detection is envisioned to be a part of a larger proactive fault management system. This is shown in Figure 6.13. After the actual prediction of an upcoming degradation, diagnosis might be required in order to find the root cause of the predicted problem. Prediction and diagnosis results can be used to schedule preventive actions and finally deploy them. The success of proactive fault management depends on all these steps each representing a whole

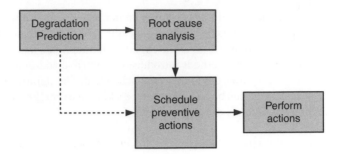

Figure 6.13 Block diagram of proactive fault management.

field of research. In the following, a closer look on the first block (i.e. degradation prediction) of this chain is taken.

From the practical viewpoint degradation prediction is a timing issue: prediction is early detection and/or diagnosis. A method that is able to predict degradations is based on detecting the very first signs of degradations. Consequently it is not possible to predict faults that have no signs manifesting themselves in performance indicators that are observed, of course. It is nevertheless reasonable to clearly separate prediction as different methods need to be applied to achieve the different tasks.

The temporal relations and general goal of prediction is depicted on Figure 6.14: at that present time the potential occurrence of degradation is to be predicted for some time in the future (lead-time, t_{lead}) based on the current and past system state (data window of length t_{data}), provided by the performance monitoring system. The prediction is valid for a certain time interval ($t_{prediction}$), which is called the prediction period. Increasing $t_{prediction}$ results in that a fault is predicted correctly. On the other hand, if the prediction period is too large, the prediction is of no practical use since it is uncertain when exactly the fault will occur. A minimal warning time is introduced ($t_{warning}$) that represents the time required to react and prepare for the predicted fault. The key requirement of a fault prediction method is that the lead-time must be greater than the minimal warning time unless there would be not enough time to perform any preparatory or preventive actions before the predicted fault actually occurs.

In the rest of the section the main approaches to degradation prediction are summarised.

6.3.6.1 Symptom Monitoring

Several approaches monitor and analyse system variables (e.g. amount of free memory) because certain errors affect the system causing degradations even before they are escalated into a serious fault. Four main method classes can be identified in this category.

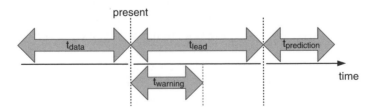

Figure 6.14 Temporal relations in degradation prediction.

1. **Function approximation** techniques model the observed system and provide expected output (target value) based on the measured system variables as input. The target function can be either probability of fault occurrence or some computing resource (Andrzejak and Silva, 2007; Hoffmann, 2006; Abraham and Grosan, 2005; Li *et al.*, 2002).
2. **Classifiers** directly analyse system variables and classify the current situation if it is failure-prone or not. The decision boundary is learned from a training data set of diagnosed cases so the classification decision is known for these reference cases (Daidone *et al.*, 2006; Turnbull and Alldrin, 2003; Hamerly and Elkan, 2001).
3. **System models** are storing knowledge about normal (i.e. faultless) operation of the observed system and are capable to produce expected output based on actual system information. The model output is then compared to the real output. If the difference is out of the acceptable range a failure is predicted (Bod'ik *et al.*, 2005; Kiciman and Fox, 2005; Chen *et al.*, 2002; Hughes *et al.*, 2002; Ward and Whitt, 2000; Singer *et al.*, 1997).
4. **Time Series Analysis** approaches handle system variables as time series thus the prediction is based on several successive samples of measurements. This method either computes the residuals of the samples and decides if the current situation is error-prone or not, or predicts the progression of the underlying variable to estimate time until a certain degradation (Meng *et al.*, 2007; Cheng *et al.*, 2005; Crowell *et al.*, 2002).

6.3.6.2 Detected Error Reporting

In contrast to symptom monitoring methods a significant number of methods work in an event based way. These methods base their fault prediction decisions on incoming notifications about errors and the expert knowledge that associates actual error status to future fault possibilities. Unlike symptom monitoring methods, where continuous input variables have to be handled, these methods usually receive binary input that reflects the existence or non-existence of a certain error.

Rule-Based Systems maintain a set of IF-THEN rules that map certain conditions (i.e. a set of errors showing up together) to certain future faults. The challenge is how to find algorithmically the near ideal number of rules to cover all possible cases while not generating too many false fault warnings (Vilalta and Ma, 2002).

To increase the efficiency of rule based systems the association between observations and future faults can be handled in some other ways. Methods in this category utilise *pattern recognition* techniques, *statistical tests* or *classifiers* to capture the association between errors and potential future faults (Salfner, 2006; Bai *et al.*, 2005; Levy and Chillarege, 2003).

6.4 Cell Outage Compensation

A cell outage occurs when a cell or multiple cells (e.g. all sectors of a base station) become unavailable or inoperational. As outlined in the previous section, the reason for a cell outage might be hardware or software malfunctions. But also a base station deliberately switched off for energy saving purposes could be considered in the same framework. Cell outage typically leads to a sudden loss of service coverage which is seen as lower network quality, lost revenue and ultimately increased churn. Therefore quick reaction to cell outage has a vital importance for the continuity of broadband mobile service. Note that coverage loss is particularly problematic, because its effects may only be partially be detectable via PM/FM data (e.g. call drops). This is the case, because UEs may not be able to connect at all any more (this is alleviated in the future by MDT (cf. Chapter 7) where UEs may log data in disconnected state and later convey it to the network).

Transmission power and antenna tilt optimisation are typical means to realise Cell Outage Compensation (COC). If adaptive antennas are used, other functionalities of adaptive antennas, such as beam-steering and beam-shaping, can be effective means to compensate outages, as proposed in (Yilmaz *et al.*, 2011). Furthermore, to solve capacity outage problems, Home-eNodeBs (HeNBs) can be used to carry macro layer traffic load and to extend Macro-eNodeB coverage as proposed in (Amirijoo *et al.*, 2011) (cf. also Chapter 10). Cell outage of a macro eNodeB could be covered by adjusting Radio Resource Management (RRM) parameters of HeNBs accordingly.

Excessive interference caused by home cells to macro cells covering the cells in outage might be reduced by adapting HeNB power and frequency band allocations.

Since the manual reaction on the changed situation is very expensive and time consuming, an automated solution with fast response time is required to compensate cell outages.

6.4.1 Activation of Cell Outage Compensation

Base station faults can be detected and diagnosed as outlined in Sections 6.2 and 6.3. In case a base station fault cannot be recovered with remote means, the BS (or an individual cell) is switched off and a trouble ticket is issued. It is also noteworthy to mention that base station or cell might be switched off also for energy saving purposes. In both situations, when it is clear that the 'outage' will persist for a significant period of time, cell outage compensation can be started for the service coverage area affected by the switched off base station or cell.

When COC is triggered, cells which are neighbours of the problematic eNodeB (e.g. those with X2 connection to base station in malfunction) become responsible for the compensation, that is, become activated for recovery of the problematic area. Operator policies on the performance trade-off between coverage and capacity might have impact on the selection or activation order of cells responsible for compensation.

6.4.2 Means of Cell Outage Compensation

COC can be done by increasing transmission power in neighbour cells or by changing antenna tilt for further coverage extension. Whether power or antenna downtilt is used depends on availability of RET-capable antennas in the system and used power levels in the base stations. It is also possible to use both, power and antenna downtilt optimisation, for instance, by first optimising power levels and then selecting appropriate antenna tilt settings.

As cell outage leads to service unavailability, it is particularly important to recover from an outage as fast as possible. Therefore, less accuracy in setting new tilt value can be tolerated as long as action is made very fast. This is why COC is typically done with single step optimisation, even if leading to less optimum result. Further optimisation after coarse optimisation by COC can be done with Coverage and Capacity Optimisation (CCO) based on Performance Management (PM) data. For more information about CCO, see Section 5.4.

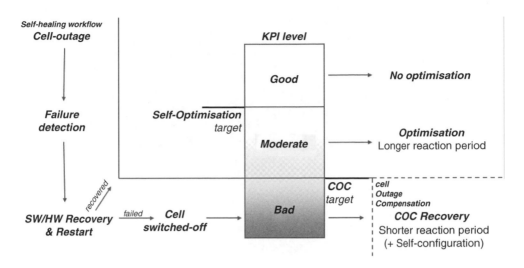

Figure 6.15   COC in SON hierarchy.

The most convenient and rapid way to compensate a cell outage is to increase transmission power at cells neighbouring to the cell outage area. On the other hand, usually when base stations are in service their transmission power is already set to the maximum available power level. Also, antenna tilt readjustment can be taken as a further action but this requires Remote Electrical Tilt (RET) antennas to be installed in the base stations.

Figure 6.15 shows a COC hierarchy in which different means to recover service quality are used based on the level of measured KPIs. In this scheme, whenever cell performance is better than the minimum service quality requirement (i.e. COC target), self-optimisation (i.e. CCO) takes place in order to optimise service quality in activated cells. Actual COC is started only when service quality goes under the minimum service quality requirement or an FM alarm is received from the malfunctioning base station. In case the service quality would be still under the minimum after COC, a further alarm would be indicated to the human operator.

A typical cell outage compensation scenario assuming a malfunctioning three-sectorised base station can be exemplified in a simulated network scenario for Helsinki downtown. The simulated LTE network planning scenario is originally a 3G WCDMA scenario adopted from (Laiho *et al.*, 2006). Users are distributed to the simulated area by taking location, for example, buildings, streets and water areas into account. Propagation is modelled in three dimensions by adding the impact of shadow fading, fast fading and antenna configurations summing with the propagation losses, calculated with ray tracing. Fast fading is modelled using a statistical sum-of-sinusoids method as described in Zheng and Xiao (2002). Modelling of antenna parameters and main simulation assumptions follow those selected by 3GPP for performance evaluations of LTE and LTE-Advanced technologies (3GPP TR36.814, 2010). The network planning modelling and simulation assumptions are described in detail in Section 5.4. While the main focus in Section 5.4 is on the extension of coverage and capacity by means of close to optimal antenna parameter selection under the SON CCO concept, in COC the main goal is to be able to provide

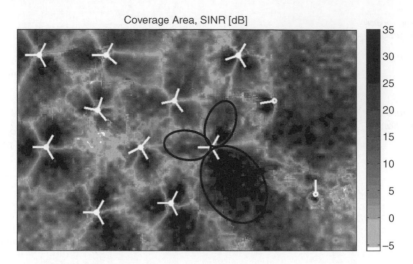

Figure 6.16 Initial network state functioning normally. Coverage area of malfunctioning base station encased with black line. Light grey are indicates weaker signal quality and dark stronger signal quality.

satisfactory service level in case of a cell-outage. Hence, the scenario is modified to introduce an eNB outage.

For the given COC scenario, where power and tilt recovery algorithms are applied, the following three steps are included:

1. In Figure 6.16, a realistic network deployment scenario is given in which network is working normally. In this scenario used power levels have been optimised to provide energy saving but base stations are not shut down. Therefore maximum power levels are not used. Coverage area of malfunction base station is outlined with black lines.
2. In Figure 6.17, one of the base stations is malfunctioning, so that cell outage, which is shown with white colour, occurs. Therefore, cell outage compensation should be triggered through the neighbouring cells. All the neighbour cells participate to compensation by power and tilt optimisation. As an example, the coverage area before compensation for a related cell is indicated with black line.
3. In Figure 6.18, transmission power and antenna tilts are optimised for the cell outage compensation and almost whole outage area has been covered by neighbouring cells. The coverage area after compensation is shown for one of the cells used for compensation.

As shown in the figures, the required service quality and continuity in the targeted LTE service coverage area could be ensured by enabling cell outage compensation functionality at self-organising networks. However, COC cannot help in all network conditions. In capacity optimised environments inter-base station distance is typically low. Thus, still continuous service coverage might be possible without changing neighbour cell configurations. Hence, optimisation, such as CCO, would be a sufficient action. On the other hand, in coverage limited scenarios inter-base station distance is usually so large that minimum or almost minimum tilt and maximum power levels are typically used, so that tilt or power level optimisation can not be used for coverage recovery.

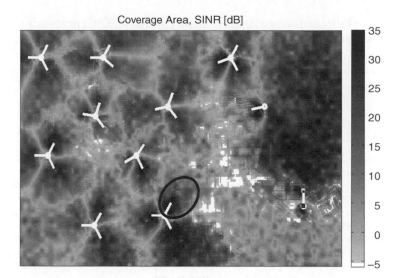

Figure 6.17 Occurrence of cell outage due to a malfunctioning base station. Coverage hole is seen as a white area. Light grey indicates an area with lower signal quality. The initial coverage area of one the neighbour cells used for compensation is encased with black line.

6.4.3 Interaction between Cell Outage Compensation and Self-Configuration Functions

In LTE networks, Physical Cell IDs (PCIs) are re-used due to their available limited number. When the area of cell in failure is covered by neighbouring cell(s), confusion or conflict of PCIs may occur as explained in Section 4.2.2. Due to that, possible PCI collision or confusion

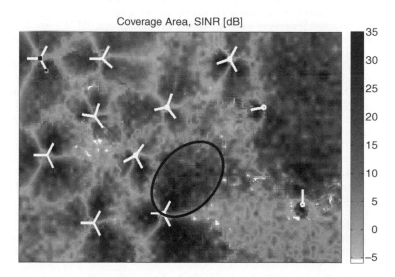

Figure 6.18 Cell outage compensation by optimisation of transmission power and tilt. The coverage area after compensation is encased with black line.

should be checked after cell outage compensation. Furthermore, it is necessary to update and check neighbour relations (e.g. via ANR) to enable reliable handovers and continuous service for mobile users. To solve all these and similar problems, self-configuration functions should be triggered at the affected base stations after the tilt change. The mechanisms to do so during SON operation are outlined in Chapter 9.

References

3GPP TS32.541 (September 2010) *Telecommunication Management; Self-Organizing Networks (SON); Self-Healing Concepts and Requirements*, 3rd Generation Partnership Project (3GPP), (Release 10).

3GPP TS32.762 (December 2010) *Evolved Universal Terrestrial Radio Access Network (E-UTRAN) Network Resource Model (NRM) Integration Reference Point (IRP); Information Service (IS)*, (Release 10).

3GPP TS32.766 (December 2010) *Evolved Universal Terrestrial Radio Access Network (E-UTRAN) Network Resource Model (NRM) Integration Reference Point (IRP); Solution Set (SS) definitions*, (Release 10).

3GPP TS32.522 (January 2011) *Self-Organizing Networks (SON) Policy Network Resource Model (NRM) Integration Reference Point (IRP); Information Service (IS)*, 3rd Generation Partnership Project (3GPP), (Release 10).

3GPP TS32.526 (January 2011) *Self-Organizing Networks (SON); Policy Network Resource Model (NRM) Integration Reference Point (IRP); Solution Set (SS) definitions*, (Release 10).

3GPP TR36.805 (January 2010) *Evolved Universal Terrestrial Radio Access (E-UTRA); Study on minimization of drive-tests in next generation networks*, 3rd Generation Partnership Project (3GPP), (Release 9).

3GPP TR36.814 (March 2010) *Evolved Universal Terrestrial Radio Access (E-UTRA); Further advancements for E-UTRA Physical layer aspects*, (Release 9).

3GPP TS32.410 (April 2011) *Telecommunication management; Key Performance Indicators (KPI) for UMTS and GSM*, 3rd Generation Partnership Project (3GPP), (Release 10).

Aamodt, A. and Plaza, E. (1994) Case-based reasoning: Foundational issues, methodological variations, and system approaches, in *Artificial Intelligence Communications*, Vol. 7 (1), IOS Press, Amsterdam. pp. 39–59.

Abraham, A. and Grosan, C. (2005) Genetic programming approach for fault modeling of electronic hardware. IEEE Proceedings Congress on Evolutionary Computation (CEC'05), Vol. 2, Edinburgh, 1563–1569.

Amirijoo, M., Jorguseski, L., Litjens, T. and Schmelz, C. (2011) Cell outage compensation in LTE networks; Algorithms and performance assessment. IEEE International Workshop on Self-Organizing Networks, May, Budapest.

Andrzejak, A. and Silva, L. (2007) Deterministic models of software aging and optimal rejuvenation schedules. 10th IEEE/IFIP International Symposium on Integrated Network Management (IM '07), pp. 159–168.

Bai, C.G., Hu, Q.P., Xie, M. and Ng, S.H. (2005) Software failure prediction based on a Markov Bayesian network model. *Journal of Systems and Software*, 74(3), 275–282.

Barco, R., Díez, L., Wille, V. and Lázaro, P. (2007) *Automatic Diagnosis of Mobile Communication Networks under Imprecise Parameters*, Series on Expert Systems with Application, Elsevier: Amsterdam.

Barco, R., Lazaro, P., Diez, L. and Wille, V. (2008) Continuous versus Discrete Model in Autodiagnosis Systems for Wireless Networks. *IEEE Transactions on Mobile Computing*, 7(6), 673–681. doi: 10.1109/TMC.2008.23.

Barco, R., Wille, V. and Diez, L. (2005) System for automated diagnosis in cellular networks based on performance indicators. *European Transactions on Telecommunications*, 16(5), 399–409.

Barreto, G.A., Mota, J.C.M., Souza, L.G.M. *et al.* (2004) A new approach to fault detection and diagnosis in cellular systems using competitive learning. Proceedings of the VII Brazilian Symposium on Neural Networks (SBRN'04).

Bod'ik, P., Friedman, G., Biewald, L. *et al.* (2005) Combining visualization and statistical analysis to improve operator confidence and efficiency for failure detection and localization. IEEE Proceedings of International Conference on Autonomic Computing (ICAC 05), pp. 89–100.

Chandola, V., Banerjee, A. and Kumar, V. (2009) Anomaly detection: A survey. *ACM Computer Survey*, 41(3), Article 15 (July 2009) 58. doi: 10.1145/1541880.1541882, http://doi.acm.org/10.1145/1541880.1541882.

Chen, M., Kiciman, E., Fratkin, E. *et al.* (2002) Pinpoint: Problem determination in large, dynamic internet services. Proceedings of 2002 International Conference on Dependable Systems and Networks (DSN), IPDS track, pp. 595–604.

Cheng, F., Wu, S., Tsai, P. *et al.* (2005) Application cluster service scheme for near-zero-downtime services. IEEE Proceedings of the International Conference on Robotics and Automation, pp. 4062–4067.

Chernogorov, F., Turkka, J., Ristaniemi, T. and Averbuch, A. (2011) Detection of sleeping cells in LTE networks using diffusion map. Proc. of the IEEE International Workshop on Self-Organizing Networks (IWSON), in conjunction with VTC 2011 Spring (May 15–18), Budapest.

Cheung, B., Kumar, G. and Rao, S. (2005) Statistical algorithms in fault detection and prediction: Toward a healthier network. *Bell Labs Technical Journal*, **9**(4), 1538–7305.

Crowell, J., Shereshevsky, M. and Cukic, B. (2002) *Using fractal analysis to model software aging*, Tech. rep. West Virginia University, Lane Department of CSEE, May, Morgantown, WV.

Daidone, A., Di Giandomenico, F., Bondavalli, A. and Chiaradonna, S. (2006) Hidden Markov models as a support for diagnosis: Formalization of the problem and synthesis of the solution. IEEE Proceedings of the 25th Symposium on Reliable Distributed Systems (SRDS 2006), Leeds.

EMC (2009) Automating Root-Cause Analysis: EMC Ionix Codebook Correlation Technology vs. Rules-based Analysis Technology Concepts and Business Considerations, A white paper from EMC (http://www.emc.com/collateral/software/white-papers/h5964-automating-root-cause-analysis-wp.pdf) [accessed 15 March 2011].

Hamerly, G. and Elkan, C. (2001) Bayesian approaches to failure prediction for disk drives, in *Proceedings of the Eighteenth International Conference on Machine Learning*, Morgan Kaufmann Publishers Inc.: Burlington, MA, pp. 202–209.

Heckerman, D., Breese, J.S. and Rommelse, K. (1995) Decision-Theoretic Troubleshooting. *Communication ACM*, **38**(3), 49–57.

Hoffmann, G.A. (2006) *Failure Prediction in Complex Computer Systems: A Probabilistic Approach*, Shaker Verlag: Aachen.

Hughes, G., Murray, J., Kreutz-Delgado, K. and Elkan, C. (2002) Improved disk-drive failure warnings. *IEEE Transactions on Reliability*, **51**(3), 350–357.

Khanafer, R.M., Solana, B., Triola, J. *et al.* (2008) Automated diagnosis for UMTS networks using bayesian network approach. *IEEE Transactions on Vehicular Technology*, **57**(4), 2451–2461. doi: 10.1109/TVT.2007.912610.

Kiciman, E. and Fox, A. (2005) Detecting application-level failures in component-based internet services. *IEEE Transactions on Neural Networks*, **16**(5), 1027–1041.

Laiho, J., Raivio, K., Lehtimaki, P. *et al.* (2005) Advanced analysis methods for 3G cellular networks. *IEEE Transactions on Wireless Communications*, **4**(3), 930–942.

Laiho, J., Wacker, A. and Novasad, T. (2006) *Radio Network Planning and Optimization for UMTS*, 2nd edn, John Wiley & Sons, Inc., New York.

Levy, D. and Chillarege, R. (2003) Early warning of failures through alarm analysis - a case study in telecom voice mail systems. ISSRE '03: Proceedings of the 14th International Symposium on Software Reliability Engineering, Washington, DC.

Li, L., Vaidyanathan, K. and Trivedi, K.S. (2002) An approach for estimation of software aging in a web server. Proceedings of the Intl. Symposium on Empirical Software Engineering, ISESE 2002. Nara, Japan.

Meng, H., Di Hou, Y. and Chen, Y. (2007) A rough wavelet network model with genetic algorithm and its application to aging forecasting of application server. IEEE Procedings of International Conference on Machine Learning and Cybernetics. Vol. **5**.

Montani, S. and Anglano, C. (2006) Case-based reasoning for autonomous service failure diagnosis and remediation in software systems. Proc. European Conference on Case-Based Reasoning (ECCBR) 2006, Lecture Notes in Artificial Intelligence, pp. 489–503.

Mueller, C., Kaschub, M., Blankenhorn, C. and Wanke, S. (2008) A cell outage detection algorithm using neighbor cell list reports. In K.A. Hummel and J.P.G. Sterbenz (eds.), *IWSOS 2008, LNCS5343*, 218–229.

Neapolitan, R.E. (2004) *Learning Bayesian Networks*, Prentice Hall, Upper Saddle River, NJ.

Nováczki, S. and Szilágyi, P. (2011) Radio Channel Degradation Detection and Diagnosis Based on Statistical Analysis. IEEE International Workshop on Self-Organizing Networks, May, Budapest.

Rao, S. (2006) Operational Fault Detection in cellular wireless base-stations. *IEEE Transactions on Network and Service Management*, **3**(2), 1–11.

Salfner, F. (2006) Modeling event-driven time series with generalized hidden semi-Markov models, Tech. Rep. 208 *Department of Computer Science*, Humboldt-Universität zu, Berlin. Available at http://edoc.hu-berlin.de/docviews/abstract.php?id=27653 [accessed 15 March 2011].

Salfner, F., Lenk, M. and Malek, M. (2010) A survey of online failure prediction methods. *ACM Computer Survey*, **42**(3), 42. doi: 10.1145/1670679.1670680, http://doi.acm.org/10.1145/1670679.1670680.

Singer, R.M., Gross, K.C., Herzog, J.P. *et al.* (1997) Model-based nuclear power plant monitoring and fault detection: Theoretical foundations. Proceedings of Intelligent System Application to Power Systems (ISAP 97). Seoul, pp. 60–65.

Turkka, J., Ristaniemi, T., David, G. and Averbuch, A. (2011) Anomaly Detection Framework for Tracing Problems in Radio Networks. Proc. of the IARIA Tenth International Conference on Networks (ICN), St. Maarten, January 23–28, Netherlands Antilles.

Turnbull, D. and Alldrin, N. (2003) *Failure Prediction in Hardware Systems*, Tech. rep, University of California, San Diego, available at http://www.cs.ucsd.edu/~dturnbul/Papers/ServerPrediction.pdf [accessed 15 March 2011].

Van den Berg, J.L., Litjens, R., Eisenblätter, A. *et al.* (2008) *SOCRATES: Self-Optimisation and self-ConfiguRATion in wirelESs networks*, COST 2100 4th MCM, February 6–8 2008, Wroclaw.

Vilalta, R. and Ma, S. (2002) Predicting rare events in temporal domains. Proceedings of the 2002 IEEE International Conference on Data Mining (ICDM'02), Washington, DC, pp. 474–482.

Varga, P. and Moldovan, L. (2007) Integration of Service-Level Monitoring with Fault Management for End-to-End Multi-Provider Ethernet Services. *IEEE Transactions on Network and Service Management*, **4**(1), 28–38.

Ward, A. and Whitt, W. (2000) Predicting response times in processor-sharing queues, in *Proc. of the Fields Institute Conf. on Comm. Networks*, (eds P.W. Glynn, D.J. MacDonald and S.J. Turner, *Fields Institute Communications 28*, American Mathematical Society, Providence, RI, pp. 1–29).

Wietgrefe, H., Tuchs, K.D., Jobmann, K. *et al.* (1997) Using neural networks for alarm correlation in cellular phone networks. International Workshop on Applications of Neural Networks to Telecommunications (IWANNT), pp. 248–255.

Yilmaz, O.N.C., Hämäläinen, S. and Hämäläinen, J. (2011) *Optimisation of Adaptive Antenna System Parameters in Self-Organizing LTE Networks* (submitted to Wireless Networks), Springer Journal.

Zanier, P., Guerzoni, R. and Soldani, D. (2006) Detection of Interference, Dominance and Coverage Problems in WCDMA Networks. IEEE 17th International Symposium on Personal, Indoor and Mobile Radio Communications, 11–14 September 2006, pp. 1–5.

Zheng, Y.R. and Xiao, C. (2002) Improved models for the generation of multiple uncorrelated Rayleigh fading waveforms. *IEEE Communications Letters*, **6**(6), 256–258.

7

Supporting Function: Minimisation of Drive Tests (MDT)

Malgorzata Tomala, Ilkka Keskitalo, Gyula Bodog and Cinzia Sartori

7.1 Introduction

7.1.1 General

Drive tests are a commonly used method by operators evolving their networks toward new technologies. The purpose of drive tests is to monitor and assess mobile network performance. The testing methodology that is currently performed comprises a series of steps and is manually controlled. Typically, when performing drive tests, test engineers need to capture measurements and course of related events of test calls performed during the movement with specially adapted equipment for this purpose terminal. In the simplest case, the testing person needs to carry a phone which supports network monitoring applications. However, the most widely practiced method and real drive tests consist of using of a test car adequately equipped with test terminal, measuring devices and additional tools (Figure 7.1). The test terminals are developed to support multiple functions and different transmission technology to provide diversity of results. They are usually integrated with a GPS receiver to get terminal coordinates and record its position. Some terminals are mounted in the car and are connected to high sensitivity antennas on the roof with different orientations, and thus different signal perceptions. In addition, the equipment includes a set of supplementary tools used for drive test control, for example, a device for measuring the network parameters, a spectrum analyser and laptops acting as local maintenance terminals.

With such test configuration, the measurement car driving along any route allows the operator to perform drive tests at different locations, for example, in typical urban and suburban sites, on campus routes or along public streets. Test calls performed during the drive route of

LTE Self-Organising Networks (SON): Network Management Automation for Operational Efficiency, First Edition.
Edited by Seppo Hämäläinen, Henning Sanneck and Cinzia Sartori.
© 2012 John Wiley & Sons, Ltd. Published 2012 by John Wiley & Sons, Ltd.

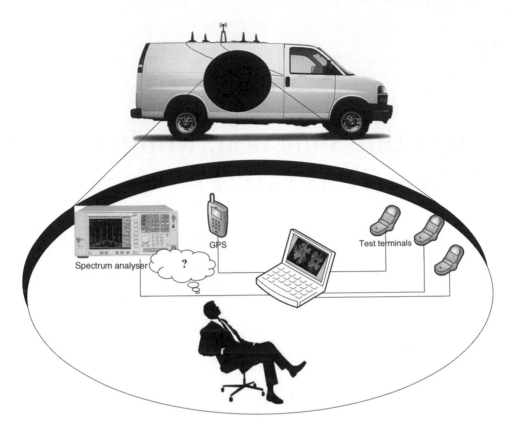

Figure 7.1 Vehicle equipped with drive testing measurement devices.

such a vehicle can detect and record a wide variety of the physical and virtual parameters of network service in a given geographical area. An intermediate result of the drive test route; that is, measured data, is stored in the database in a log file, commonly representing a tremendous amount of data which requires further analysis. The process of studying the results and associating each drive test measurement with a transmitting cell may be a laborious task, but definitely is an inseparable and expected effect of drive testing. The root cause analysis helps to determine the course of corrective action to be taken to tackle the problem.

Thus, drive test steps allow the operator to collect appropriate statistics, perform network design verification and further identify and categorise network, its coverage and quality problems. While measuring what the network conditions are in any specific area that is visited by the test car, it is likely to get realistic data comparable to actual user experience results. Hence, the possibility of getting an accurate map of base station service areas and signal strengths from the perspective of the subscriber is very useful as well as greatly desired by operators. By exploring the network conditions, the operator can get knowledge about the weak points of the network and make direct changes to the network configuration to provide better coverage and service to their customers.

Although operators continuously monitoring, developing and enhancing network quality strongly rely on drive tests, the traditional and professional way of carrying out drive tests brings about some uneasiness. Apart from expensive highly specialised measurement equipment, the process of execution implies additional expense on contractors, drive test service and solutions providers or licences for tools. Moreover, the procedure involves human resources and the effort required for drive testing itself. Further post processing might also take a long time. As a result, the overall operation cost operators considerable amount of money and strenuous effort.

Drive tests are also used for performing competitive analysis amongst operators. Hence, this technique advancement will really benefit the operators. It will help to drive down network exploring costs. For this reason, several facilities to the traditional drive tests method are being adopted and offered to operators, for instance, through the installation of an automatic measuring device in the taxi. The measuring device, with ability to automate data collection provides operators with results that are post-processed manually. As taxis frequently drive across the service areas of the operators, test engineers do not have to physically carry out the drive testing and audit the field. However, the simplification does not actually resolve all the inconveniences. The drive test business remains still costly and causes additional CO_2 emissions.

Another, essential aspect is that existing practices for drive tests are usually done on roads, without the possibility of providing operators with information on users' experiences within indoor coverage.

Consequently, this has been driving the need for continued innovations. The motives behind drive test evolution have been also considered in 3GPP standardisation. With the advent of LTE and its large-scale commercial deployment, it becames desirable to develop automated solutions including involvement of commercial UEs. This approach allows operators to provide measurement data for radio network fault detection and optimisation in all possible locations covered by the network. In the wake of the ultimate objective: to reduce cost of managing the networks, the idea followed by introducing a new feature in Rel10.

The focus of this chapter is on the technologies defined for this evolution in the 3GPP standards, that is, on the Minimisation of Drive Tests (MDT) functionality. The chapter gives an overview on the MDT concept, use cases, procedures and measurements in UMTS/HSPA and LTE technologies, followed by the progress and plans toward Release 11.

7.1.2 History and Background

Before entering into a detailed discussion on technical realisation, it is probably worth highlighting some of the driving factors giving the origin of further working assumptions and understanding where the technical requirements and solutions are coming from.

Looking back to when 3GPP activities on the minimisation of drive tests started, it begun with early discussions amongst RAN (Radio Access Network) and SA (System Architecture) technical specifications groups. Since the automation of drive tests became operators' requirement in the NGMN forum (Lehser, 2008) and therefore a top priority for their networks' deployments, operators taking part in 3GPP took the subject there also. 3GPP groups acknowledged the need and established a framework in order to define use cases, requirements and solutions for minimisation of drive tests.

The high level concept assumed that a solution will utilise measurement capabilities in UEs in order to increase the information available at the O&M on the status of the network, and thus reduce an operator's dependence on manual drive tests. As far as the end user involvement and his field measurements delivery to O&M system were apparent, the transport mechanism between these two end-points was indistinct at the starting point of discussions.

The clarity on user participation in the foreseen evolved resolution for drive tests data collection resulted in a study item creation in 3GPP TSG RAN, the groups having adequate insight into the radio interface and the UE functionalities. The work was initiated to:

- identify use cases and requirements to minimise drive tests;
- study the necessity of defining new UE measurements logging and reporting capabilities;
- analyse the impact on the UE and recommend the development of new capabilities.

At one time during the study phase, there were two alternative architecture proposals:

- user plane architecture (see Figure 7.2);
- control plane architecture (see Figure 7.3).

The two architectures were studied in 3GPP TSG SA WG5 in 3GPP TR32.827 (2010).

In user plane architecture the configuration of the UE is done by utilising the OMA defined device management feature. The reporting of the UE measurement data is done by establishing a normal data connection to a specific file server. In this architectural model, the access network is not involved in a policy control of UE measurement collection. The configuration and

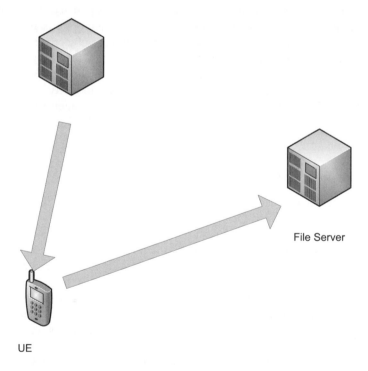

File Server

UE

Figure 7.2 User plane architecture.

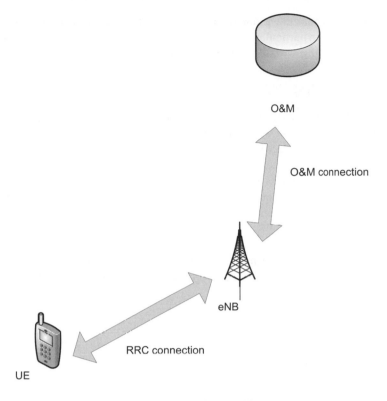

Figure 7.3 Control plane architecture.

reporting procedures are terminated directly between a file server and UE, thus transport of UE reports is treated as user plane data by access network. The transport of the measurements is transparent to the radio access and core network.

In control plane architecture the policy for measurement data collection and reporting from the operators' O&M system is targeted; the eNB/RNC and then eNB/RNC configures the UE via RRC connection. The UE reports back the measurements to the eNB/RNC via RRC and the UE measurement data is transferred onwards from the eNB/RNC to O&M via the O&M connection. This approach allows the collection of the UE measurements in the eNB/RNC, possibly combining them with results already available in the eNB/RNC and forwarding the combined measurements to the O&M system.

The studies done under the study item phase have shown that for efficient network optimisation it would be beneficial to use UE measurements per physical location together with information available in the radio access network. Examinations on transport mechanisms determined control plane solutions as feasible mean to acquire the information from devices and possibility to add network characteristics. These conclusions were documented in a 3GPP technical report, 3GPP TR36.805 (2009) and approved with a justification for further work on a concrete solution based on Control Plane Architecture.

Thus, after the initial requirements stage, where targets and objectives of MDT were settled, the 3GPP TSG RAN and SA working groups established two parallel work items. The TSG

RAN group started to examine a functioning for MDT to be performed in the UE and lower layers of the networks; in particular, the air interface. At the same time the 3GPP TSG SA group launched a corresponding work on the MDT management.

7.2 Relation to SON

Self-Organising Networks, as the name suggests, focus on adaptive network maintenance, and manual intervention reduction while optimising the network. SON operations are definitely crucial for achieving most effective network performance, especially, in the face of increasing complexity in the current networks.

Both SON and MDT address the same objectives, that is to reduce operational efforts, increase network performance, quality and, at the same time, decrease the maintenance costs; both techniques are very promising for network optimisation and can be used independently one from the other.

Behind SON mechanisms there is the general image that the network itself, based on certain algorithms, takes decisions to optimise and auto-tune its own settings. Inputs to trigger those decisions are predefined thresholds based on operator preferences with the ultimate objective to reduce human intervention.

The main differences between SON and MDT are:

- SON is aiming at instant/automated reaction on short- to middle-term network issues, MDT is more about collecting measurements for further analysis and processing (either manually or automated).
- Use case applicability: SON includes self-configuration, self-optimisation and self-healing, while MDT mainly focuses on optimisation.

Different SON use cases are relevant at different stages of network operation, for example, during roll-out and early phases of operation, or in high loaded and mature networks. In general, the use cases related to self-configuration and coverage are most important in earlier phases, whereas quality and capacity based use cases become more relevant when network usage increases.

While in the initial stages of the network, for example, pre-launch optimisation and during first rollouts of new infrastructures (e.g. 3G or LTE), the need for drive testing campaigns becomes especially augmented. It is evident, that operators offering a new technology need to be able to check services of a new wireless communication network and run trials to verify RF design before commercial launch. Currently, in order to accept data throughput and trans-mission quality in a rollout network, deployment drive tests are performed in an iterative manner so that it is possible to do some modifications if necessary. Network optimisation engineers take decisions on remedial solutions that are implemented in the network. After the implementation of changes there is another iteration of the monitoring period to verify if the issue has been resolved. A decisional flowchart of the process is presented in Figure 7.4. Consequently, based on drive test deliverables in the phase between the planning and new infrastructure acceptance it is feasible to introduce an initially optimised and fully functional network.

With MDT procedures, the above process is enhanced by automatic input delivered from *'normal'* UEs camping and connected to the network; measurements from UEs, which

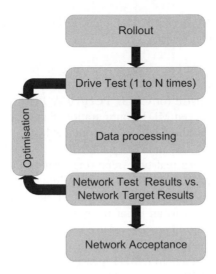

Figure 7.4 Network acceptance process.

experience problems, are then collected and (automatically) post-processed for further optimisation.

The above process can also be used to optimise a mature network (e.g. to verify the field parameter changes, to solve problems in a particular area or in a swap projects) so that MDT brings benefits not only at pre-operational phases of network deployment, but can nicely complement configuration and healing stages as well.

Since the UE reported results through MDT may be utilised for radio optimisation purposes, MDT does not include any automatism as SON does. In addition SON and MDT are independent one from the other, for example, the operator may decide not to deploy SON features, but would still like to reduce manual testing.

Even though the two solutions may work independently, the combination of the two is a powerful method to enhance end user performance, improving overall network operability.

7.3 Requirements

Prior to elaboration of possibilities to replace some of the drive tests by new measurements and data logging carried out at the UE, it was essential to establish high level requirements for MDT. The future prospect of minimising drive tests by using ordinary UEs, where a large part of the value would come from the use of GNSS (e.g. GPS or Galileo) positions, gave remarkable consideration to consequent UE impacts. It was seen as important to secure the feasibility of UE implementation also for the lowest cost categories. Therefore, all direct and indirect impacts from the new feature would be taken into account while defining the measurements and related procedures. It was felt that list of requirements should cover at least issues related to UE complexity, power consumption and the end user experience. Consequently, 3GPP agreed on a set of general principles and requirements guiding the definition of functions for MDT, which to large extent, was a trade-off between the feature efficiency and UE impact.

The requirements specified in 3GPP TS37.320 (2011) as a basis for the development of MDT, are divided into several areas:

- **MDT (reporting) mode:** The requirement introduces a split with regard to UE state. There are two modes of reporting for the MDT measurements: non-real-time or immediate reporting. A UE in RRC_IDLE (or CELL_PCH/URA_PCH in UTRA) mode of operation can be configured with Logged MDT configuration, that is, after having initiated Logged MDT the UE will not support immediate reporting but instead will apply specific non-real-time reporting triggers. A UE in RRC_CONNECTED mode can be configured with Immediate MDT that implies real-time reporting, that is, the most measurement results are communicated by the UE as MDT information immediately.
- **UE measurement configuration:** With the use of air interface resources, the UE typically provides radio measurements as an input for radio resource management (RRM). RRM functions are commonly utilised procedures to ensure the efficient use of the available radio resources. MDT measurement must fit in to normal UE operations. Thus, the requirement states that it shall be able to configure MDT measurements independently from the network configurations for normal RRM purposes. Hovever, the availability of measurement results is conditionally dependent on the UE RRM performance.
- **UE measurement collection and reporting:** MDT measurement reports may consist of multiple events and measurements taken over time. While standardising MDT solutions the requirement underwent different interpretations conditioned by the MDT reporting mode:
  - For *Immediate MDT* it is possible to configure several instances of a measurement trigger. Consequently, when the preconfigured triggers are met, the UE immediately reports the measured radio conditions accordingly as a real-time report.
  - For *Logged MDT* it is possible to configure periodical trigger only. When the predefined periodical timer expires, the UE stores the measured radio conditions available at this point of time for further reporting. The time interval for logged measurement collection and reporting is decoupled in order to limit the impact on the UE battery consumption and network signalling load.
- **Geographical scope:** Of measurement logging drive testing of a targeted geographical area (e.g. problem area) is part of the usual activities for operators. With MDT configuration, it shall be possible to follow this approach and configure the geographical area where the defined set of measurements shall be collected.
- **Location information:** A large part of the manual drive test value comes from the use of GPS position and all drive tests used to rely on satellite based positioning, for example, GPS. Hence, the solutions for minimising drive tests also requires the measurements to be linked with the available location information or other measurements that can be used to derive the actual location of the UE. Since the existence of GPS receiver as such does not necessarily guarantee that accurate GSP position for MDT will be always available at the UE and for all anticipated measurements, there are no absolute accuracy requirements on the adopted location methods. Bearing in mind the availability of the UE location in practical situations, it is assumed that measurement results for MDT, will be tagged by location information in a best effort way, that is, in the absence of GPS position, cell identity may be used as a rough indicator for finding problem areas (see Section 7.7.5 for further details).
- **Time information:** The time information is important in order to examine MDT reports in post processing operations. Generating and storing a timestamp indicating the time when

each measurement is taken enables to track back the measurement data and make a reference to traffic conditions possible. Therefore, the measurements performed for MDT purposes shall be linked with time information. For non-real-time reporting, the measurements in measurement logs shall be tagged with a time stamp that is generated by the UE. For real-time reporting, the time information shall be included by network.

- **UE capability information:** The network needs to be able to choose suitable terminals for performing MDT measurements. Thus, the network shall request the terminal capabilities, which allow the network to carefully select the right terminals for MDT measurements. Possible indications include GNSS positioning or logging feature support. Since real-time reporting does not apply to MDT functionality exclusively, no specific capability is required to indicate immediate MDT support.
- **Dependency on SON:** As mentioned in Section 7.2, there is a synergy between MDT and SON. Information collected for MDT may also be used by SON functions as well as by other optimisation entities with defined interfaces. Nevertheless, the solutions for MDT shall be able to work independently from SON support in the network. At the UE, the relation between measurements for MDT and UE side SON functions shall be established in a way that re-use of functions is achieved where possible.
- **Dependency on TRACE:** The existing trace functionality was agreed as a feasible mean for MDT feature management. Thus, two trace techniques are to be reused and extended to serve MDT. If the MDT is initiated toward to a specific UE (e.g. based on IMSI, IMEI-SV), the signalling based trace procedure is used, otherwise the management based trace procedure is utilised.

While defining new requirements, an especially cautious approach was considered with regard to user and terminal impact. Hence, solutions for MDT processes should take into account the following constraints 3GPP TS37.320 (2011) over and above the list of requirements, with respect to:

- **UE measurements logging:** The UE measurement logging mechanism is an optional feature. In order to limit processing and the impact on UE power consumption, MDT shall utilise the measurements that are performed in the UE according to RRM enforced by the network. Consequently, measurement logging functionality should rely on the available measurements as much as possible.
- **Location information:** The availability of location information is subject to UE capability and/or UE implementation. It should be considered that to be able to have valid position information at the time when the measurement trigger happens, the GNSS receiver should be typically started in advance and be active for sufficiently long time. This may become a severe power consumption issue. Hence, solutions requiring detailed location information shall take into account power consumption of the UE due to the need to run its positioning components.

7.4 Use Cases

In essence, one important driver for MDT is collection of information about radio network performance by the UE autonomously. However, in the process of specifying functional requirements, also the list of use cases was identified to understand a set of possible scenarios, in which the foreseen solutions help to minimise the drive tests. Subsequent sections list use cases where MDT can have potential to provide useful information for network deployment verification and optimisation. In most of the cases the actual MDT solution is open. The solutions developed and specified in 3GPP for Release 10 (3GPP TR32.827, 2010; 3GPP TS36.331, 2011) only focus on the coverage optimisation use case. Remaining use cases became eventual subject for evaluation in future releases.

7.4.1 Operator Scenarios

The main triggers for operators to perform drive tests outlined in 3GPP TR36.805 (2009) are:

- **Deployment of new base stations** Constructing new cell sites is the typical operation when drive test needs to be performed in order to verify service activation. When the network is built or expanded, drive test ensures that new base station installation is realised at an optimal level of efficiency. In practice, deployment of a new base station means that initially the transmission is activated in a 'test mode'. Drive test intends to collect downlink and uplink coverage measurements of the new cell and neighbour cells in the tested area. Based on the first testing, a coarse area tuning is performed (e.g. physical changes of antenna tilts and azimuths or some parameter settings, like transmission (TX power)) with the scope of optimising the coverage and the quality of the network. Commercial service with the new cell can be started after such initial tuning. However, after the launch, networks are 'alive', always changing due traffic conditions, addition of new sites, new software upgrades. Thus, drive tests have been continued to collect more extensive data of coverage/throughput measurements in the intended area to ensure that good service is being provided and allow performing a fine tuning of the network (i.e. parameter based).

    New base station deployment is a continuous and long lasting process aiming to keep the high level of performance. This process is becoming more important now with new deployment scenarios or new RAN elements (e.g. femtos, picos, remote radio heads) which bring additional factors affecting fine tune coverage of the operator's network.

- **Construction of new highways/railways/major buildings** This is an additional stage of network monitoring, when drive test that is performed for a particular area in an event driven manner.

    Construction of new highways or major buildings forecasts areas where network traffic will increase. Also, such large obstacles normally introduce weak signal areas as they become new obstacles causing additional shadowing in the radio network. Therefore, whenever new highways/railways/office buildings are constructed, it is important to provide a snapshot of surrounding network performance. Operators perform drive tests in the relevant areas to see if downlink/uplink coverage or throughput requires improvement due to changes in volume of network traffic or if there is a need to enhance in-building traffic. If coverage improvement is deemed necessary, operators may take action to add extra capacity in high-traffic areas, deploy new cells, adjust the parameter settings of existing cells, install new cell sites, and so on. To check whether coverage has been actually improved in the area to an adequate level after such action, additional verification has to be performed again and another campaign of drive test should follow.

- **Customer's complaint** This is another trigger for network monitoring, when drive test is performed for a particular cell in an event driven manner. An area where drive tests need to be performed is determined by a customer's complaint against network service. When a customer experiences bad quality of the voice or data and indicates these concerns to the operator, the operator performs drive tests in the relevant place to observe the downlink coverage and service quality. Based on the information from customer, operators can understand where problems lie, why and how customers are affected. If the coverage improvement is deemed necessary, operators take corrective actions. The response in such reactive manner helps in providing high-quality service to the end user.

- **Periodic drive tests** Continuous monitoring of a mature network carrying traffic is also necessary and a commonly utilised practice by operators (e.g. to verify on the field parameter changes, to solve problems in a particular area and in a swap project amongst others). Drive testing that is performed for a particular cell in a regular manner reflects the actual performance of the network and identifies areas for improvement. Thus, as previously mentioned, drive tests also can be used once sites are fully operational. It is important for operators to periodically perform drive tests in order to update their understanding of the coverage and throughput levels provided in their networks. Any issues detected have to be investigated and corrected.

Given the variety of operators' scenarios performed during different network operational stages, it seems apparent that drive tests currently performed can be reduced within the framework of MDT if measurements collected for these purposes are carried out by commercial devices. This will also significantly reduce network maintenance costs. Furthermore, it will enable operators to collect measurements from areas which are difficult to access by traditional drive tests (e.g. narrow roads, non-public areas, private houses or other indoor areas).

Based on the above triggers, several use cases that are discussed in the following sections have been identified within the scope of MDT solutions. The use cases generalise different scenarios into assorted optimisation purposes.

7.4.2 Coverage Optimisation

Coverage is one of the fundamental aspects of network performance and end-user experience. In principle, coverage is the key assessment rating network service and major criterion that a customer considers when comparing service provided by different operators. Accordingly, it is very important for network providers to be aware of the downlink and uplink coverage, as well as to find the origin of the coverage related problems. Radio coverage related measurements are a great value for operators for the general improvement of network performance, such as network planning, network optimisation and RRM parameter optimisation. Therefore, the coverage optimisation use case, as the most crucial in terms of network maintenance, became top priority when defining MDT solutions.

Operators' main interest in a set of MDT measurements for coverage optimisation generally means getting reliable coverage maps for optimisation needs. However, statistics collected for the purpose of coverage monitoring may be also used to detect coverage issues that have to be solved. Thus, relevant situations where coverage will be measured and verified are categorised into the following sub-use cases:

- **Coverage mapping** Coverage maps based upon the signal strength provide operators with a valuable insight into the services levels that can be provided. For the purpose of

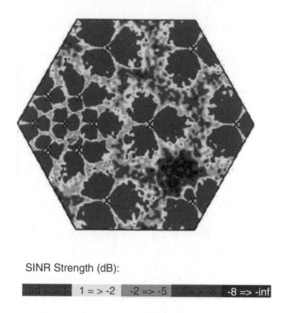

SINR Strength (dB):

Inf => -1   1 = > -2   -2 => -5   -5 > -8   -8 => -inf

Figure 7.5 Coverage map.

coverage maps visualisation, there should be insight into the signal levels per physical location in the cell areas. This means also that there should be measurements collected in all parts of the network and not just in the areas where there are potential coverage issues. Hence, the operator should be able to collect statistics from the deployed network about the signal levels and/or qualities that the UEs experience throughout the network. The statistics can be collected with different cell/network load situations. With all such information, there will be knowledge about the distribution of service levels that can be provided within a cell and the overall network. An example of a coverage map is shown in Figure 7.5.

- **Coverage hole detection**   Despite network planning, there could be areas where no sufficiently high downlink signal is received by the UEs obviously causing disconnection of services. An area where the signal level of both serving and allowed neighbour cells is below the level needed to maintain basic service is usually caused by physical obstructions such as new buildings, hills, by unsuitable antenna parameters, or by inadequate RF planning. UE in coverage hole will suffer from call drop and radio link failure. It is essential that such areas can be identified in order to be able to re-plan the network and make required corrections. Clear holes in the cellular coverage shall be discovered and eventually removed to provide seamless services throughout the network. An example of a coverage hole is shown in Figure 7.6.
- **Identification of weak coverage**   Weak coverage occurs when the signal level of serving cell is below the level needed to maintain a planned performance requirement. Areas where the signal levels are the weakest and where the probability of experiencing problems to maintain the connection or executing handovers is highest can be identified by UEs receiving signal level below certain threshold. This information should reveal the

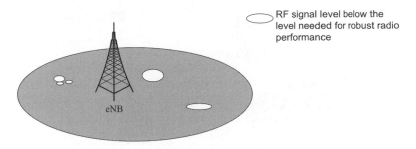

Figure 7.6 Coverage hole.

locations where signal levels are the weakest and there is a risk for handover failures, call drops, degradation of user data throughputs, and so on. Hence, related statistics should be limited to areas where the serving cell signal level is low to prevent unnecessary involvement of UEs to collect data mainly from relevant areas only. An example of weak coverage is shown in Figure 7.7.

- **Detection of excessive interference** Excessive interference may be caused by large overlap of cell coverage areas or unexpected signal propagation between the cells which has not been predicted during the network planning. The experienced interference will degrade the network capacity especially when the traffic load is high. Typically in this situation UEs may experience high serving cell signal level simultaneously with high interference from the neighbouring cell(s) resulting in poor connection quality and user experience. The main target for data collection is to get samples from the areas where the problem is present. An example of pilot pollution is shown in Figure 7.8.
- **Overshoot coverage detection** The overshoot coverage appears when coverage of a cell reaches far beyond what is planned. The overshoot coverage can is perceived as 'island' coverage of a neighbouring cell, that is, a situation where a strong neighbour cell signal is received within the serving cell coverage area. The neighbour cell signal may come from adjacent cell but also from a more distant cell. At that particular position the UE will experiences similar situation as with the cell overlap discussed above. Detection of overshoot coverage or island coverage, where the UE experiences strong neighbour cell signal within the serving cell coverage area shall be effective after collecting sufficient statistics from

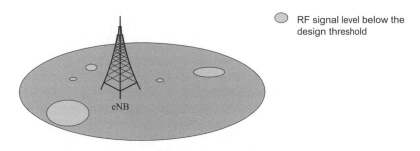

Figure 7.7 Weak coverage.

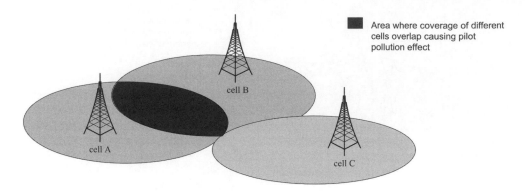

Figure 7.8 Pilot pollution.

relevant areas about serving cell level and interfering signals of neighbouring cells. An example of overshoot coverage is shown in Figure 7.9.

- **Uplink coverage verification** Uplink performance is also noticeable for the users and depends on many factors discussed in Chapter 8 in Holma et al. (2010). Poor uplink communication might impact user experience in terms of call setup failure, call drop, uplink voice quality, and so on. The coverage should be balanced between uplink and downlink connections. Ideally, the uplink coverage area should equal the downlink coverage area. However, base stations installed in field may experience unequal uplink and downlink link budgets. The downlink may have better coverage area than is there in the uplink direction. For the purpose of uplink coverage prediction, monitoring uplink transmit power levels at the UE is crucial.

Uplink coverage optimisation is not only about adapting the cellular coverage by changing the site configuration (antennas) but also about readjusting parameter settings in the way that they allow optimised usage of available uplink TX power in different environments. The performance of the uplink receiver can also be improved so that the uplink and downlink coverage

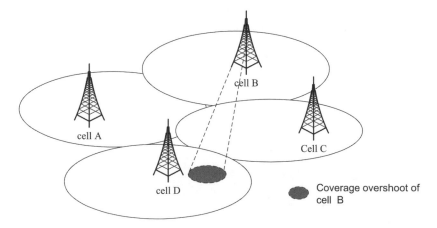

Figure 7.9 Overshoot/island coverage.

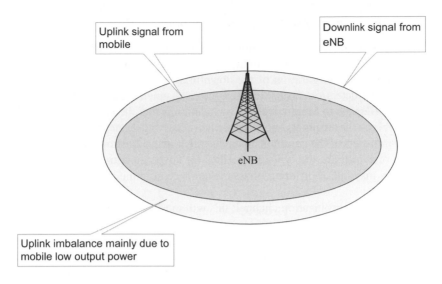

Figure 7.10 Imbalanced coverage between uplink and downlink.

areas become more equal. An example of imbalanced coverage between the uplink and downlink is shown in Figure 7.10.

7.4.3 Mobility Optimisation

The solution for drive test minimisation can also provide information for network to optimise mobility performance. By collection of radio measurements related to mobility events one can get an insight to places where performance degradation exists (e.g. cells with high handover failure rate). Besides measurement results of the source cell, the target cell and also other neighbour cells, provision of location information can be especially helpful to localise service disruption accurately and correlate the failure with specific network topology. Triggering such information by problems or failures in the mobility support allows identifying areas with the need to adapt the network parameters setting (e.g. to improve the handover success rate). In effect, operators can get essential ground on users' mobility patterns, use it to achieve terminal seamless mobility through handovers and hence cope with its mobility at the network in an optimised way. Normally the prerequisite for mobility optimisation is that the radio network coverage has been sufficiently verified so that the mobility issues are not mainly caused by poor radio coverage.

7.4.4 Capacity Optimisation

The main objective of the use case is to improve strategies in network capacity planning, verification and optimisation. Proper network capacity monitoring helps to, for example, determine placement of new cells and optimise other capacity related network parameters. For this purpose, average load on the network as a whole is important, but operator also needs to understand how potential traffic flow relates to the location of users. Thus, appropriate statistics should allow detection of locations where the traffic is unevenly distributed or the user throughput is low.

7.4.5 Parameterisation for Common Channels

Common channel parameters configuration may have impacts to system performance. Non-optimum parameter setting, in addition to the coverage issues, may affect the failure rate of the connection setup. The configuration of the downlink and uplink common channels shall be defined so that the detection probability of those channels match with the planned cell coverage. The difference to basic coverage verification is that the network may not be aware of potential problems the terminals encounters when accessing the network. Also the target is to optimise the parameters of the common channels rather than adjusting the antenna pattern, TX power, and so on which are the typical remedies for coverage optimisation.

The reception may fail at different phases when accessing the network:

- Reception of the synchronisation channel; this will most likely indicate lack of coverage in some areas of the network.
- Decoding of the PBCH (Physical Broadcast Channel); contains MIB (Master Information Block). MIB is transmitted in the first sub-frame in each frame. MIB periodicity is 40 ms, that is, the transmission is repeated four times allowing (soft) combining of the received signal to improve the decoding performance.
- Decoding of PDCCH (Physical Downlink Common Channel). The PDCCH carries information about the allocation of different channels on the PDSCH (Physical Downlink Shared Channel). This information points to the resources allocated for paging (indicated by P-RNTI, Paging Radio Network Temporary Identity), system information (S-RNTI) or dedicated connections (C-RNTI, Connection RNTI)
- Decoding of SIB1 (System Information Block 1). SIB1 is sent in a fixed location (fifth sub-frame) in every other frame (20 ms interval).
- Decoding of the other SIB information, SIB2 to SIB12. The transmission of these SIBs is scheduled dynamically on the PDSCH.

While in idle the UE shall also monitor the paging channel to be able to react to incoming calls. The paging reception failure may be due to following reasons:

- Decoding of PDCCH at the paging occasions fail.
- Decoding of the paging information fails from the allocated radio resources.

The problem is similar to detection of the broadcast channels with the exception that the network knows when to expect response from the terminal. However, the problem evaluation cannot be done on the cell basis as the paging will be sent throughout the location, routing or tracking area is aware of instantaneous paging decoding failures at the UE reception. Inter-cell coordination may therefore be needed.

7.4.6 QoS Verification

User experienced quality of service (QoS) is comprised of several aspects such as the data rate and delay, service response time, loss, interrupts, and so on. The network operators like to verify the service level that can be offered to the users in different parts of the network. The experienced QoS per user is affected not only by the cell coverage and interference conditions but also the choices the operator has done regarding the radio resource management methods and principles. In the packet type connections the scheduling principle in particular adopted by the operator will affect the QoS that the user may receive in different parts of the network.

Current network capabilities can provide much information, which gives an indication about the user experienced QoS. For example, the user data throughput can be traced in the network side and the instantaneous radio resource scheduling could be tracked. However, due to interests to be able the verify the service quality in the network, further elaboration is expected in 3GPP standardisation for MDT solutions beyond Release 10 to find out what would be the essential and most beneficial information that should be collected either from the network nodes and/or from the terminal for the QoS verification.

The fairness factor of the used scheduling method will affect the throughput that can be offered to a UE in particular radio conditions. Low fairness emphasises the cell throughput but cell edge users (with poor radio conditions) will not achieve satisfactory QoS. When the fairness increases the offered service quality will become more even. Therefore, there seem to be a lot of factors should be taken into account while assessing the optimality of the network operation.

For MDT, the user throughput could be mapped to the service level on the location basis. Hence, the network operator would for example be able to identify potential problematic areas regarding the provided service level that can be offered in different parts of the radio network.

7.5 Overall Architecture

Based on the requirements summarised in Section 7.3 parallel solutions have been developed for MDT:

- *Area based MDT* (Figure 7.11), where the UE measurement data is collected in a certain area where the area is determined by a set of UTRAN/E-UTRAN cells or a set of Location Area/ Routing Area/Tracking Area.
- *Subscription based MDT* (Figure 7.12), where the UE measurement data is collected for one specific subscriber or equipment.

Cell trace function is used for managing area based MDT. Activation of the MDT functionality occurs in the management system. The management system communicates with eNB/RNC directly in the cells where the MDT needs to be activated. The radio nodes are using the RRC protocol over the Uu interface to communicate further with UEs to start the measurement collections in the UE.

The subscriber and equipment trace function is used to manage the subscription based MDT. The MDT activation is started in this case also from the management system but as it targets one specific subscriber identified by IMSI or one specific user equipment identified by IMEI(SV), the activation goes first to the Home Subscriber Server (HSS) database. The HSS propagates the parameters of the session further to the radio network via the core network entities (MME/ SGSN). The main benefit of using the subscriber and equipment trace function for the subscription based MDT is that the mobility of the subscriber is considered that is, the MDT data is collected wherever the specified subscriber or user equipment is located.

In both area based MDT and subscription based MDT the data collection can be done while the UE is in RRC Connected mode or in RRC Idle mode: for Immediate MDT and Logged MDT respectively (see Section 7.7 for further details).

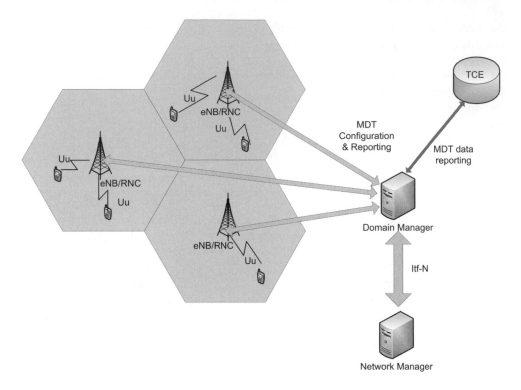

Figure 7.11   Architecture of area based MDT.

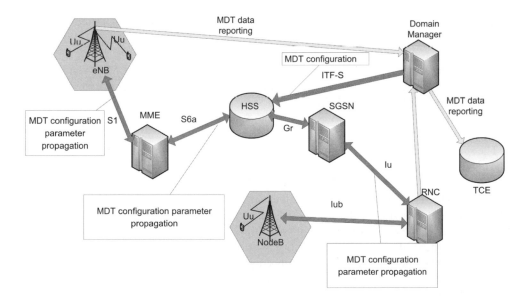

Figure 7.12   Architecture of subscription based MDT.

7.6 Managing MDT

7.6.1 Subscriber and Equipment Trace

The subscriber and equipment trace feature is used for configuring and reporting MDT, therefore it is essential to introduce shortly the feature in order to better understand the following sections.

Subscriber and equipment trace is used mainly for troubleshooting in a wireless network. With the feature it is possible to follow all activities made by a certain subscriber (IMSI) or equipment (IMEISV). Depending on the configuration of the feature different level of information can be collected.

There are two ways of using the subscriber and equipment trace feature:

1. Signalling based activation;
2. Management based activation.

In case of signalling based activation the configuration is always started in the core network and the core network initiates the Trace Session to the radio network during an activity for example, during a call setup. The signalling based activation ensures also the tracking of user mobility, that is, whenever the user is moving to a new area/cell the network follows the user.

In case of management based activation the configuration is always targeting one specific network element. Typically this is an RNC or eNB. In this scenario if the user is moving outside the area of the RNC/eNB that has been configured, the trace will not follow and the tracking is restricted only in the specified area.

The so-called cell trace feature is a specific subset of the management based trace. In this scenario the trace is activated to a specific cell or a group of cell and all users in the specified area is traced.

More details on the trace functionality can be found in 3GPP TS32.422 (2011).

7.6.2 MDT Configuration Parameters

In order to achieve a maximum flexibility in the MDT usage several configuration possibilities are defined. In this section the configuration parameters that is used for MDT is described.

7.6.2.1 List of Measurement

The parameter specifies what measurements should be collected by the UE in an immediate MDT session. In UMTS the following measurements can be configured:

- M1 for FDD mode: CPICH RSCP and CPICH Ec/No measurement by UE.
- M1 for 1.28 Mcps TDD mode: P-CCPCH RSCP and Timeslot ISCP measurement by UE.
- M2 SIR/SIRerror measurement taken by NodeB.

In LTE the following measurements can be collected:

- M1: RSRP and RSRQ by UE.
- M2: Power Headroom (PH) by UE.

7.6.2.2 Reporting Trigger

The parameter specifies the trigger when to make a measurement report in an immediate MDT session. Two triggers have been specified in Release 10 for immediate MDT, namely periodical and event based trigger (when serving cell becomes worse than threshold, that is, A2 event for LTE, 1F for UMTS FDD and 1L for UMTS TDD).

7.6.2.3 Report Interval

The parameter specifies the interval between the periodical measurements in an immediate MDT session. In UMTS the following report intervals are available: (the values are in milliseconds) 250, 500, 1000, 2000, 3000, 4000, 6000, 8000, 12 000, 16 000, 20 000, 24 000, 28 000, 32 000, 64 000.

In LTE the following report intervals are available: (the values are in milliseconds) 120, 240, 480, 640, 1024, 2048, 5120, 10 240, 60 000, 360 000, 720 000, 1 800 000, 3 600 000.

7.6.2.4 Report Amount

The parameter specifies the number of measurement reports that should be taken for periodical reporting in an immediate MDT session. The configurable report amounts for both LTE and UMTS are as follows: 1, 2, 4, 8, 16, 32, 64 and infinity.

7.6.2.5 Event Threshold

The parameter specifies the threshold for an event based reporting in an immediate MDT session (serving cell becomes worse than threshold). The available thresholds are 0–97 for RSRP and 0–34 for RSRQ.

7.6.2.6 Logging Interval

The parameter defines the periodicity for logging the measurements in logged MDT. The following logging intervals are available for both UMTS and LTE: 1.28, 2.56, 5.12, 10.24, 20.48, 30.72, 40.96, 61.44.

7.6.2.7 Logging Duration

This parameter determines how long the MDT configuration is valid at the UE. The timer starts when the MDT configuration is received. The timer is not stopped neither interrupted during UE state transitions or RAT/PLMN changes. The following logging durations are available for both UMTS and LTE (the values are in seconds): 600, 1200, 2400, 3600, 5400, 7200.

7.6.2.8 Area Scope

The parameter specifies geographical area in terms of cells or tracking/routing/location area where the MDT data collection should take place. In UMTS the area scope can be a list of cells or location/routing areas. In LTE the area scope can be given by list of cells or tracking areas. Maximum values that can be configured are: 32 cells or 8 tracking/location/routing areas. If the parameter is not present the MDT data should be collected in the whole RPLMN of the UE where the logged measurement configuration was received. The parameter can be present for both management based trace and for signalling based trace.

In case of signalling based trace the MDT is collected for a specific IMSI or IMEI when it is roaming in the specified area only.

7.6.2.9 Trace Reference

The parameter is composed as follows: MCC + MNC + Trace ID, where the Mobile Country Code and Mobile Network Code identifying one PLMN, which contains the Network Manager configuring MDT, and Trace ID is a 3 byte octet string. The Trace Reference parameter is globally unique and is used to uniquely identify the MDT/trace session.

7.6.2.10 Trace Recording Session Reference

The parameter identifies the trace recording session within a trace session. This parameter is a 2 byte octet string. The parameter together with the Trace Reference can be used to correlate the data gathered by one UE at the Trace Collection Entity: *TCE IP.*

This parameter specifies IP address of the Trace Collection Entity to which the Trace Records should be transferred. It can be signalled in a form of IPv4 or IPv6 address(es).

7.6.2.11 TCE ID

This is an identification parameter of the TCE that is sent to the UE instead of the IP address. There is a mapping table at eNB/RNC that provides the mapping between the TCE ID and TCE IP address.

7.6.3 Subscription Based MDT

The subscription based MDT targets one specific subscriber or user equipment for MDT data collection. The signalling based trace feature is used for managing the subscription based MDT, therefore the parameters used for subscriber and equipment trace is extended with the MDT parameters. The MDT activation starts from the management system. The management system directs the Trace Session activation to the HSS where the subscriber data is stored for the specific subscriber/equipment. As the MDT is targeted a specific subscriber user consent checking is required to ensure user privacy. If the specific subscriber/user has not given consent beforehand MDT cannot be started therefore HSS will reject the MDT activation, otherwise the MDT parameters are stored in the HSS. If the subscriber is not yet registered to the network, the Trace Session is not propagated further, that is, the Trace Session is kept within the HSS. Once the subscriber is registering to the network with the attach procedure, the Trace Session is propagated further in the network.

The detailed procedure in E-UTRAN is shown in Figure 7.13 which is adapted from 3GPP TS32.422 (2011).

The HSS activates the Trace Session to the MME by sending the trace control and configuration parameters together with the MDT parameters in the *S6a-UPDATE LOCATION ANSWER* message. MME stores the received trace control and configuration parameters together with the MDT parameters.

If the UE is already registered to the network, the Trace Session is activated to the MME by using the Insert Subscriber Data procedure.

When the UE gets connected again for example, initiates a session setup, the MME first checks if area scope is given for the Trace Session. If area scope is also given, it means that the

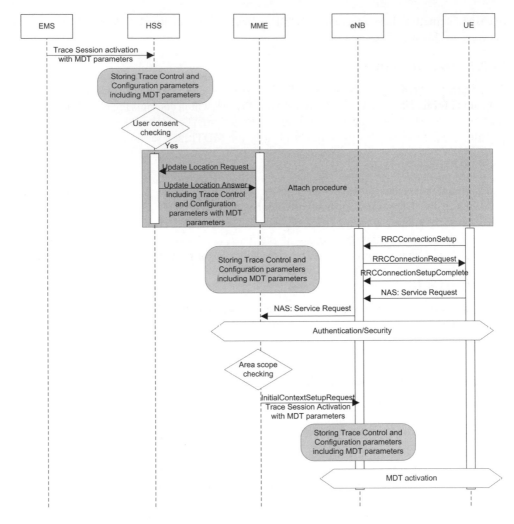

Figure 7.13 MDT activation in E-UTRAN (3GPP TS32.422, 2011). Adapted with permission from 3GPP.

MDT should be performed for the specific subscriber or equipment only if it is roaming in the specified area. If the subscriber or equipment is roaming in the specified area, the MME sends the trace control and configuration parameters with the MDT parameters to the eNB in the *S1-INITIAL CONTEXT SETUP REQUEST* message.

If the UE has already an active connection, the MME can propagate the MDT activation via the *S1-TRACE START* message.

At this point the eNB can configure the UE for MDT data collection according to the received MDT parameters. The method used for UE configuration depends on the type of MDT (immediate or logged). In case of immediate MDT the configuration is done by using existing RRC procedures and messages. In case of logged MDT the configuration is done by using a new RRC message the *LoggedMeasurementConfiguration* message (see Section 7.7 for further details).

It should be noted that for signalling based logged MDT it is not required to send to the UE the trace reference, trace recording session reference and TCE ID parameters. In signalling based MDT the trace session is activated to the eNB from the MME. Therefore when the UE sends the collected logs (as described in Section 7.6.5) the eNB will receive the MDT parameters, including the trace reference and the IP address of the Trace Collection Entity), from the MME.

In order to avoid unnecessary logged MDT configuration to the UE over the air interface in signalling based MDT, the eNB has to check whether it has received the logged availability indication beforehand when it received the trace session activation from the MME.

The procedure used in UTRAN is very similar to the E-UTRAN case and shown in Figure 7.14 (adapted from 3GPP TS32.422, 2011).

If the UE is not registered to the network the MDT activation is sent as part of the attach procedure similarly to the E-UTRAN case. The MDT parameters are sent to the SGSN in the

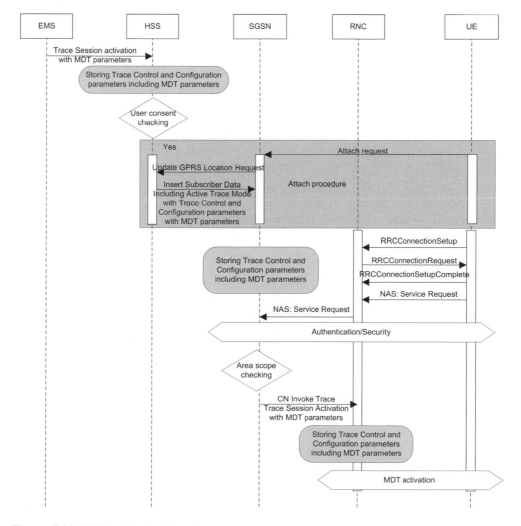

Figure 7.14 MDT activation in UTRAN via PS domain/(3GPP TS32.422, 2011). Adapted with permission from 3GPP.

MAP-INSERT SUBSCRIBER DATA message, which includes the *MAP ACTIVATE TRACE MODE* procedure.

If the UE is already registered to the network the MDT parameters can be sent to the SGSN in a standalone *MAP-INSERT SUBSCRIBER DATA* message. Once the SGSN received the MDT parameters it stores in its database similarly to the MME. If area scope is also given for the subscription based MDT, SGSN is responsible to check if the user is roaming in the specified area or not. The MDT activation is sent to the RNC over the Iu interface by using the *RANAP CN INVOKE TRACE* message.

When RNC receives the MDT parameters it is responsible for activating the MDT functionality to the UE. The method used towards the UE depends on the MDT mode (immediate MDT or logged MDT) (see Section 7.7 for further details).

It should be noted that the procedures describe above for UTRAN taken the example of a PS connection, but the MDT can also be activated through CS domain. In this case the MDT activation goes through the MSC Server and MSC Server is responsible to do the same functionality from MDT point of view as described here for SGSN.

7.6.3.1 Mobility Handling

In subscription based immediate MDT when the signalling based trace procedures are used the mobility of the user is followed. It means that whenever the user moves to a new location that is, to a new cell, the network is responsible for keeping the MDT context and forwarding the MDT parameters to the target node during handover. An X2 based handover scenario is shown in Figure 7.15.

The handover procedure is started with the handover preparation. During the handover preparation the source eNB sends an *X2 Handover Request* message to the target eNB. If the traget eNB belongs to the same PLMN the source eNB is responsible to forward the trace and MDT parameters to the target eNB. Therefore the trace session and the MDT can continue in the target side as well.

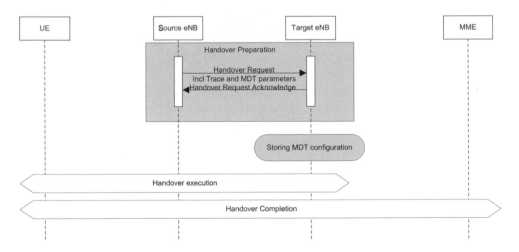

Figure 7.15   X2 based handover with MDT activation (3GPP TS23.401, 2011). Adapted with permission from 3GPP.

If the target eNB belongs to a different PLMN, that is, the handover is an inter-PLMN handover the source eNB shall not transfer the MDT context to avoid MDT data collection in a foreign PLMN. Even though the exact point of deactivation to the UE is up to RAN implementation, the MDT functionality should be not continued.

If there is no X2 interface deployed between the source eNB and the target eNB, the handover is made via the S1 interface.

If the handover is made via the S1 interface, the procedure is controlled by the MME. If area scope is also defined for the MDT, that is, the MDT should be carried out for a specific IMSI in a specific area, the source MME needs to check whether the target cell belongs to the specified area. If the target cell belongs to the specified area for MDT data collection the source MME is responsible for forwarding the MDT parameters to the target eNB. If the target eNB connects to another MME, the source MME sends the MDT parameters to the target eNB via the target MME as shown in Figure 7.16.

If there is an inter-PLMN handover via the S1 interface, the MME deactivates the MDT in the source eNB. In this case the source cells do not forward the MDT context to the target cell. Otherwise the removal of the MDT measurement configuration is left for RAN implementation. If the UE later on moves back to the original PLMN, again the MME is responsible to re-activate the MDT.

For logged MDT it is not required to pass the MDT context to the target eNB as the source eNB already configured the UE for logged MDT and in logged MDT the measurements are done only when the UE is in IDLE mode.

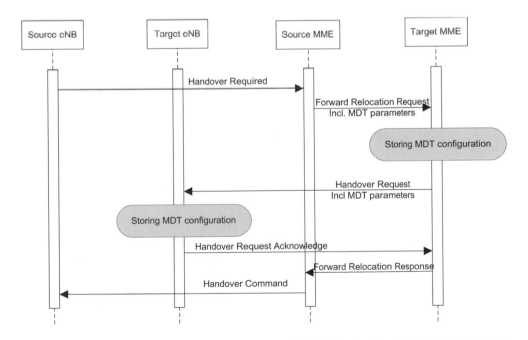

Figure 7.16 S1 based handover (with MME change) with MDT activation (3GPP TS23.401, 2011). Adapted with permission from 3GPP.

7.6.4 Area Based MDT

The area based MDT targets one specific area. The area is defined as either one or a list of UTRAN or E-UTRAN cell or tracking area, location area or routing area. For one MDT session maximum 32 cells or eight tracking/location/routing areas can be given. It should be noted that area scope can also be defined in subscription based MDT. In such scenario the MDT is carried out only the area defined by the area scope for the specified IMSI or IMEI number. Once the MDT is activated for an area, without selecting particular user (by IMSI or IMEI), all MDT capable UEs in the specified area are configured for MDT. In order to ensure user privacy it is required to collect user consent from the users before initiating the MDT in the area. Only such users can participate in the MDT data collection that has given consent beforehand. The mechanism to store and deliver the user consent to the radio network (eNB/RNC) is the same as in subscription based MDT and shown in Figure 7.17.

The method and the way in which users can give consent to the service providers are not determined and it is up to each service provider what kind of solution they deploy. The user consent is stored in the HSS database as part of the subscription data and propagated further to the MME/SGSN/MSC Server when the subscription data is delivered or updated. The core network elements (MME SGSN or MSC Server) check the roaming status of the user. If the user is an inbound roamer the user consent is not delivered to the eNB or RNC to exclude the inbound roamers from the MDT data collection. The main reason for excluding inbound roamers is because different privacy regulations are applicable in different countries and users usually gives consent only to their home operators. Therefore, only their home operators are allowed to collect the sensitive data. On the other hand, MDT data can contain sensitive data from the service provider's point of view, which results in that MDT data collected in one

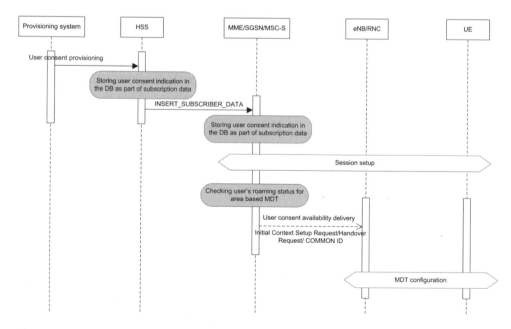

Figure 7.17 User consent delivery (3GPP TS32.422, 2011). Adapted with permission from 3GPP.

PLMN should be used and delivered only to a TCE that is located in the same PLMN. If the core network entity determines that the user is a home user and there is user consent available for the user, the user consent availability indicator is sent to the eNB/RNC, which is will use during the UE selection procedure.

Similarly to the subscription based MDT, the MDT activation is started from the management system. In area based MDT the activation is targeting the eNB or RNC which controls the specified cells. When eNB or RNC activates the MDT to the UE the same procedure is used here as for subscription based MDT:

- for immediate MDT the *RRCConnectionReconfiguration* message is used;
- for logged MDT the new *LoggedMeasurementConfiguration* message is used.

7.6.5 Supporting Functionality in the Management System

As already described, the subscriber and equipment trace feature is used for managing the MDT functionality. Therefore the supporting feature in Itf-N for MDT is the Trace IRP, which defines the operations used in Itf-N (see 3GPP TS32.441, 2011 and 3GPP TS32.421, 2011).

ActivateTraceJob operation is used to activate the trace functionality, MDT respectively. The operation is extended with the attributes required for MDT:

- *Area scope:* defining a specific area in terms of cell where the MDT should be performed.
- *Job type:* defines whether the requested job is for MDT only, trace only or a combined MDT and trace.
- All parameters as defined in Section 7.6.2.

DeactivateTraceJob operation is used to stop the MDT. The MDT session is identified by the trace reference, which is an already existing parameter of the operation.

The *ListTraceJob* operation is used to query the parameters of an existing trace session, MDT session respectively. This operation can provide the actual value of the parameters of an activated MDT session.

7.6.6 MDT Reporting

The overall MDT operation aims at delivering MDT reports to a data repository. All the collected MDT measurements are transferred to a specified file server which is called as Trace Collection Entity.

7.6.6.1 Immediate MDT Reporting

The procedure used for immediate MDT is shown in Figure 7.18.

Once the UE is configured with immediate MDT, the measurement report will be sent to RAN when the conditions of the configured reporting trigger are met. Radio interface reporting procedure is described in more detail in Section 7.2. Further, to complete the reporting operation data are processed in the eNB/RNC (e.g. time stamped, network measurements attached) and forwarded onwards to the TCE. The method when the MDT data is sent from the eNB/RNC to the TCE depends on each vendor. There are certain criteria defined by each

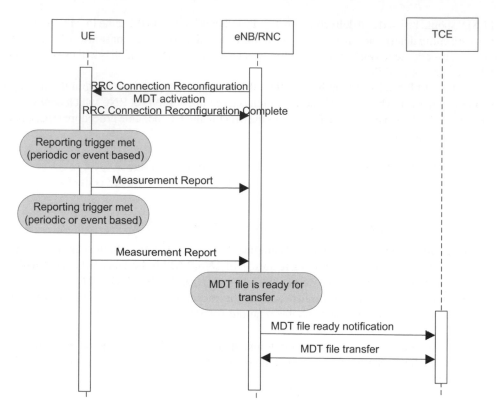

Figure 7.18 Immediate MDT configuration and reporting procedures (3GPP TS32.422, 2011). Adapted with permission from 3GPP.

vendor when the file is sent. The file format is defined as an XML file and when the file closing criteria is met the eNB/RNC sends a notification to the TCE informing that the file is ready. When the TCE receives the notification it can start downloading the file with a file transfer protocol.

7.6.6.2 Logged MDT Reporting

The reporting mechanism in logged MDT case is a little bit different from the immediate MDT. The procedure is shown in Figure 7.19.

Section 7.23 gives a more detailed description of the mechanism in context of radio procedures. However, the reporting is likewise finalised by RAN node. Similar to the immediate MDT reporting the criteria when the MDT log is sent to the TCE by the eNB/RNC is not specified by 3GPP and it is up to each vendor. When the criteria for closing the file are fulfilled a notification is sent to the TCE and TCE can download the MDT log by using a file transfer protocol.

The main difference in logged MDT reports delivery compared to immediate MDT is that before the MDT data is transferred to TCE, the eNB/RNC is not aware of trace relevant configuration (MDT context is released after RRC connection release). Hence, trace

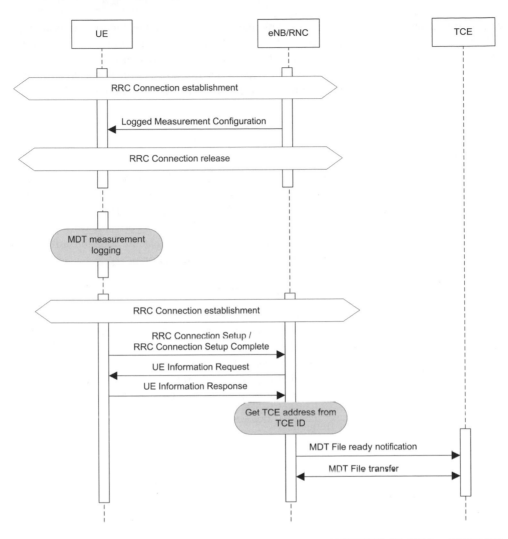

Figure 7.19 Logged MDT configuration and reporting procedures (3GPP TS32.422, 2011 and TS37.320, 2011 and TS37.320, 2011). Adapted with permission from 3GPP.

relevant parameters (trace reference, trace recording session, TCE ID), memorised by the UE, are reported back to the network and attached by the UE to MDT log. Trace reference and trace recording session reference is used to correlate the data at the TCE that belongs to the same trace (MDT) session. TCE ID is used to get the correct IP address of the TCE, where the data should be finally passed.

7.7 MDT Radio Interface Procedures

There are two MDT operational modes how the measurement collection can be done, namely Immediate MDT and Logged MDT. The former one uses the normal RRC measurement configuration and reporting principles with the exception that the reported measurement data may include the UE location information at the time of measurement results are obtained. The latter one is a new mechanism for idle state UEs (and PCH state UEs in UTRA) to store the radio measurement results to be reported later when the connection is set up next time. Details of the configuration, measurements and reporting for the two modes are described in subsequent chapters.

7.7.1 Immediate MDT

The immediate MDT mode is applicable for both UTRA and E-UTRA. However, the UTRA specifications already have all the required features before the Release 10. Hence, also legacy terminals can be used for MDT measurements in UTRAN. In this section the immediate MDT function is described using E-UTRA as an example. There are corresponding signalling procedures and messages also for UTRA with the exception that the location information request is a message separate from the measurement configuration.

7.7.1.1 Immediate MDT Configuration in E-UTRAN

Figure 7.20 illustrates Immediate MDT configuration in E-UTRAN. Immediate MDT uses existing RRC signalling procedures and messages to configure the MDT reporting and sending the measurement reports to RAN. In E-UTRAN eNB translates the MDT configuration parameters received from Network Management system (see Section 7.6.2) into the *RRCConnectionReconfiguration* and sends the message to the UE to initiate MDT measurements and reporting. The configuration message includes amongst others following information:

- Measurement object: carrier frequency.
- Reporting configuration.

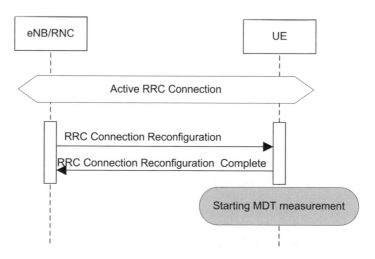

Figure 7.20   Immediate MDT configuration in E-UTRAN (3GPP TS32.422, 2011 and TS37.320, 2011). Adapted with permission from 3GPP.

The MDT reporting configuration specifies on what conditions the report shall be sent and what will be the measured parameters to be included in the report:

- Reporting trigger: periodical and A2 event (serving cell becomes worse than threshold).
- Trigger quantity; either RSRP or RSRQ.
- Reporting quantity; either same as trigger quantity or both.
- Max reported cells excluding the serving cell; up to eight cells.
- Reporting interval; 120 ms, 240 ms, . . ., 1 h.
- Reporting amount (for periodical reporting); 1, 2, 4, . . ., 64, infinity.
- Request to include available location information.

The periodical reporting can also be started with A2 event that is, as event based periodical reporting. This will be useful when collecting measurement results in the problematic coverage areas. 'Reporting amount' can be used to define the measurement 'window' for the reported data.

7.7.1.2 Immediate MDT Reporting

Once the UE is configured with immediate MDT, the measurement report will be sent when the conditions of the configured reporting trigger are met: either expiration of reporting period or A2 event (1F in UMTS FDD or 1L in UMTS TDD). Figure 7.21 illustrates reporting based on an A2 event.

The UE shall send *MeasurementReport* message to RAN including latest measurement results for the parameters defined in the configuration message, as shown in Figure 7.22.

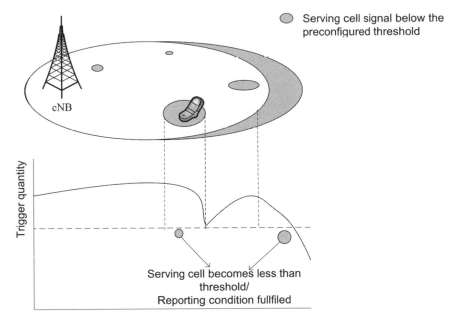

Figure 7.21 A2 reporting trigger.

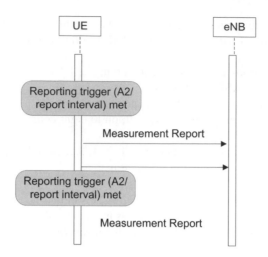

Figure 7.22 Immediate MDT reporting procedures in E-UTRAN.

The UE includes the location information to the report whenever it is available at the time when the measurement was done. The location information can be either cell ID, RF fingerprint (neighbour cell measurement results) or accurate location information from the stand-alone positioning function. The accurate location information will come from the GNSS (Global Navigation Satellite System) circuitry. GNSS can be for example, GPS, Galileo, Glonass or alike.

The report is sent on Signalling Radio Bearer 2, SRB2 (SRB4 in UTRA). The usage of SRB2 has been extended to be used for MDT reporting in addition to NAS messages. The priority of the SRB2 is lower than with SRB1 which carries the RRC messages as well as NAS messages prior to SRB2 establishment.

7.7.2 Logged MDT

Logged MDT will utilise non-active UEs for the measurement data collection. In E-UTRA terminals may be configured to log measurement data while in idle state. In UTRA logging can be done also when the UE is in Cell_PCH and URA_PCH states. Logged MDT allows less frequent reporting and extended terminal usage. The 'cost' for this mode comes with the memory requirement (storage for logged data) and somewhat increased complexity due to new procedures.

The overall principle of the logged MDT operation in E-UTRAN is illustrated in Figure 7.23. The configuration will be done when the UE is in connected state. The measurement logging is started when the UE goes to idle state. The logged data will be reported with the request from the network when the connection has been set up again. Note that the UE may be moving around the network (not illustrated here) and the reporting can take place in cell of the same PLMN.

7.7.2.1 Logged MDT Configuration

Logged MDT configuration will use the new signalling procedure illustrated in Figure 7.24. The selected UE in connected mode will be configured to collect idle mode (or Cell/URA-PCH mode in UTRAN) measurement results. When the UE goes to those modes, it starts logging of the measurement results according to logged MDT configuration.

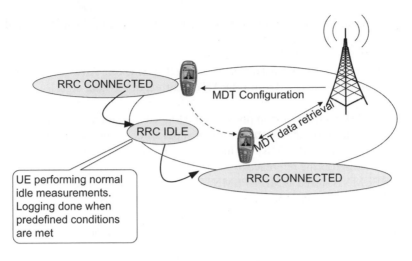

Figure 7.23 Logged MDT operation in E-UTRAN.

The RRC configuration is created based on the MDT configuration parameters received from Network Management system (see Section 7.6.2) and sent on *LoggedMeasurement-Configuration* (LOGGED MEASURMENTS CONFIGURATION in UTRA) message. The following parameters are included in the configuration to define the measurements and logging criteria:

- Absolute time information; used as the reference for the relative time stamp.
- Logging trigger; currently only periodical logging.

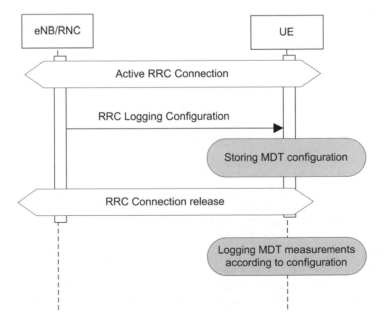

Figure 7.24 Logged MDT configuration in (E-)UTRAN.

- Logging duration; 10 min, 20 min, 40 min, 1 h, 1 h 30 min or 2 h. Counting started from the reception of the logged MDT configuration.
- Logging interval; 1.28 s, 2.56 s, ..., 61.44 s.
- Area configuration (optional); list of cell IDs (max 32), Tracking Area (TA, in E-UTRAN) or Location/Routing Area (LA/RA in UTRAN) identifications (max eight).

The logged MDT configuration implies that the logged measurement results will be tagged with the available location information without separate configuration for it.

The registered PLMN (RPLMN), at the time when the logged MDT configuration was received, shall be used as the reference to determine when the logging and reporting is allowed or not. This RPLMN is stored as MDT-PLMN. The MDT log shall not be collected nor reported (if requested) if current RPLMN is not equal to the MDT-PLMN.

In addition, the logged MDT configuration includes following parameters in order to ensure that the collected MDT data is transferred to the correct file server:

- Trace reference.
- Trace session recording reference.
- TCE ID.

Based on these parameters the RAN node shall be able to determine correct TCE address where the MDT report should be forwarded.

Trace reference and trace recording session reference is used to correlate the data at the TCE that belongs to the same trace (MDT) session. TCE ID is used to get the correct IP address of the TCE. The IP address can be obtained in two ways: either by using a DNS resolution or based on a table that is configured to the eNB/RNC by the OAM. The TCE ID is a 1-byte-long integer number and that is used towards the UE instead of the IP address to avoid any security issues, by sending an IP address of a network node to the UE. Further details on Trace relevant parameters are discussed in Section 7.6.

7.7.2.2 Logged MDT Effectiveness

Preliminary condition for Logged MDT configuration performance is UE state. Measurement logging shall take place when the UE is in idle state (UTRA and E-UTRA) or in Cell_PCH or URA-PCH (in UTRAN).

Furthermore when started, UE shall continue MDT logging until:

- the timer for the logging duration expires;
- the storage of the MDT log becomes full (the storage size shall be at least 64 kB);
- UE returns to RRC connected state (Cell_DCH or Cell_FACH in UTRA);
- UE moves to other RAT (Radio Access Technology) or roams to another network (PLMN).

In the last two cases, MDT logging will resume when the UE returns to idle state (or Cell/URA_PCH in UTRA) or comes back to the RAT that configured the MDT logging provided that the first two conditions have not been reached, and, provided that the PLMN is the same as the configuring PLMN. The logging configuration shall survive through multiple state transitions.

At the UE side at one time only one logged MDT configuration (i.e. one MDT context) is stored, therefore if the UE receives a new logged MDT configuration, the previously collected MDT logs are cleared together with the MDT configuration and the new MDT configuration is stored.

7.7.2.3 Logged MDT Data

Once the logging trigger is met (i.e. periodical timer expires) the UE stores MDT record, which consists of the available cell measurement results attached with a time stamp and available accurate positioning information.

Radio Measurements
Every MDT log contains measurement results for the serving cell (RSRP/RSRQ in E-UTRA, CPICH RSCP and Ec/N0 in UTRA). In addition, the cell measurements may include available neighbour cell results (see Section 7.7.4.1). The neighbour cell results can be used also as the RF fingerprint for the location estimate if the accurate location is not available. The neighbour cell results are not duplicated for the UE positioning.

Time Stamp
To be able to correlate the reported measurement results with other information, the logged results will be associated with timing information. In the configuration message the network sends an absolute time value which the UE stores and includes later in the report. When the configuration is received the UE starts a timer. Each time a measurement sample is logged, the timer value will be attached to the log entry. Using the absolute time information and the relative time per log entry, the absolute time for each log entries can derived.

Location Information
At the time when the measurement result is logged, the UE checks if new accurate location information is available from the stand alone positioning function (GNSS). If not, the UE location will be estimated from the neighbour cell results (RF fingerprinting) or from the cell ID. Accurate location information is added on best effort basis that is, if the GNSS is activated by another function or application.

7.7.2.4 Logged MDT Reporting

When the UE returns to connected state in the same RAT as when it received the MDT configuration (i.e. if the configuration is made in LTE and the UE connects to the network in LTE) it indicates the log availability with a one bit indicator. The indication can be sent in following messages:

UTRA:

- RRC CONNECTION SETUP: message.
- At SRNC relocation in UTRAN MOBILITY INFORMATION CONFIRM: message.
- CELL/URA UPDATE: messages.
- In MEASUREMENT REPORT: message in case of seamless transition between CELL_PCH and CELL_FACH states.

E-UTRA:

- *RRCConnectionSetupComplete* message.
- *RRCConnectionReestablishmentComplete* message.
- After handover in *RRCConnectionReconfigurationComplete* message, possibly updated.

When the log availability indication is received the RAN node may request, at a suitable time, the UE to report the log by using the *UEInformationRequest* message. The log will be sent to RAN in *UEInformationResponse* message. This message is sent using SRB2 (Signalling Radio Bearer) in E-UTRA or SRB4 in UTRA.

The log indication is sent also in case where the timer for the logging duration has not expired. The network may request a partial log from the UE and the UE will continue MDT logging until the stopping criterion is met. The UE can delete the log data that is already reported. In case the RAN node does not request the log, the UE should maintain non-retrieved logged measurements for 48 hours. The logged MDT reporting procedure is shown in Figure 7.25.

If the log size is large, UE shall split the log in segments and send the log in several messages as shown in Figure 7.26. This will improve the probability for the whole report to be successfully sent over the radio interface as the SRB do not have re-transmission possibility on the RRC layer. The losses of the report will affect separately each segment. The MDT report (and the segment) will be self-explanatory which means that the network can utilise the reported data on report/segment basis. It implies every MDT report conveyed on *UEInformationResponse* message, apart from logged MDT data, needs to contain trace reference, trace recording

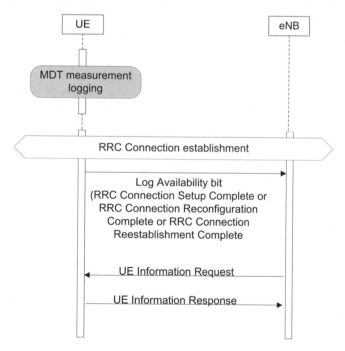

Figure 7.25 Logged MDT reporting procedure in E-UTRAN (3GPP TS32.422, 2011 and TS37.320, 2011). Adapted with permission from 3GPP.

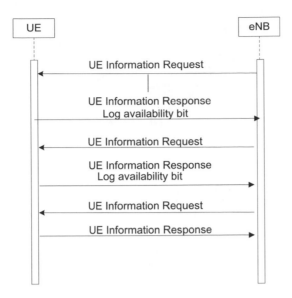

Figure 7.26 Logged MDT reporting for log that does not match one RRC message.

session reference and TCE ID to provide network with necessary information on final log destination (see Section 7.6.6).

7.7.3 RLF Reporting

In addition to immediate MDT and logged MDT, in E-UTRAN UE may send also a separate report from the connection failure situations. The failure can be either radio link failure (RLF) or handover failure (HOF). The information reported from the connection failures will provide useful information especially about problematic areas in the network coverage. The failure reporting will be effective by limiting the reporting only to the situations where UE experiences problems.

The information sent with the RLF report can be used for coverage optimisation (as MDT data) but it will provide relevant information to identify issues with the mobility behaviour (support for SON Mobility Robustness Optimisation).

From Figure 7.27 it can be seen that once the eNB receives RLF report, it may forward the data directly to TCE within the framework of MDT feature or it may use it to exchange information along to neighbouring eNBs for SON purposes.

There is no separate configuration for the RLF reporting but the UEs supporting this feature can collect the specified RLF data and send it to the network if requested. There will be an indication about the availability of RLF data and the network may retrieve that with specific signalling procedure.

The UE shall collect following information at the failure and preceding that:

- Cell measurement results at the time of the failure; include serving cell as well as available neighbour cell results and cell IDs (physical cell identification).

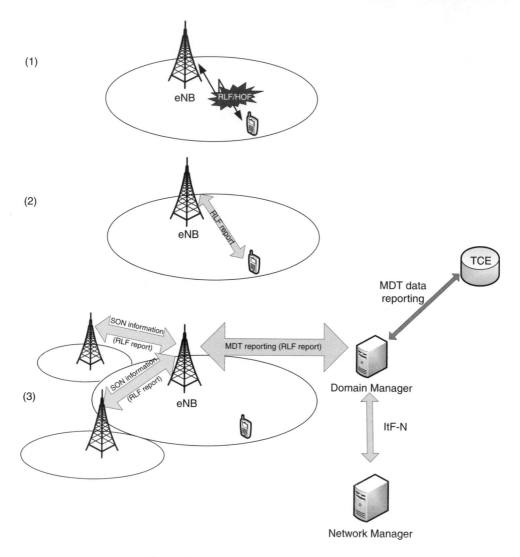

Figure 7.27 RLF reporting in SON and MDT.

- Available accurate location information (from stand alone GNSS function) at the time of the failure.
- Cell identification (ID) for following cells, either global cell ID, E-CGI, or physical cell ID, PCI:
  - Cell ID where the failure happened, either E-CGI (RLF) or PCI (HOF).
  - Cell ID of the cell where the last HO command was received (E-CGI).
  - Cell ID of the cell to which UE attempted establish the connection after the failure (E-CGI).
- Time between last successful HO initialisation and the failure.
- Indicator on connection failure type, either RLF or HOF.

The same way as with logged MDT reporting, the UE shall first send an indication about the existence of the RLF report. This will be sent after successful connection re-establishment (*RRCConnectionReestablishmentComplete* message) or, if the re-establishment fails or is rejected, after successful connection set up. In the latter case the messages where the indication can be sent can be either *RRCConnectionSetupComplete*, or, after handover, *RRCConnectionReconfigurationComplete*.

The RLF report is sent using the *UEInformationRequest/Reponse* signalling procedure, the same way as the logged MDT report. UE shall also check whether the requesting PLMN is the same as the RPLMN at the time when the failure happened. The UE shall not sent the report if the PLMN IDs are not the same.

The RLF report shall be kept stored for 48 hours if it has not been sent to the network. It shall also survive state transitions and RAT changes. The RLF information can be deleted when it has been reported or when the 48 hour timer expires. The RLF reporting procedure is shown in Figure 7.28.

7.7.4 Measurement Parameters

This section discusses the radio measurements that can be provided to the MDT data collection both from the UE and RAN nodes. The reporting mechanisms for the UE are specified in the RRC specification 3GPP TS25.331 (2011) and 3GPP TS36.331 (2011) for UTRA and E-UTRA, respectively.

7.7.4.1 Downlink Measurements

Immediate MDT
For immediate MDT the UE shall measure following parameters:

UTRAN
- CPICH RSCP and CPICH Ec/No measurement by UE.
- P-CCPCH RSCP and Timeslot ISCP for UTRA 1.28 TDD.

CPICH RSCP, Common PIlot CHannel Received Signal Code Power: The parameter represents the signal level of the primary CPICH that the UE can receive from the measured cell.

CPICH Ec/No, chip energy per noise spectral density: This is indicating the signal quality of the received signal. 'No' represents the total received power including thermal noise and interference.

P-CCPCH RSCP, Primary Common Control Physical Channel (carrying BCH transport channel) RSCP.

Timeslot ISCP, Interference Signal Code Power: Given only interference power is received, the RRC filtered mean power of the received signal after despreading to the code and combining. Equivalent to the RSCP value but now only interference is received instead of signal.

E-UTRAN
- RSRP and RSRQ measurement by UE.

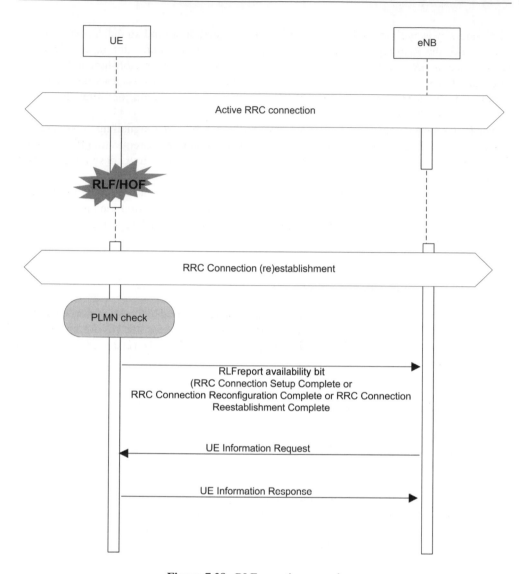

Figure 7.28 RLF reporting procedure.

RSRP, Reference Signal Received Power: This is the measured power of the reference signals distributed in time and frequency domain in the downlink symbols.

RSRQ, Reference Signal Received Quality: The RSRQ is the RSRP divided by the carrier RSSI (Received Signal Strength Indication) which comprises the linear average of the total received power (in [W]) observed only in OFDM symbols containing reference symbols.

During active connection UE may send also Power Headroom Report (PHR) which is used for uplink scheduling and power control purposes. Power headroom is the margin between the used TX power and the maximum allowed UL TX power in the cell. PHR is not on via MAC signalling and it is not repeated as an RRC measurement report for MDT purposes.

The PHR has been considered to be useful when analysing the root cause for potential coverage problem and, whether the problem is on the UL connection or in DL. In E-UTRAN the PHR is sent on the MAC layer and it will not be repeated in the RRC messages. It is up to eNB implementation to forward the information to relevant function inside the eNB and to include that later in the reports forwarded to TCE.

In E-UTRAN the power headroom is measured using the actual TX power (for all allocated physical resource blocks, PRBs) but it is seems beneficial to normalise the power headroom per PRB as it will more indicative for the actual power reserve UE may have.

Because the PHR is not sent on RRC and there is no MDT extensions specified for MAC reporting, the location information, if wanted for MDT data collection, can be obtained by activating DL MDT measurements and reporting, for example periodical DL measurement reporting.

Other downlink measurements that could be used for MDT:

CQI (Channel Quality Indication) information will be sent for DL scheduling purposes but it will provide also useful information how the UE 'sees' the DL connection quality and what would be the appropriate signal format for the DL transmission. The CQI can be either wideband CQI or sub-band CQI. In the latter case UE may report n sub-bands with best CQI value, or the network may request CQIs for particular sub-bands.

There are several ways how the PHR and CQI can be utilised for MDT data analysis but the methods and algorithms are left for network implementation.

Logged MDT:

The UE shall log measurement results that will be available from normal cell (re-)selection process. The logged MDT measurement parameters are not configured. The measurement quantity consists of both RSRP and RSRQ for EUTRA, both RSCP and Ec/No for UTRA, P-CCPCH RSCP for UTRA 1.28 TDD, and Rxlev (received signal level) for GERAN (as inter-RAT neighbour cell measurement).

The number of neighbouring cells to be logged is limited by a fixed upper limit per frequency for each category. The maximum number of the neighbouring cells, that UE should log the measurement results is defined in Table 7.1:

Table 7.1 Upper limits for logged measurement results of neighbouring cells

Category	Number of cells
Intra-frequency neighbouring cells	6
Inter-frequency neighbouring cells	3
GERAN neighbouring cells	3
UTRAN (non-serving) neighbouring cells	3
E-UTRAN (non-serving) neighbouring cells	3
CDMA2000 (if serving is E-UTRA) neighbouring cells	3

The measurement reports for neighbour cells consist of:

- Physical cell identity of the logged cell;
- Carrier frequency;
- RSRP and RSRQ for EUTRA;
- RSCP and Ec/No for UTRA;
- P-CCPCH RSCP for UTRA 1.28 TDD; and
- Rxlev for GERAN;
- Pilot Pn Phase and Pilot Strength for CDMA2000.

7.7.4.2 Uplink Measurements

RRC controlled radio measurements for MDT allow to utilise uplink measurements results in addition to the measurement reports coming from the UE. The data collection may combine both UL and DL measurement results in order to make the data analysis more reliable while trying to identify potential radio network problems or optimising the network performance.

UL measurements have been considered useful for MDT purposes:

- UTRAN: UL SIR and SIR error measured by NodeB.
- E-UTRAN: SINR and Uplink signal strength measured by eNB.

There will be freedom left for vendor specific implementation how the UL measurements will be used and not all will be necessarily standardised. MDT data record, however, can contain vendor specific measurements. This possibility facilitates any available uplink measurement delivery from eNB/RNC to MDT server.

7.7.5 Location Information

The UE location information will be included in the MDT reports in best-effort manner. There are three ways how the UE location may be determined:

1. Cell ID.
2. RF fingerprint (using neighbour cell measurements).
3. Using stand-alone GNSS (Global Navigation Satellite System) positioning function.

The cell ID will be available always known with immediate reporting (= serving) cell and the serving cell ID will be stored with logged MDT. The neighbour cell measurement results will be included in the report/log whenever available. As the UE may not be able to detect all the time neighbour cell signals (e.g. when close to the serving cell base station), the positioning with RF fingerprint is not guaranteed in all locations.

The accurate location (physical coordinates) from the GNSS circuitry can be obtained if the satellite positioning has been activated by another function or application. Even though active, the GNSS may not be able to provide position information continuously when poor signals received from the satellites. This is particularly the case when indoors or is some locations in urban areas.

The MDT measurements and GNSS function are normally independent functions and therefore also the timing, when the measurement results and GNSS coordinates become available, can be random. The MDT function at the UE shall tag the measured result with the latest location information. A certain location sample shall be used only once in the MDT report/log. The next accurate location information shall be included first when new coordinates are provided by the GNSS.

To check the validity of the location information, the MDT report shall include also the GNSS time information which the network may use to assess what is the time difference between the time instants of the measurements and the positioning. If the time offset is large the actual location of the UE at the time of measurements may deviate too much to be valid for MDT positioning. The resulted error depends on the UE velocity and it will be up to the network to decide whether to use the location information or not.

7.8 Conclusion

This chapter covered drive test method evolution developed in 3GPP. Release 10 defined MDT functionality which provides automated solutions collecting feedback from the end user and hence reducing operator's dependence on manual drive tests. By using MDT, the operator can explore the network from subscriber perspective and get the overall picture of the actual network performance with the minimised operational effort and cost. The network performance can be assessed via key radio measurements collected by ordinary UEs in the field. For collecting the measurements UEs in connected are involved, as well as new capability to log idle mode UE measurements is supported. Specific Trace procedures enhanced for MDT purposes allow complementing the measurement by eNB/RNC input and delivering the MDT reports to a centralised entity in Network Management, where data can be post processed to identify the problem areas with respect to radio coverage.

Since it is especially valuable in drive tests to identify possible 'pain points' for radio coverage, most essential enhancement to the existing measurement procedures is the addition of location information. It easily allows creating comprehensible reports on network performance per physical location, and hence localising potential network faults or coverage issues. In other respects, MDT utilises standard radio measurements and procedures to a great extent. The integration of MDT functionality with the Realease 10 terminals is therefore expected to be smooth, making it possbile to have a large volume of MDT capable terminals in the market. While devices support the MDT operation, a clear increase in statistcs volume and greater insight to network conditions than in case of manual data collection is provided. Then the MDT technique will able to really benefit the operators to drive down costs of network audits.

Although, the focus of Release 10 solutions for MDT was on coverage optimisation, no doubt the evolution of MDT will continue beyond the steps decribed in this chapter. Already during the course of Release 10 work, several improvements were found. Thus, the foundation of MDT is being further developed in Release 11. Besides adding further enhancements for coverage optimisation on the top of the Release 10 MDT feature, a number of enhancements mainly related to QoS verification use case will be considered. It is obvious that possible techniques raised in 3GPP for MDT evolution will include further improvements with the ultimate goal to enhance end user performance and improve overall network operability.

In summary, MDT represents important improvement in general network optimisation process. The use of MDT fits well with the use of SON, as the collected MDT measurements may be used for a wide scope of different use cases aiming at a consistently high level of network coverage, capacity and quality.

References

3GPP TR36.805 (2009) Technical Specification Group Radio Access Network, *Evolved Universal Terrestrial Radio Access (E-UTRA); Study on Minimization of Drive-Tests in Nest Generation Networks*, ver.9.0.0., Release 9, 5 January 2010. Available from http://www.3gpp.org/ftp/Specs/archive/36_series/36.805/36805-900.zip [accessed 30 June 2011].

3GPP TR32.827 (2010) Technical Specification Group System Architecture, *Telecommunication Management; Integration of Device Management Information with Itf-N*, ver.10.1.0, Release 10, 22 June 2010. Available from http://www.3gpp.org/ftp/Specs/archive/32_series/32.827/32827-a10.zip [accessed 30 June 2011].

3GPP TS25.331 (2011) Technical Specification Group Radio Access Network, *Radio Resource Control (RRC); Protocol specification*, ver.10.3.1., Release 10, 11 April 2011. Available from http://www.3gpp.org/ftp/Specs/ archive/25_series/25.331/25331-a31.zip [accessed 30 June 2011].

3GPP TS32.401 (2011) Technical Specification Group Services and System Aspects, *General Packet Radio Service (GPRS) enhancements for Evolved Universal Terrestrial Radio Access Network (E-UTRAN) access*, ver.10.4.0, Release 10, 21 June 2011. Available from http://www.3gpp.org/ftp/Specs/latest/Rel-10/23_series/23401-a40.zip [accessed 30 June 2011].

3GPP TS32.422 (2011) Technical Specification Group System Architecture, *Telecommunication Management; Subscriber and Equipment Trace; Trace Control and Configuration Management*, ver.10.2.0., Release 10, 30 December 2011. Available from http://www.3gpp.org/ftp/Specs/archive/32_series/32.422/32422-a20.zip [accessed 30 June 2011].

3GPP TS32.421 (2011) Technical Specification Group System Architecture, *Telecommunication Management; Subscriber and Equipment Trace; Trace Concepts and Requirements*, ver.10.2.0., Release 10, 01 April 2011. Available from http://www.3gpp.org/ftp/Specs/archive/32_series/32.421/32421-a20.zip [accessed 30 June 2011].

3GPP TS32.441 (2011) Technical Specification Group System Architecture, *Telecommunication management; Trace Management Integration Reference Point (IRP): Requirements*, ver.10.1.0., Release 10, 4 April 2011. Available from http://www.3gpp.org/ftp/Specs/archive/32_series/32.441/32441-a10.zip [accessed 30 June 2011].

3GPP TS36.331 (2011) *Evolved Universal Terrestrial Radio Access (E-UTRA); Radio Resource Control (RRC); Protocol Specification*, ver.10.1.0.

3GPP TS37.320 (2011) Technical Specification Group Radio Access Network, *Universal Terrestrial Radio Access (UTRA) and Evolved Universal Terrestrial Radio Access (E-UTRA); Radio Measurement Collection for Minimization of Drive Tests (MDT)*, Overall description; Stage 2, ver.10.1.0., Release 10, 5 April 2011. Available from http://www.3gpp.org/ftp/Specs/archive/37_series/37.320/37320-a10.zip [accessed 30 June 2011].

Holma, H., Toskala A. (2010) *WCDMA for UMTS: HSPA Evolution and LTE*, 5th edn, John Wiley & Sons, Ltd., Chichester.

Lehser (2008) *Review of Project-SON achievements and agree proposals for ongoing work in Release 8/9/10*, NGMN SON Workshop, October.

8

SON for Core Networks

Anssi Juppi, Risto Kauppinen and Seppo Hämäläinen

8.1 Introduction

Self-Organising Network principles are feasible also for automation of Core Networks management. Core Networks-related 3GPP standardised SON use cases and functionalities are related to Evolved Packet System (EPS). Some of the EPS related use cases are such that they span over radio and core networks and require functionality from both networks. An example is the ANR function that is based on detection of new neighbours based on UE reports in the base stations. However, in order to create actual X2 links between base stations, EPS support is needed to discover IP addresses of new neighbours. In addition, SON functionality that is not based on standard signalling or interfaces can be identified.

As for any SON, it can be concluded that SON use cases for core networks are not limited to LTE networks only, but many of them can be already taken into use in 2G/3G networks.

8.2 SON for Packet Core Networks

8.2.1 Packet Core Element Auto-Configuration

Auto-configuration is particularly useful for configuration of Enhanced Packet Core. In LTE networks tens of thousands of connections between eNodeBs and MMEs may occur. In addition, configuration of those connections may change frequently. Thus, their manual configuration would be very laborious and auto-configuration therefore particularly beneficial. The LTE and 2G/3G networks are different in a sense that in 2G/3G networks the number of interfaces is much smaller due to controllers that act as mediators between the base stations and the core network elements. Therefore part of the configuration challenge that was earlier with configuration of BSC/RNC has moved to configuration of MME.

LTE Self-Organising Networks (SON): Network Management Automation for Operational Efficiency, First Edition.
Edited by Seppo Hämäläinen, Henning Sanneck and Cinzia Sartori.
© 2012 John Wiley & Sons, Ltd. Published 2012 by John Wiley & Sons, Ltd.

Configuration of Packet Core network elements is a rather complex task due to many interfaces with neighbouring network elements. Thus, in the following chapter, a couple of use case examples are given on how the Packet Core element configuration tasks can be partially automated.

8.2.1.1   Auto-Configuration with S1 Setup

3GPP has specified auto-configuration of S1 setup (3GPP TS36.413, 2011) which is a part of self-configuration process for a new base station. Auto-configuration with S1 setup reduces manual work related to configuration of eNodeBs and MMEs. Especially the MME config-uration effort is reduced as the number of eNodeBs is typically large in a commercial LTE network.

The auto-configuration process is initiated by the eNodeB that sends its eNodeB identifier and the Tracking Areas (TAs) it supports to the MME. By doing so, MME does not need to be configured manually with a large matrix of different eNodeBs and TAs, but instead the MME learns eNodeB and TA configuration during runtime from eNodeB. eNodeB identifier and TA configuration for eNodeB is created as a part of network planning.

The MME replies to the eNodeB with its relative capacity and Globally Unique MME Identifiers (GUMMEIs). Through these steps, eNodeB avoids configuration of the GUMMEI(s) that are needed to be configured only in the MME. Auto-configuration of S1 setup is described in Figure 8.1.

8.2.1.2   Automatic Software Update and Upgrade

The operational aspects of core network elements can be improved through automation of element software updates and upgrades. The automatic software update and upgrade function provides software download checks from the external software repository where the new software locates. The software repository server is typically a part of the Domain Management System. The trigger to start software upgrade is usually manual but subsequent actions are automated. Once a new software build becomes available, the element performs pre-condition checks to ensure that the new software can be uploaded. Once the actual software has

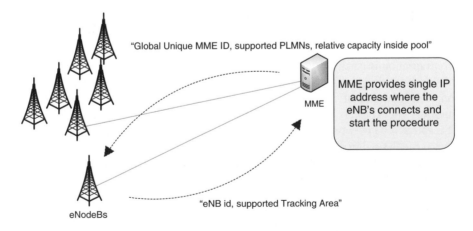

Figure 8.1   Auto-configuration of S1 setup.

been upgraded to each of the element units, a post-condition check is made to ensure that element is fully operational in the network environment where it was operating previously. As the core elements are complex, benefit from this is the reduced amount of manual work and therefore reduced number of human errors that are typical when core network are elements configured manually.

8.2.2 Automatic Neighbour Relation

MME provides the 3GPP specified support (3GPP TS23.401, 2011) for automatic configuration of neighbours and X2 interfaces between neighbouring cells in the radio network. As wrong neighbour relation configurations cause handover failures or dropped calls, relations between neighbour cells need to be carefully known and planned. On the other hand the number of neighbour relations is large in a typical LTE network. Therefore manual neighbour planning is time consuming and expensive.

ANR is an eNodeB function with MME extension to relay information. The MMEs role consists of relaying the neighbouring cell relay requests from the source eNodeB to the target eNodeB so that X2 connection and eNodeB neighbouring cell relation table can be automatically updated in both eNodeBs. This MME supporting role is kown as Transport Network Layer (TNL) address discovery of candidate eNodeB via S1 interface or Configuration Transfer Procedure.

Automatic Neighbour Relation function is described in Section 4.2.3. ANR provides operator significant savings as it can be used to make automatic configuration of X2 connections for each eNodeB and neighbour relation table.

8.2.2.1 Configuration Transfer Procedure

The Configuration Transfer Procedure is a 3GPP standardised functionality (3GPP TS23.401, 2011) whose purpose is to enable transfer of MME transparent information between applications belonging to two eNodeBs. Configuration transfer provides a generic mechanism for the exchange information at any time via S1 interface and the Core Network. Configuration Transfer for ANR is used only for intra-RAT E-UTRAN information exchange. The information is transferred between two eNodeBs via the MME core network node(s). In order to make the information transparent for the Core Network, the information is included in an E-UTRAN transparent container that includes source and target eNodeB addresses, which allows the Core Network nodes to route the messages.

Each Configuration Transfer message carrying the E-UTRAN transparent container is routed and relayed independently by the core network node(s). Any relation between messages is transparent for the MME. A request/response exchange between applications, for example SON applications, is routed and relayed as two independent messages by the MME.

When an eNodeB detects a new neighbour to which the X2 link needs to be created, it initiates the configuration transfer procedure by sending a Configuration Transfer message to the MME that forwards message transparently to the target eNodeB. The message contains a request for X2 TNL IP address, the global cell identifier of the target eNodeB and transport address of the source eNodeB. Once receiving the configuration transfer message from the MME, the target eNodeB transfers back its transport layer address. After receiving X2 transport layer address from the target eNodeB, the source eNodeB initiates the X2 TNL establishment by sending an X2 Setup request message to the target eNodeB. As a result from this, X2

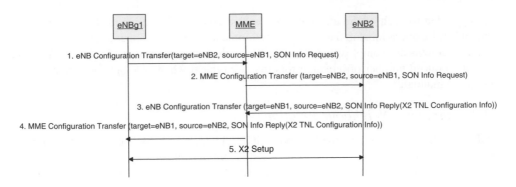

Figure 8.2 Configuration transfer procedure signalling.

interface is established between two eNodeBs. The signalling diagram for the Configuration Transfer Procedure is shown in Figure 8.2.

8.2.3 S1 Flex (MME Pooling)

MME pooling with MME offload can be seen as a SON use case for packet core when SON scope is widely thought. MME pooling involves intelligence in the communications between the eNodeB and MME in terms of reporting and selecting the less loaded MME from a pool of MMEs. Additional extension of MME pooling is smart offload of the subscribers without user session breaks using 3GPP specified signalling procedures. Therefore MME pooling facilitates self-healing and self-optimisation of MME network elements. 3GPP has specified MME pooling related 3GPP functionality in several specifications, see (3GPP TS36.413, 2011; 3GPP TS23.401, 2011 and 3GPP TS23.060, 2011).

The S1-Flex, also known as Multipoint S1 or MME Pooling, enables the connection from each eNodeBs to all EPC nodes within a pool area. This feature removes the strict hierarchy in which one eNodeB is allowed to connect to one MME only. Instead, a group of MMEs serve an MME pool area, consisting of several eNodeBs. An MME pool area is defined as an area within which a UE may be served without the need to change the serving MME (3GPP TS23.401, 2011). MME pool areas are a collection of complete tracking areas and they may overlap each other. Pool areas must be configured to E-UTRAN and this information is also needed in the Domain Name System (DNS). Figure 8.3 gives an example of S1-Flex implementation.

S1 Flex provides geographical resilience: using multiple MMEs within a single pool area increases the service availability, as other MMEs can provide services for an eNodeB that primary MME is not available. It also enlarges the served area compared to the service area of only one MME.

S1 Flex provides MME capacity optimisation in terms of balancing load between multiple MMEs. The eNodeB selects the MME from the pool area on the basis of MME relative available capacity. Relative capacity can be modified by a specific algorithm based on load measurements. Actual load balancing can be made based on the MME configuration update procedure for changing relative capacity informed to the eNodeBs or alternatively measuring the MME elements load status in the defined pool and using the MME offloading functionality in a controlled manner to move subscribers to other less loaded network elements in the pool.

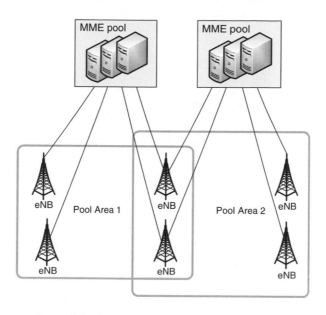

Figure 8.3 An example of S1-Flex implementation.

Thus, there is some load distribution of the signalling traffic between MME nodes. These measurements and offloading functionality are automated to facilitate MME load balancing in pool area.

S1 Flex optimisation has an additional effect on the network signalling load. With the S1-Flex feature, there is reduced signalling load on the S6a interface between MME and Home Subscriber Server (HSS), as subscribers do not move as much between MMEs.

S1-Flex related MME offloading, which is used for load balancing between the MMEs, enables operator to move subscribers between the MMEs that are in the same MME pool without breaking the ongoing user sessions. As seen in Figure 8.4, MME offloading starts with setting MME relative capacity first to zero to inform the eNodeBs that new UEs should not be allocated to the MME that is being offloaded. A command is then performed at the MME which initiates the MME to start releasing the current S1 connections on active subscribers with a cause code 'Load balancing TAU required' sent to UE. This triggers a Tracking Area Update (TAU) procedure at the UE, and as a consequence the UEs are moved to a new MME. Moving of inactive subscribers is done in such a way that they are first changed to active state using the paging procedure and after that applying 'Load balancing TAU required' procedure. MME offloading can be partial for load balancing reasons, also the duration in which the offloading takes place is configurable. For load balancing purposes this procedure or part of the procedure is performed to achieve optimal load in each of the MMEs in pool.

8.2.4 Signalling Optimisation

In following a couple of different use cases demonstrating how the LTE specification provides capabilities for signalling optimisation in the network are given.

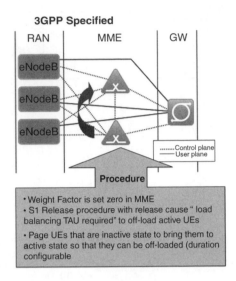

3GPP Specified

Figure 8.4   Automated MME Off load Procedure.

8.2.4.1   Paging Optimisation

A key idea in paging optimisation is that MME pages UEs step-wise to minimise number of paging messages. In the LTE environment the paging load may grow as the number of eNodeBs is high. The ways to minimise number of paging messages sent out from MME can be optimised with a step-wise paging algorithm. The SON functionality can select most suitable paging strategy for certain type of terminals from the following list of example algorithms:

- Page first the eNodeB the paged UE was last connected to.
- Maintain a history list of latest eNodeBs the paged UE has been connected to and page those first.
- Page the latest Tracking Area (TA) first where the UE was located.
- Maintain a history list of latest TAs and page those first.
- Only in case there is no response the entire registered TA list is paged.

In addition to the number of eNodeBs, other things that impact paging load are, for example; TA size, the way the registered TA list is built and UE type. The paging optimisation greatly reduces signalling load in the network. Reduced signalling in the network also leads to increased terminal battery life.

8.2.4.2   Tracking Area List for Idle Mode Signalling

The smart provisioning of Tracking Area list for UEs reduces remarkably UE idle mode signalling in the network. The UE can move within given tracking areas in a UE specific TA list without causing any additional signalling towards the network. The UE needs to do signalling to the network only when leaving tracking areas specified for the UE. On the other hand, the size of TA list should be small enough to reduce paging signalling load.

Therefore optimisation of TA list benefits operator through optimised idle mode signalling load and paging load. Reduced signalling in the network also leads to increased terminal battery life. Optimisation of tracking area lists is made by creating TA lists dynamically for the UE, based on UE behaviour instead of static configuration. Tracking area lists are automatically created based on TAU history for the UE. Shorter or longer TA lists can be created based on UE's mobility pattern. The UE will be provided with a TA List which best matches its frequently visited tracking areas.

8.2.5 Latency Optimisation

When a UE registers to the evolved packet system network, it must be assigned with the appropriate Serving Gateways (S-GW) and Packet data network Gateways (P-GW). The gateway selection is done automatically in the MME with the help of Domain Name System. DNS framework is a hierarchical system composed of DNS servers, which form a distributed database for domain names and IP addresses. DNS clients, also known as DNS resolvers, perform queries to that database for finding out the mapping between a particular domain name and an IP address.

Latency optimisation as a SON use case is mainly related to the selection of S-GW. In addition to S-GW, also P-GW needs to be allocated for a user. Today P-GW selection is not regarded as SON functionality, but as regular telecom functionality. Bases of P-GW selection is obtained list of P-GW names and their priorities based on Access Point Name (APN) which is given by HSS. In 3GPP there are standard initiatives to enhance optimisation of P-GW selection (3GPP TS23.401, 2011 and 3GPP TS29.303, 2011).

Gateway selection allows the selection of separate or geographically collocated Serving Gateways: the automatic S- and P-GW selection from nearest core sites, or from the site where MME locates creates savings in the transport costs and optimised latency for latency sensitive features like VoIP.

In S-GW selection, the initial DNS query is performed for obtaining a list of S-GW identities (names) and their priorities that match with the given Tracking Area Identifier (TAI). Thus the S-GW selection is based on subscriber location, but not the subscriber identity itself. MME chooses the target S-GW based on the priority information. It is preferable to choose the S-GW closest to the user location, but other locations are also possible, depending on the DNS configuration. A second DNS query is performed thereafter to obtain the valid IP address for the desired S-GW. If several IP addresses are returned by the DNS, the MME uses a round-robin mechanism to balance the load between interfaces of the selected S-GW. After the S-GW has been chosen for the initial default bearer, it is used also for additional default and dedicated bearers, meaning that there is only one S-GW for an attached LTE subscriber.

In the gateway selection the automation can be done on configuring the best TAI and S-GW latency information mapping automatically to DNS based on the topology information managed by the network management system. In this case it is assumed that the topologically closest TAI and S-GW combination provides the best latency. Further the S-GW could measure the latency towards eNodeB and the related Tracking Areas IDs and provide the list of mapping towards management system. Management system would then utilise the mapping list to configure appropriate settings to the DNS referring the current operational capabilities of the network.

8.2.6   Fast Gateway Convergence with Bidirectional Forward Detection

For self-healing purposes, a network element application needs to know if a link is failing in order to be able to trigger switchover to a redundant link. Bidirectional Forward Detection (BFD) can be used for automated fast recovery of network element functionality. BFD is an IETF specified (Katz, 2010) network protocol used to detect faults between two forwarding engines connected by a link. It provides fast low-overhead detection of faults even on physical media that do not support failure detection of any kind, such as Ethernet, virtual circuits, tunnels, Multiprotocol Label Switching (MPLS) Label Switched Paths and routers not supporting Open Shortest Path First (OSPF) fast hello.

In an SON context the BFD is used to detect link failures and when the failure is detected then the network element application triggers automatic switchover in the element, that is, re-routing is made. Identification of link failures takes place over one or more configured LAN/Router hops (sending and receiving of L2 layer 'hello' messages).

BFD support would be recommended to be a separate process to minimise impact on other functionality of element like for example, routing.

8.2.7   Dynamic IP Pool Allocation

While Packet Core Networks use IPv4 addresses, the shortage of IPv4 addresses may become an issue in the network. Automating IP address pool allocation and making it dynamic will bring benefits to Connection Service Provider (CSP) as it enables full control of the whole IP address pool and fast address allocation to S-GWs when more IP addresses are needed. Full control of the IP address pool gives also possibility to re-allocate IP address of an unavailable S-GW to other S-GWs automatically when S-GW outage is detected.

Dynamic IP pool allocation use case can be implemented with a Process Automation Solution, as shown in Figure 8.5. The process automation solution is responsible for collecting

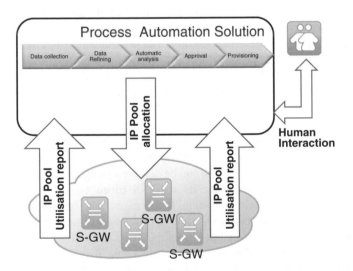

Figure 8.5   Dynamic IP Pool Allocation Solution with Process Automation.

reports, monitoring S-GWs availability, allocating IP address pools to S-GWs and keeping track of IP address space usage. The collected reports can include information on IP pool utilisation or IP address utilisation levels at S-GWs Based on set threshold levels, the Process Automation Solution can allocate more IP addresses to a network element. Counter wise, IP address pools can be de-allocated from S-GWs if there's less traffic.

8.2.8 Energy Saving

Dynamic power consumption introduces an intelligent way to lower the network element power consumption. The main idea is to lower the network element's processing power during the time traffic is lower. This could be done by shutting down certain CPUs or lowering the clock frequency, for example.

Energy saving features raise the energy efficiency and helps drive environmentally sustainable business. Lower energy consumption also brings cost benefits on the necessary cooling systems.

8.3 SON for Voice Core Networks

8.3.1 Voice Over IP Quality Monitoring and Management

In LTE networks, VoIP is planned to be used widely for voice services. Quality of Service (QoS) in VoIP calls becomes an important issue while moving from traditional Circuit Switched calls to VoIP. One mechanism to ensure QoS and to make VoIP successful in LTE networks is to monitor end-to-end voice quality and conduct actions when finding problems with it.

Quality of service data can be collected from various different levels in LTE networks; on a connection, element or domain level. However, to create a complete picture of the end-to-end voice quality, it is important to collect data from end user voice clients. 3GPP has defined standard mechanisms, for this purpose based on quality of experience reports (3GPP TS26.114, 2010). The quality of experience reports from VoIP clients include for instance frame rate and jitter measurements.

Figure 8.6 shows the VoIP quality monitoring and management solution based on process automation. The VoIP quality monitoring and management use case can be split into two steps: the first step is to collect data and create a holistic, end-to-end quality view of VoIP calls in the network. When these reports are combined with the location information that is, origin and destination location, it is possible to analyse and create a view of VoIP call quality in different parts of the network. The location information can be determined for example from the IP addresses of the calling parties. The analysis may include also data collection from network elements or OSS. Based on the analysis it is possible to drill-down to the poor voice quality areas and find places for improvement.

The second phase in VoIP quality monitoring and management is about changing the network resource usage. These actions may be for instance proactive care actions or automated route re-selections to improve voice quality.

VoIP call quality monitoring data can be also used for integrating to customer care process and fault localisation, but this is outside of SON scope.

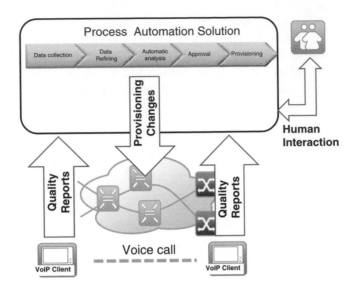

Figure 8.6 VoIP Quality Monitoring and Management Solution with Process Automation.

8.3.2 Resource Optimisation in Voice Core Network

LTE voice core networks are based on either IP Multimedia Subsystem (IMS) or with Circuit Switched (CS) Fallback (CSFB), which basically means connecting voice calls over existing 2G or 3G network, see Section 2.1.7. In IMS, control plane (control) and user plane (voice traffic) are separated and this is also the case in today's CS core networks. The focus of resource optimisation is on efficient usage of control plane resources and load balancing between the control plane elements based on pre-defined parameters. Control plane elements can be pooled and a control element (node selection) function is used to choose the most suitable element from the pool to serve a user. In IMS network, the control element is Serving-Call Session Control Function (S-CSCF) and the selection function resides in Proxy-CSCF (P-CSCF). When a subscriber registers to IMS based Voice Core network, P-CSCF is responsible for selecting an S-CSCF to serve the subscriber. In CS core network (Rel-4 based), Mobile Switching Centre (MSC) server is the control element and the network NAS Node Selection Function (NNSF) may reside either in Media Gateway (MGW) or in RAN nodes (3GPP TS23.236, 2009). When a subscriber registers to CS based Voice Core network, NNSF function will select MSC server to serve the subscriber.

To determine the right balance between the control plane elements, it's necessary to collect data of the utilisation level of the control plane elements or of the resources these elements use. This data may be for instance Visitor Location Register (VLR) utilisation level, CPU load or link capacity utilisation. From the collected data it is possible to analyse and decide the right balance of allocating new subscribers to pooled control plane elements. For this, the SON function shall adjust the load balancing parameters at the NNSF. By implementing this automated process, it is possible to achieve evenly balanced resource utilisation between the control plane elements in the pool and avoid congestion situations in different parts of voice core network. Figure 8.7 shows an example for balancing resources of a pooled MSC server.

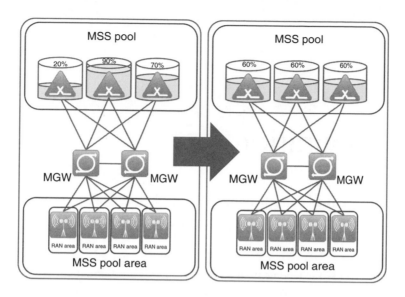

Figure 8.7 Balancing pooled MSC Server for optimal resource usage.

References

3GPP TS23.236 (2009) Technical Specification Group System Architecture, *Intra-domain connection of Radio Access Network (RAN) nodes to multiple Core Network (CN) nodes*, ver.9.0.0., Release 9, 14 December 2009. Available from http://www.3gpp.org/ftp/Specs/archive/23_series/23.236/23236-900.zip [accessed 30 June 2011].

3GPP TS26.114 (2010) Technical Specification Group System Architecture *IP Multimedia Subsystem (IMS); Multimedia telephony; Media handling and interaction*, ver.9.3.0., Release 9, 18 June 2010. Available from http://www.3gpp.org/ftp/Specs/archive/26_series/26.114/26114-930.zip [accessed 30 June 2011].

3GPP TS23.060 (2011) Technical Specification Group System Architecture *General Packet Radio Service (GPRS); Service description; Stage 2*, ver.9.8.0., Release 9, 28 March, 2011. Available from http://www.3gpp.org/ftp/Specs/archive/23_series/23.060/23060-980.zip [accessed 30 June 2011].

3GPP TS23.401 (2011) Technical Specification Group System Architecture, *General Packet Radio Service (GPRS) enhancements for Evolved Universal Terrestrial Radio Access Network (E-UTRAN) access*, ver.9.8.0., Release 9, 28 March 2011. Available from http://www.3gpp.org/ftp/Specs/archive/23_series/23.401/23401-980.zip [accessed 30 June 2011].

3GPP TS36.413 (2011) Technical Specification Group Radio Access Network, *Evolved Universal Terrestrial Radio Access Network (E-UTRAN); S1 Application Protocol (S1AP)*, ver.9.6.1., Release 9, 6 May 2011. Available from http://www.3gpp.org/ftp/Specs/archive/36_series/36.413/36413-961.zip [accessed 30 June 2011].

3GPP TS29.303 (2011) Technical Specification Group Core Network and Terminals *Domain Name System Procedures; Stage 3*, ver.9.4.0., Release 9, 15 June 2011. Available from http://www.3gpp.org/ftp/Specs/archive/29_series/29.303/29303-940.zip [accessed 30 June 2011].

Katz, D. (2010) Bidirectional Forwarding Detection (BFD), IETF RFC 5880.

9

SON Operation

Tobias Bandh, Haitao Tang, Henning Sanneck and Christoph Schmelz

In traditional network management (e.g. cf. Laiho *et al.*, 2006), network operation and optimisation are coordinated through interactions of the human network operators between themselves and with the system. The network operation personnel need to have a good overview over ongoing configuration, optimisation and troubleshooting processes within the network. This overview combined with their operational experience and their ability to assess the importance of a detected situation (and to trigger the required corresponding actions) allows avoiding the effects of negative interaction between management actions and thus to reach a stable operational state.

Additionally, network management has the requirement to be highly efficient. Detected failures shall obviously be handled with minimal delay. Moreover, the optimisation of the network has almost equal importance. In case management actions do not interact with each other they can be executed in parallel, while potentially conflicting actions have to be handled in a way that the conflict is either avoided (e.g. by serialisation), or does not have a negative impact on the system performance.

In the operation of a SON-enabled system, both challenges need to be addressed at the level of the automated SON function execution (rather than the human level), see Section 3.2 and (SOCRATES, 2008). In the previous chapters which have targeted the individual SON areas of self-configuration, -optimisation and -healing, some of the potentially negative interactions between SON functions have been mentioned. Also, those challenges have been identified and explored by earlier research: (Doettling and Viering, 2009; Baliosian *et al.*, 2006; Strassner *et al.*, 2006; Calder *et al.*, 2002; Cleaveland and Smolka, 1996). Therefore, it is now widely accepted that automatic methods are required to govern the execution of individual SON functions in a wider system-level context (SOCRATES, 2011; Schmelz *et al.*, 2011; Barth and Kuehn, 2010; Sanneck *et al.*, 2010). These methods aim at reaching a similar or even higher degree of stability, efficiency and thus, in the end, performance when compared with the traditional network management approach.

LTE Self-Organising Networks (SON): Network Management Automation for Operational Efficiency, First Edition.
Edited by Seppo Hämäläinen, Henning Sanneck and Cinzia Sartori.
© 2012 John Wiley & Sons, Ltd. Published 2012 by John Wiley & Sons, Ltd.

The first section of this chapter gives an introduction of the nature of SON function interactions. Examples of some specific SON function interactions are given. In Section 9.2, actual methods for handling SON interactions are introduced. The focus is on how interactions between SON functions can be handled (even with a large number of deployed interacting or even conflicting SON functions) while still maintaining the overall efficiency of the system.

Besides the handling of SON function interactions, there are a number of other aspects with regards to the operation of SON functions (e.g. tracking of SON function actions (reporting), SON function target setting derived from operator business policies). These aspects have basically not been addressed in the SON research community so far, however.

The actual integration of SON functions into the operational processes (cf. Section 3.2.1) is highly operator-specific and therefore considered to be beyond the scope of the chapter.

9.1 SON Function Interactions

To describe SON function interactions, the following definitions are required:

- **SON Use Case:** A description of 'who' can do 'what' with a conceptual self-organising functionality inside a system.
- **SON Function:** A specific realisation of a conceptual self-organising functionality described by a SON use case.
- **SON Function Instance:** A specific run-time process that instantiates a SON function in a given system (e.g. network) environment. The SON function instance contains a monitoring capability (that may be always active when the SON system is in operation), a data collection, data analysis, decision making and action execution capability.

A SON interaction refers to the situation that one SON function instance (for example, Instance A) impacts another SON function instance (Instance B) such that the originally intended operation from Instance B is affected and the related system performance may be different from what has been intended to be (operator policy). Such an interaction can only appear when there is a dependency between the two SON function instances with regards to spatial and temporal dimensions (cf. Section 3.4).

SON function interactions may decrease the network system performance. We describe such 'negative interactions' as 'SON function conflicts'. The impact of a SON function conflict caused by Instance A can be:

- to modify the change that has been made by an action of Instance B;
- to spuriously trigger an Instance B, since the measurements that trigger Instance B have been influenced;
- to spuriously influence the decision making of a running Instance B due to affected measurements and/or input parameters;
- to block the execution of Instance B;
- to cancel an action that Instance B intends to take;
- to delete or diminish the performance gain achievable by Instance B.

Some of these impacts may be specific to an implementation of the SON system and functions, for example, that a SON function instance is blocked or deleted if corresponding lock or deletion mechanisms are implemented.

9.1.1 Spatial Characteristic

Every SON function instance is triggered for a single or a set of NE(s), for example, a group of cells within a network. The changes performed by the function instance directly affect those cells. Those cells form the *function area* of a SON function instance. Apart from the function area, there are often additional network elements or cells respectively which are important for the SON functions and the function coordination in particular. Either because they provide directly input to the SON function or because the function will compute erroneous results if those network elements or cells are changed during its run-time. The combination of function area and those additional NEs or cells is called the *impact area* of a SON function instance (cf. the discussion of the 'spatial scope' of a SON function in Section 3.4).

Figure 9.1 shows an example for conflicts due to the spatial characteristics of SON functions. SON Function Instance A modifies NE I. These changes impact NEs II and III, amongst others. SON Function Instance B modifies NE II, thereby causing a conflict. Furthermore NE IV is modified through SON Function C which causes a conflict at NE III.

9.1.2 Temporal Characteristic

A second SON function characteristic that is important for the function coordination is the *impact time*. A SON function instance is incorporated into coordination decisions from the

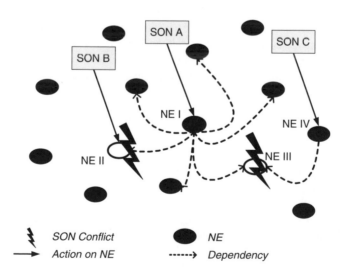

Figure 9.1 SON function instance conflict: spatial characteristic.

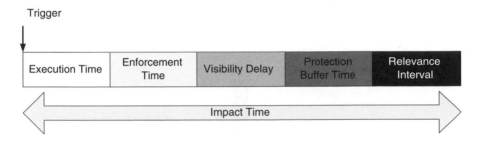

Figure 9.2 Temporal Characteristic: Impact Time Components.

moment it has been started until the end of its impact time (during this interval the instance is considered to be 'active'). After this point in time a function instance does not have any effects on subsequently requested SON function instances anymore. Note that for the discussion of spatial characteristics above, an overlapping impact time of all the mentioned function instances has been assumed.

The impact time (cf. the discussion of timing requirements in Section 3.4.) consists of several components. The length of each of the components is dependent on the particular SON function. In the following, an introduction and reasoning for the impact time components is given (see also Figure 9.2):

- **Execution Time:** For a given SON function, this component is defined as the period after the SON function instance has been started, during which the SON algorithm is running to compute new parameter values.
- **Enforcement Time:** It can take some time until a requested configuration change is actually enforced on the target NE.
- **Visibility Delay:** In case changes performed by a SON function instance become visible in measurement values or KPIs, some time is required until all changes are fully reflected in these measurement values or KPIs. This delay between an applied configuration change and its visibility in the KPIs is called 'visibility delay'.
- **Protection Buffer Time:** Some SON functions use PM data which is collected over a certain time interval as input. The protection buffer time is used to assure that the used input reflects the actual state of the network and all previously performed changes are already visible, that is, the visibility delay of the previous SON function has expired. Dependent on the requirements of the specific SON function, the protection buffer time is adapted. Note that the length of the protection buffer time interval depends on the SON function the currently executed SON function interacts with.
- **Relevance Interval:** Some changes performed by SON function instances are relevant for subsequent SON function instances over a rather long time. This fact can be expressed with the relevance interval, especially in order to prevent oscillating reconfigurations. Therefore the relevance interval has a particular importance for the coordination of subsequent instances of the same SON function.

Note that the different time components are SON function-specific and also context specific, for instance, depending on the NE type. For some SON functions these intervals

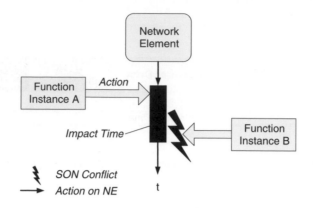

Figure 9.3 SON function instance conflict: temporal characteristic (cf. Bandh *et al.*, 2011).

may also be zero or infinite. For example, the change of a PCI is instantly visible (visibility delay is zero).

Furthermore, it is important to understand that a SON function interaction is always directional, for example, from A to B as introduced above. An interaction from A to B does not imply that there also exists an interaction from B to A. In cases there exists an interaction also from B to A this interaction is to be treated separately.

In Figure 9.3 the temporal characteristic of a SON function conflict is shown. SON Function B modifies the same (set of) parameters as SON Function A during the impact time of A (Conflict Type A2 according to Table 9.1, see next Section).

9.1.3 Categories of SON Conflicts

Table 9.1 lists the different categories of SON conflicts with corresponding examples:

9.1.3.1 Example for Conflict Type A2

Two SON Function Instances F1 and F2 may compete over their shared parameters (see Figure 9.4), where Function Instance F1 requests to increase the handover parameter 'Qoffset', while Function Instance F2 requests to decrease this parameter after a short time interval. Without coordination this could lead to an oscillation effect for this parameter which is obviously not the desired behaviour.

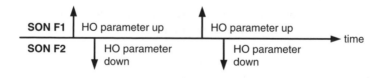

Figure 9.4 Parameter competition between different functions instances.

Table 9.1 SON Conflict Categories

Cat.	Conflict	Description	Example
A	Configuration Conflict	A conflict induced by changes to a parameter.	
A1	Input Parameter Conflict	SON functions that deal with parameters whose values are dependent on the values of other parameters can suffer from an input parameter conflict, as they rely on the stability of the values of read parameters to compute the new configuration settings. In case the values of these read parameters change during computation the new configuration settings may be wrong.	PCI allocation (cf. Section 4.2.1): the PCI configurations of neighbour cells are gathered; based on this information the PCI of the target cell is allocated. If a neighbour PCI is changed during the runtime of the PCI allocation function instance the resulting configuration can be erroneous.
A2	Output parameter conflict	When a SON function instance tries to modify a configuration parameter within the impact time of another function instance (i.e. that configuration has been/is manipulated by that other function instance), an output parameter conflict occurs.	See example following table
B	Measurement Conflict	A conflict induced by the change to a measurement	
B1	Measurement Conflict	Measurements may on the one hand trigger SON function algorithm execution, and on the other hand serve as input to the SON function algorithm to evaluate the current state of the system and deduce appropriate configurations/actions to reach the intended target. Parameter changes made to the system hence influence the measurements, but it takes some time until these changes show effect in the measurements. This delay may lead to conflicts in case a SON function is either triggered or computes new parameter values based on actually 'outdated' measurements.	See example following table

Table 9.1 (*continued*)

Cat.	Conflict	Description	Example
C	Characteristics Conflict	A conflict induced by the change of a cell's characteristics	
		A characteristic is defined as a property of a cell which is difficult to measure or even not measurable at all, as for example the cell size.	
		A KPI is a metric which is calculated from one or several measurements.	
C1	Direct Characteristic Conflict	Two SON function that modify different parameters aim at changing different metrics (KPIs) of a cell, but they may influence the same cell characteristic. Thus a conflict cannot be detected as a configuration conflict but only through the target metrics the functions want to modify. However, this requires identification of the characteristics that are associated with a metric during design of the SON functions.	The modification of both, downlink transmission power and electrical antenna tilt, influence the cell size.
C2	Logical Dependency Conflict	This conflict appears if there is a logical dependency between the metrics influenced respectively used by a SON function.	See example following table

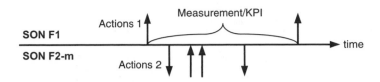

Figure 9.5 Measurement conflict between different function instances.

9.1.3.2 Example for Conflict Type B1

Two or more SON function instances (F1 and F2-m) may compete over different time scales where the actions of Function Instance F2-m may distort the measurements needed for F1 to make a correct decision, as shown in Figure 9.5.

9.1.3.3 Example for Conflict Type C2

Two function instances may cancel the performance gains made by each other, as shown in Figure 9.6. For example, F2 performing CCO modifies downlink transmission power and electrical tilt and therefore has an impact on cell size, and instance F1 (e.g. MRO or MLB) modifies the handover hysteresis and therefore the handover trigger threshold. In combination F2 may cancel the effects of the improved handover settings made by F1, as by F1 changing the cell size, the overlap area between the two neighbouring cells changes, thus having an impact on the load of the cell targeted by F1 (since the handover procedures may start earlier or later). The effect is that the performance gain achieved with F1 is cancelled through the modifications done by F2.

Another similar example has been introduced in Section 6.4 where a Cell Outage Compensation (COC) function changes the cell size which may invalidate the assumptions under which the PCI of the cell (and those of neighbouring cells) have been computed.

9.1.4 Network Parameters Related to SON Functions

There are many network parameters (a few hundred, yet the majority of them rarely change) related to SON functions in a self-organising LTE network. Amongst them, there are a small number of crucial parameters, for example, cell-related IDs, radio transmission power and other antenna parameters, radio channel parameters, neighbour cell parameters (cf. Section 4.2), mobility parameters (cf. Section 5.1), or cell activity status (due to energy saving, cf. Section 5.3.5), and so on. Figure 9.7 shows some of the network parameters that are directly related to two or more different SON functions, with the parameters being shared inputs, shared

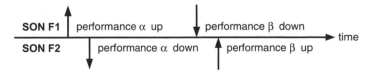

Figure 9.6 Logical dependency conflict between two function instances.

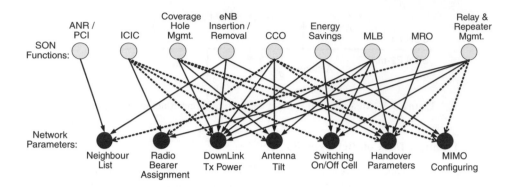

Figure 9.7 Shared configuration parameters between different SON functions.

outputs, or both. In the figure, a solid arrow line denotes that the parameter is a major parameter for the SON function, while the dashed arrow line denotes that the parameter is a secondary parameter for the SON function. The figure depicts the possibility that one SON function may interact with another SON function through their directly shared network parameters.

In fact, the situation is even more complicated than what is shown in Figure 9.7, since not only identical parameters have to be taken into account, but also different parameters both influencing the same physical characteristic of a cell (like TX power and tilt influencing cell coverage, cf. Conflict Category C in Table 9.1). One SON function may thus also interact with another SON function through the shared physical characteristics of some different parameters used by the SON functions.

9.1.5 Examples for Conflicts between SON Functions

In this section some examples for conflicts between SON functions are given, to explain the conflict categories introduced above in more detail.

9.1.5.1 SON Conflict Between PCI Function Instances

There can be two or more instances of the Physical Cell ID (PCI) allocation function active in the same network or network area at the same time. They are triggered by the insertion of new cells to assign a new ID to the inserted cell and potentially also to surrounding cells in case cell ID conflicts exist.

There exists the possibility of a conflict (cell ID collision or confusion) between the results of two or more simultaneously running, but spatially separated instances of the PCI function. As shown in Figure 9.8, Cell X is 'confused' (cf. Section 4.2.2) by two of its neighbour cells that have the same Physical Cell ID 'Cell A' after the PCI Instance II assigns the same physical cell ID to the latest inserted cell, if PCI Instances I and II are not coordinated. According to Table 9.1 this is an Input Parameter Conflict (Category A1).

The impact time of the conflict is 'short' to 'medium', depending on the execution duration of the PCI algorithm. The conflict probability may be low due to the relatively rare cases of macro cell insertions, or the need to update a PCI, respectively. However, this may be different for scenarios with many small cells (cf. Chapter 10), including femto cells which can be

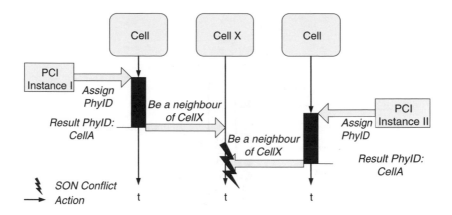

Figure 9.8   A conflict between two PCI instances.

switched on and off rather often. In case a cell ID collision or confusion conflict actually occurs, the impact is severe as a cell may become inoperable or the handover to a cell may become impossible. There is thus the need of coordination at run-time, so that a potential conflict can be prevented or resolved.

9.1.5.2   SON Conflict Between an MRO Function Instance and an MLB Function Instance

Assume the Mobility Robustness Optimisation (MRO) function and the Mobility Load Balancing (MLB) function are designed separately and not coordinated at run-time. An MLB function instance and an MRO function instance may conflict heavily with each other by cancelling the handover performance that would otherwise be improved by each of them, as shown in Figure 9.9. In fact this is a simplified view on MRO and MLB interactions as described in Section 5.2.7. According to Table 9.1 this is an Output Parameter Conflict (Category A2).

The probability that such a conflict occurs in case the MRO and MLB functions run in an uncoordinated manner is high, and the impact of the conflict is severe. There is thus the need of

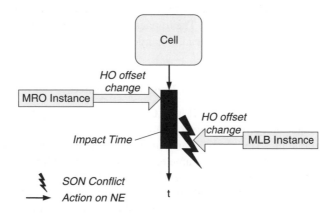

Figure 9.9   A conflict between MRO and MLB instances.

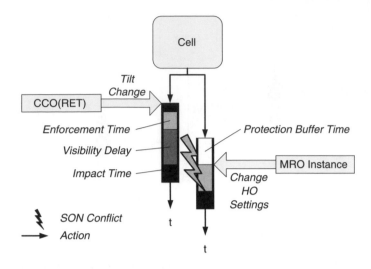

Figure 9.10   A conflict between CCO and MRO instances.

coordination between them in run-time or the two SON functions need to be designed together such that a potential conflict can be prevented.

9.1.5.3   SON Conflict Between an MRO Function Instance and a CCO Function Instance

An instance of the CCO function changes the tilt settings of the cell to achieve the optimum computed by the CCO algorithm. This SON function has a rather long visibility delay time, that is, the time until the modifications can be seen from the measurements. During the visibility delay time of the CCO function, an MRO function is triggered, which uses measurements collected in the time before the trigger to calculate new HO settings (see Figure 9.10). However, these measurements are influenced by the changes of CCO as the tilt modifications change the cell size and hence the hysteresis of the cell. Because the MRO and CCO functions are not coordinated, the MRO function hence uses wrong measurements as basis for the calculation. This type of conflict is a Measurement Conflict (Category B) according to Table 9.1.

The conflict probability is high in case CCO and MRO are not coordinated, as the visibility delay of the CCO function is rather long. The impact is also significant as a cell working with non-optimal handover parameters may lead to handover problems such as handover failures and handover ping-pong.

9.1.5.4   SON Conflict Between two CCO Function Instances

Except the key parameters CQI and RSRP, a CCO function optimising the antenna tilt (CCO (RET)) and a CCO function optimising the transmission power (CCO(TXP), cf. Section 5.4.3) share little input parameters. Their triggers are partly different, but their output parameters are clearly different. However, the output parameters impact similar radio characteristics of the affected cells since changes in tilt and TX power have similar effects. For example, downtilting and decreasing the TX power both shrink the coverage area of the cell, while simultaneous uptilting and decreasing TX power may neutralise each other. The situation may be even more complex in case CCO(RET) performs horizontal beam-steering while CCO(TXP) changes the

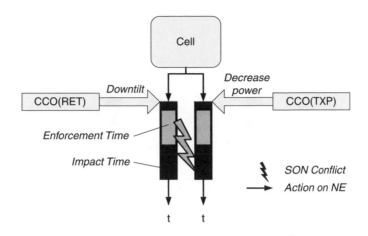

Figure 9.11 A conflict between two CCO instances.

TX power of the cell. Therefore, given the fact that CCO(RET) and CCO(TXP) are implemented and act as independent (non-coordinated) SON functions within one cell, they may heavily conflict with each other even though they share only a few key parameters. It is to be noted that changes to TX power and tilt in general have a rather long impact time as the corresponding visibility delay and relevance intervals (cf. Section 9.1.1) and hence the impact time of the SON functions is rather long.

In the case of overlapping SON function areas, the conflicts between CCO(RET) and CCO (TXP) can happen if CCO(RET) and CCO(TXP) are triggered simultaneously (see Figure 9.11) and change the respective radio parameters of the same cell. They do not change the same parameters (tilt versus power), but they impact the same characteristic of the cell, in this case the cell size. This is a direct characteristic conflict (Category C1) according to Table 9.1.

9.1.5.5 SON Conflict Between CCO and MLB Instances

An instance of the CCO function in Cell A modifies TX power and electrical tilt and therefore has an impact on the cell size. The cell size is a characteristic which is difficult to measure and may hence not easily be visible from measurements. An MLB function simultaneously running on a neighbouring Cell B modifies the handover hysteresis of Cell B, to achieve a better load distribution between Cell A and Cell B. However, the size of Cell A modified by the CCO function clearly has an impact on the handover hysteresis as the overlap area with the neighbouring Cell B changes. This causes the MLB parameter changes in Cell B to be potentially incorrect, for example, such that the handover hysteresis modified by the MLB function and hence the handover trigger threshold does not match with the cell overlap that has been changed by CCO in Cell A (see also Figure 9.12). In case the CCO and MLB functions on neighbouring cells are not coordinated with each other, one SON function may cancel the effects of the other one. However, this is difficult to measure and may hence not be recognised by the SON functions. This type of conflict is a Logical Characteristic Conflict (Category C2) according to Table 9.1.

The conflict probability is high in case CCO and MLB functions are not coordinated, as the visibility delay of the CCO(RET) function is rather long, and the MLB function may not

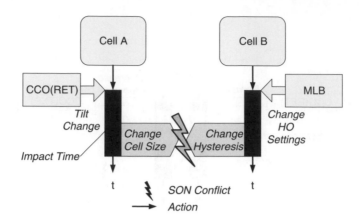

Figure 9.12   A Conflict between CCO and MLB functions.

recognise the changes. The impact is also significant as a cell working with non-optimal handover parameters may lead to handover problems such as handover failures and handover ping-pong.

9.2   Coordination of SON Functions

9.2.1   Basic Options for SON Coordination

On a conceptual level, there are two ways to reach the required scheme of operation, where the desired SON function interactions are enabled and the SON function conflicts are prevented or resolved. SON function co-design and SON function coordination both target this issue but follow different approaches.

In the following, the basics of both concepts are introduced and it is shown that none of those concepts is sufficient by itself but both will be required and used in a complementary way.

9.2.1.1   SON Function Co-Design

Detecting and handling conflicting SON function instances at 'run-time' can be hard to accomplish. If possible, conflicting SON function interactions should already be prevented at 'design-time' (when SON functions are created or prepared for operation, operational work-flows and policies are created, cf. Section 3.6). Section 9.1 has shown that conflicts between SON functions often have their roots in the fact that they operate on a non-disjoint set of parameters or change the same characteristics.

SON function co-design aims to provide SON functions which will not conflict at run-time by design.

Co-design of SON functions has to obey the following guidelines:

- Reduce the number of shared parameters by pre-assigning clear responsibilities for parameters to SON functions to the maximal possible extent. In the best case SON functions operate on fully disjoint parameter sets.

- If required, combine the functionality of several SON functions into a single function, for example, to create an optimisation function that optimises several target parameters at the same time instead of having several single target parameter SON functions.
- Exclude a SON conflict a priori by co-designing SON functions as a 'function group'. Within a function group, there is a defined run-time interaction between the individual SON functions which assures that there is no conflicting behaviour. The co-design of the MLB and MRO SON functions (cf. Section 5.2.7) is a good example of this type of conflict prevention. Uncoordinated function execution of those functions could cancel the intended handover performance gain as shown in Section 9.1.5.2. After the co-design MLB will set limiting values within which MRO is allowed to operate.
- Enable required interactions between SON functions to the extent possible. It has to be assured that those interactions are not excluded by some other constraints.

The result of SON function co-design is visualised in Figures 9.13 and 9.14. Figure 9.13 shows a set of SON functions. The target parameters of the SON functions are illustrated by different shapes. There are multiple functions with common target parameters. Figure 9.14 shows the results of the co-design. The number of shared parameters has been reduced and the functions can be separated into three groups:

1. A function group with defined run-time interactions. Those functions target the same parameters and need to interact with each other to avoid conflicts.
2. Two functions have been merged into a single function, which performs the tasks of both functions.
3. A group of functions that could be changed in a way that they use a fully disjoint set of target parameters.

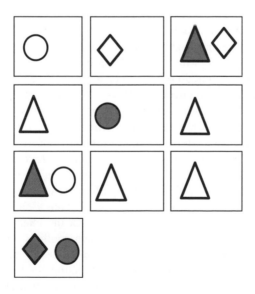

Figure 9.13 Non co-designed SON functions.

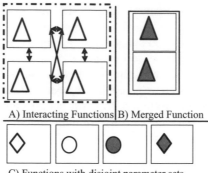

A) Interacting Functions | B) Merged Function

C) Functions with disjoint parameter sets

Figure 9.14 Co-designed SON functions.

The benefit of SON function co-design is that it is often a simple and foremost feasible approach when there are only a few SON functions in the self-organising network.

SON function co-design as a method to avoid run-time conflicts reaches its limits as soon as the overall number of deployed SON functions increases and especially if the set of deployed functions is changed at run-time. The introduction of additional functions could require a re-design of already deployed functions, which increases the overall overhead and possibly creates new dependencies between functions. Even if it is theoretically possible to design all SON functions with disjoint parameter sets, the major problem persists, co-design of SON functions from different parties (which did only marginally or not at all align their work) is not possible.

There are additional properties and requirements of SON functions and SON in general which can hardly be covered by co-design. SON function co-design relies mainly on the systematic analysis of the parameters, KPIs and measurements used or targeted by the functions, to design functions that do not conflict in this domain. The existence of characteristic conflicts (category C, Table 9.1) which cannot be resolved is a major issue for co-design. In that case either very different functions have to be merged into a single function or extensive communication overhead is introduced to allow a conflict free operation. Both solutions contradict the basic SON function paradigm of independently operating functions.

There are SON functions that target network optimisation and functions for failure recovery, which operate on identical parameters. There is a high probability that those functions are in conflict. For example Cell Outage Compensation (Section 6.4) functions will try to re-establish full coverage as fast as possible, which is conflicting with functions that target an equal load distribution within all cells. If an optimisation is performed when some cells fail, it is obvious that the conflict will occur but precedence needs to be given to failure recovery.

A parameter analysis would reveal this conflict, but the existence of these conflicting functions is fully acceptable. To be able to solve such conflicts with co-design a function designer has to know that those 'conflicts' on parameter level cannot be resolved. Either the decision must be to create a combined SON function for both, failure recovery and optimisation, or to integrate run-time interaction capabilities to allow the functions to solve the conflicts.

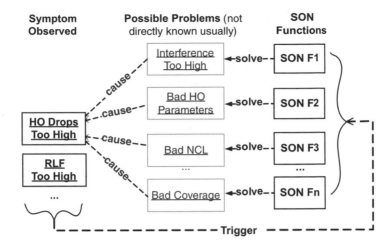

Figure 9.15 An example of SON function conflicts due to their shared triggering conditions.

But even if SON function co-design is not applicable as the sole conflict prevention solution, the principles can be used to significantly reduce the amount of potential SON conflicts. This reduces the number of conflicts that need to be handled separately and facilitates run-time conflict resolution by SON function coordination.

9.2.1.2 Co-Design Example: SON Function Conflicts due to Shared Triggering Conditions

Figure 9.15 shows a typical scenario where co-design is a first important step to counteract potential conflicts. The figure shows a number of SON functions F1 to Fn, which target different problems that may occur in the network. The major issue is that all those different problems result in the same effect: they all cause a high number of handover (HO) drops. Hence, at the design-time of all of these SON functions this observed symptom is used as a trigger for the SON function execution. If the function-designer is not aware of this fact and uses the handover drop rate as the sole triggering parameter for each of the SON functions, all functions will become active as soon as an increased HO drop rate is observed. Each of the functions will try to solve its problem independently if it has been the root cause for the observed symptom.

In such a situation it is very likely that the SON functions will not solve the actual problem but make the actual solution harder to reach or longer to take effect. As a first step towards the resolution of this problem a careful co-design should be performed, resulting in more than a single observed symptom as a trigger. After the co-design most of the SON functions will have independent triggers, for those where it is not possible to separate the triggers the conflict has to be resolved by SON function coordination which depends on the specific operation strategy of a given network operator.

9.2.1.3 SON Function Coordination

Coordination of SON function execution is the logical next step if co-design is not capable to provide a conflict free SON system, but it is more than a bare concept for conflict resolution.

Two basic ideas drive SON function coordination: On the one hand to perform a low-level coordination of autonomically operating SON functions in order to prevent or resolve SON function conflicts and on the other hand at the same time give control over the SON-enabled system to the human operator, to allow a governance of the SON function behaviour based on high level operator requirements without the need for continuous interaction between human operator and management system.

The operator requirements specify guidelines for SON function execution. There is a wide array of possibilities for the operator to express those guidelines, for example a KPI might have to be interpreted differently for different deployment scenarios and different thresholds might apply only in rural or urban areas. A respective operator requirement is then used as input to the function coordination to determine wanted or unwanted network behaviour and enforce different function behaviour for the individual deployment scenarios.

All decisions are taken based on the operator requirements with respect to the current context of the network. The function coordination can determine how to deal with conflicting SON functions, whether, for example, the optimisation of a set of cells should be continued or the changes of cell sizes is related to failure recovery and should therefore be prioritised.

SON function coordination is constructed around the question how potential interactions and more importantly conflicts between SON functions can be detected and prevented (respectively resolved) at run-time. The detection and resolution approaches have to be based on the context of the SON functions and the network entities they operate on.

The main concept of SON function coordination is based on the following principles:

- SON function instance execution is under full control of a coordination function. Here, 'coordination function' refers to an entity which enforces the coordination logic. It does not imply any centralised versus distributed choice of implementation.
- The execution of a SON function instance or the part of a SON function instance that potentially causes conflicts needs to be approved by the coordination function.
- The coordination function must be able to prioritise a SON function instance by pre-empting already running SON function instances.
- SON function coordination decisions are based on operator requirements (e.g. operational policies, SLAs, etc.), the state of the network and the characteristics of the requested SON functions.

SON function coordination provides a very powerful tool to deal with function conflicts that could not be resolved through co-design and provides additional means for run-time control of function behaviour. Though for a simple SON network with a small number of deployed SON functions an additional coordination function may introduce too much overhead.

9.2.2 Goals of SON Function Coordination

Based on the previous sections, the tasks of a coordination function have been identified as coordination of the SON function execution in order to:

- Prevent undesired SON function conflicts.
- Support of required SON function interactions.

Function coordination is also the best place to enforce of management goals as well as operational policies and in addition assure a very high efficiency for the overall SON system. Essentially, SON function coordination enables a fully automated network management system by linking different parts of the system together and allowing to express many different operator constraints on the functions.

- *Active protection* is the most obvious task that is performed. Running SON function instances need to be protected against negative impacts caused by other SON function instances.

  For each SON function instance execution request the coordination function evaluates properties of the requested SON function instance and the current network context. Only in case the analysis does not reveal any negative impacts from the requested function to any other currently active SON function instances, the execution request is acknowledged.
- *Pro-active protection* is performed to avoid future conflicts. A requested function instance can be negatively affected by already active function instances. In addition, pro-active protection targets also future SON function instances. As a result of changes performed by a SON function instance often some particular function instances need to be executed subsequently to guarantee the overall consistency of the network configuration. Conflicts between currently executed and future functions (cf. Figure 9.16) can prevent the successful execution of the future functions. A coordination decision based on knowledge about such dependencies allows to proactively protect subsequent function instances. Deadlocks or priority-based function termination due to increased impact areas are typical examples for future conflicts. If it is known that a currently running function will trigger another function with a larger impact area which conflicts with the requested or one of the subsequent function instances, this future conflict can be omitted through an appropriate coordination decision. Future conflicts are obviously much harder to assess at run-time than conflicts between active and requested SON functions instances. To be able to use the coordination function to perform this kind of proactive protection, knowledge about typical sequences of functions and their characteristics has to be used as input to the design-process of the coordination logic.
- *Enforcement of management goals:* Each network operator has high level management goals which govern the overall network operation, for example: 'prioritise full coverage availability over load optimisation.' Such a general policy is also required to resolve conflicts between failure recovery and optimisation functions. The coordination function is used to enforce those high level management goals by giving priority to functions in a way that those management goals are met. But also other operator policies like maximum change values for certain parameters can be enforced via the coordination function.

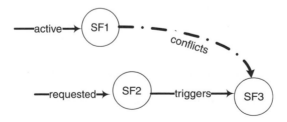

Figure 9.16 Conflict between active and subsequently triggered SON function.

- A *high efficiency of the overall SON system* is partly also reached through SON function coordination. Several individual parts of the coordination contribute to the efficiency of the system. A first important part is *parallelisation*: the coordination function acts as a kind of smart scheduler. In a network with thousands of network elements, spread over a wide area, several potentially conflicting functions can safely be executed in parallel as long as their impact areas do not intersect. When there are no potential conflicts detected, parallel SON function execution requests should be acknowledged to increase the efficiency of network operations. A second important part that determines the efficiency of the system is whether a SON function has all intended effects or if the effects are reduced or even eliminated through conflicting functions. Several types of conflicts have been introduced in Section 9.1.3. Active and proactive protection together with the enforcement of operator policies serves several goals:

  - Protect the effects already reached by preventing the execution of functions which counteract previously performed changes.
  - Reject functions whose actions will not be enforced or will be undone by other (future) functions. In case it can be foreseen that another function will block the action or even revert a performed configuration due to a higher priority it is better to reject a function execution request and thus to reach a higher overall efficiency. Figure 9.16 shows such a constellation when SF2 is requested, SF1 is active and will still be active when SF3 is triggered by SF2. The execution of SF3 is required as a result of the changes performed by SF2. In order to avoid conflicting behaviour between SF1 and SF3 the coordination function will not immediately acknowledge the execution of SF2. The request is either rejected completely or rescheduled to a later point in time where there is no conflict between SF1 and SF3.

In summary it can be stated that a careful design of the coordination logic which is enforced by the coordination function gives the operator of a network control over the execution of the SON functions without the need to tightly interact with the system. The coordination function will automatically acknowledge as many parallel SON function instances as possible, deal with conflicting functions and enforce operator policies. Functions that do not conflict with other functions or need to be coordinated according to operator policies are not affected by the coordination function.

9.2.3 SON Coordination Function Concept

In order to be able reach the goals outlined in the previous section, it is important to map those goals as requirements into a general coordination concept. This section will introduce such a general SON function coordination concept step-by-step. First, some important characteristics and requirements of SON functions and the way they are executed are introduced. This knowledge is in a subsequent step used to show how SON functions are analysed and how from the results of the analysis the coordination logic can be derived and mapped into a solution-agnostic representation which in a last step can be used for a concrete implementation of a coordination function.

9.2.3.1 Information Requirements and Effects on Efficiency

Efficient and reliable SON function coordination is a central challenge. In large networks with large numbers of deployed network elements and therefore a large number of SON function execution requests it is important that the coordination function does not become the bottleneck that causes service quality degradations, which could happen if a large amount of information needs to be stored and analysed for each decision. Therefore one of the main questions that determine the presented coordination concept is: 'What is the minimal amount of information required for reliable and efficient SON function coordination?'

The answer to this question is highly dependent on the SON functions, the input they require and the actions they perform. To allow for function coordination, the interactions of each SON function have to be analysed already at design-time. This analysis should reveal possible interactions and conflicts between SON functions.

The most important characteristics of the SON functions that are needed at design-time to specify the decision logic are the spatial and temporal characteristics of a SON function as introduced in Section 3.4. Based on this, definitions for impact area and impact time have been introduced in Section 9.1.

To take a decision the coordination function needs to consider at first the spatial characteristics of a SON function and whether there is an overlap with another SON function independently from the question whether the functions are in a conflict according to Table 9.1. Therefore the information of the spatial scope of a SON function is the most important information for coordination decisions.

The next important information that is required is the temporal characteristics as defined by the impact time (cf. Section 9.1). The impact time is closely connected to potential conflicts as described in Table 9.1. The minimal information required to take a coordination decision consists therefore of impact area, impact time and the potential conflicts between SON functions.

9.2.3.2 SON Function Interaction and Decision Logic

The SON function interaction and therefore also the decision logic of the coordination function, as already shown in the previous section is based on:

- Impact area;
- Impact time;
- Input parameters;
- Output parameters;
- Used measurements;
- Affected measurements.

A detailed per-function design-time analysis is required to determine conflicts and interactions according to the classification in Table 9.1. The analysis has to be done in a way to reveal all potential interactions between SON functions and allow in combination with information on impact time and area the decision logic to be established.

After the design of a SON function an interaction and conflict analysis towards all other already deployed SON functions has to be performed. Category A and B conflicts (cf. Table 9.1) are mostly detectable by an automatic analysis of the used or affected parameters and measurement values. For conflicts of Category C operational experience is required for detection. In the future, semantically enhanced data and information models may allow to even perform an automatic analysis of Category C conflicts.

After the identification of potential conflicts the impact times for each conflicting function have to be defined (cf. Section 9.1). While execution, enforcement and visibility times are given by the already deployed functions, the protection and relevance intervals have to be defined for the new function. There is not a single impact time for a new function but it has to be pair wise defined for each function the new function is in conflict with. This is obvious, since protection buffer time is dependent on the actual conflict. If it is not a Category B but Category A conflict, the protection buffer time is not required. The same applies for the relevance part of the impact time. Only if, for example, oscillation detection is required a relevance interval is included in the impact time.

After these steps all information required to design the coordination logic are available. The SON function designer looks at the conflicts and decides on how each conflict should be handled. Often these decisions are already pre-defined due to the priority of SON functions that are given by operational goals or logical and sequential dependencies.

9.2.3.3 Representation of Decision Logic

Decision trees are a solution-agnostic way to express the decision logic that needs to be applied to respond to a SON function instance execution request. They capture dependencies and interactions of a given function with other deployed functions. For each SON function they provide a sequence of conditions that need to be evaluated to take a coordination decision. The great benefit of a solution-agnostic representation of decision logic is its independence of the technology to be used for SON function coordination.

A coordination function will use the decision trees and combine them with impact area and time for the given SON function instances to come to a coordination decision. Whenever a SON function execution request is received the coordination function has to determine the impact area of the requested function instance and then traverse the decision tree. Figure 9.17 gives a basic example of a decision tree. The root of the tree represents the SON function execution

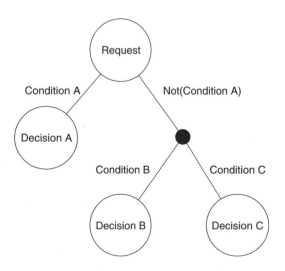

Figure 9.17 Generic decision tree.

request. The decision logic is annotated at the edges between the nodes. Typically the decision logic forces the coordination function to perform different types of evaluations:

- Check whether one or more SON functions are currently active within the impact area of the requested SON function.
- Evaluate the requested changes whether they are in conflict with previously executed changes to the targeted network elements.

The information required to evaluate the condition is rather limited. There is no need to keep extensive state information. It is sufficient if the coordination function has access to the information on the currently active SON functions affecting the impact area of the requested function and their impact times. For some SON functions, additional information on performed changes is required. This information is only available until the end of the impact time. In case there is no information about SON functions within the impact area of the requested function, the SON function execution request is instantaneously acknowledged.

The leaves of the decision trees contain the result of the decision process. This gives a good indication on how decision trees should be constructed to support efficient decision making. The more information is required for a particular decision the lower it should be placed within the tree. Decisions that lead to an instantaneous decision should be placed in the upper part of the tree. For example, if there is a conflicting SON function, which, if present, leads to a direct rejection of the requested SON function execution, it should be placed close to the root of the decision tree. Such a tree organisation facilitates the understanding of the decision logic and speeds up the overall decision process.

Independently from the used technology, the availability of decision trees simplifies the actual implementation of the decision making. On the occurrence of a SON function instance execution request, the coordination function only has to traverse the respective decision tree until it reaches a leaf. The contained decision is then enforced.

As it has been shown, in order to allow successful SON function coordination, the construction of the decision trees has to be an integral part of the SON function design process. In case the analysis of the new SON function reveals conflicts with already deployed functions the decision trees of those functions have to be re-evaluated if there is a need for adaptation. This approach brings several benefits most importantly:

The decision logic of all SON functions without interactions stays untouched.

All information that is required for the decision making is automatically provided within the coordination request. No or only little additional information about the SON functions needs to be acquired from the OAM system, which reduces the overall communication overhead. For example, information that was contained in a previous coordination request is re-used for further coordination decisions until the impact time of the previous function expired.

The information that is required for SON function coordination can even be further reduced by aggregating and grouping information about individual SON functions with identical decision logic. This information compression facilitates the introduction of new functions and increases the coordination efficiency.

The coordination logic needs to be designed carefully to meet all operational requirements, preventing conflicts but also supporting the required SON function interactions. The benefit is that this approach is useable even for systems with a large number of deployed SON functions. Figure 9.18 shows an example based on the set of SON functions introduced in Figure 9.13.

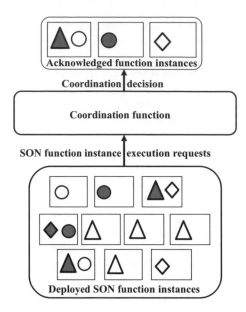

Figure 9.18 SON function coordination.

The SON functions request permission to execute from the coordination function. In case a conflict free execution is possible the coordination function acknowledges the execution request and the SON functions are executed.

9.2.3.4 Coordination Decisions

SON function coordination is used to either prevent or resolve SON function conflicts. At the end of a coordination process there are single or multiple coordination decisions that affect the following execution of SON function instances. This section provides an overview over possible coordination results and a reasoning as to why they are required. Depending on the system setup and the assumptions that are made on monitoring and SON functions in general, only a subset of the presented decisions or even additional decisions may be required. The selection shown here is seen as a good balance between simplification of the decision logic and decisions required to increase the operational efficiency.

- **Acknowledge SON function request:** Obviously this decision is taken if no conflicting SON functions have been detected. An acknowledged request leads to the execution of the requested SON function.
- **Reject:** Multiple reasons can cause a rejection of a SON function request. In general it requires another SON function to be active on the same target or intersecting impact areas of requested and active function. The requested SON function must not be executed. For a full rejection of a request often a higher prioritised SON function is required to be active.
- **Reschedule:** A 'softer' variant of a rejection is the rescheduling decision. The coordination function has determined that due to the current context of the network the SON function should not be executed, but it is reasonable to execute the function at a later point in time,

when the impact time of the currently active conflicting SON functions has timed out. Such a decision may be taken when the impact time is close to the end of the protection time, especially if the used monitoring intervals are rather long. Rescheduling is seen as a mean to increase the efficiency of the network operations. Rescheduling is also used in combination with the pre-emption decision.

- **Pre-emption:** A requested SON function can have a higher priority compared to the SON function currently active on the target. In this case it needs to be possible to terminate a running SON function and give precedence to the requested one. A typical example is an optimisation function that is active when some failure recovery is required. In such a case the execution of the optimisation function could be terminated in order to let the failure recovery function execute. Pre-emption will often be combined with rescheduling, the pre-empted functions will be re-executed after the higher priority function's execution has finished.
- **Rollback:** Is used in case the actions performed by a previously executed SON function have a negative effect on the requested function or the expected result of the function. In this case, the coordination function can combine an acknowledgement with a rollback decision. Rollbacks have to be enforced before the requested SON function may be executed. Combinations of acknowledgements and rollbacks are for example used if different functions can be used to reach the similar effect. For example, both tilting down the antenna and reducing the output power leads to a reduction of the cell size. Depending on the context of the network one or the other function can be beneficial. In case either one of them has been used without the expected effects or even with unexpected side-effects a rollback followed by the other function can be used to reach the desired effect.

Depending on the capabilities of a coordination function in a concrete implementation additional coordination decisions are conceivable. For example in the case of two conflicting SON functions, instead of acknowledging one and rejecting the other function both functions could be rejected and a third function is triggered that performs the tasks of both functions in a combined way. Another solution to this issue could be to integrate some sort of mediation functionality into the coordination function that changes the behaviour of the functions in a way that a non-conflicting result is reached.

Example Use Case: Coverage and Capacity Optimisation
This example shows how the decision logic for a well known SON function is mapped into a decision tree.

Coverage and capacity optimisation (CCO) is one of the key use cases defined by the 3GPP (cf. Section 3.2.3). CCO for a particular cell can be done with the help of two SON functions that influence the cell performance:

- CCO(RET): Changes the electrical tilt of the antenna.
- CCO(TXP): Changes the transmit power.

A decision tree that is used to coordinate those SON functions could look like the one shown in Figure 9.19. It contains the conditions that have to be evaluated whenever a CCO(TXP) function execution request is received. In case there is no concurrent CCO(RET) function on the same target cell the TXP request is immediately acknowledged. If a CCO(RET) function is currently active on the same target cell it has to be evaluated whether both function request

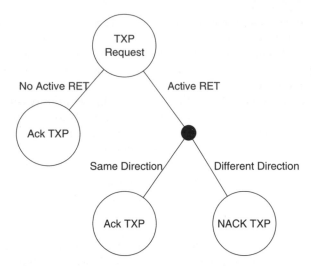

Figure 9.19 Decision tree for CCO(TXP) coordination.

changes in the same direction. The TXP requests to increase the transmission power and the RET function tilts up the antenna for the cell. Depending on the results the CCO(TXP) request is either acknowledged or rejected.

9.2.4 Coordination Schemes

Depending on the SON functions, the conflicts between SON functions and the operational goals in a network, different requirements are imposed on the SON function coordination.

Coordination schemes describe how a given SON function is coordinated. In particular, it is specified in which phase of the SON function instance execution it makes sense to take coordination decisions for a function instance. This sub-section will give an introduction to three generic types of coordination schemes and highlight their advantages and challenges. They are rooted in the conceptual considerations on the setup of SON functions and impact time and area that have been presented in the previous sub-sections.

It will be shown why some SON functions should be coordinated using a particular scheme while other functions benefit if a different coordination scheme is applied.

Conceptually, a SON function can be split into three operational phases:

- **Monitoring or Perception Phase:** For each SON function a trigger is defined. During the monitoring phase the network is monitored, if the set of observed parameters or measurement values indicate the requirement that the SON algorithm execution has to be requested. Monitoring activity is use case specific. It ranges from continuous monitoring over monitoring in predefined intervals to a bare on-demand activity.
- **Algorithm Execution Phase:** In this phase the algorithm of the SON function, which computes new configuration values, is executed. This phase could also contain additional data acquisition steps. Acquired data are either more measurement values or parameter configurations that are required to compute the new settings.

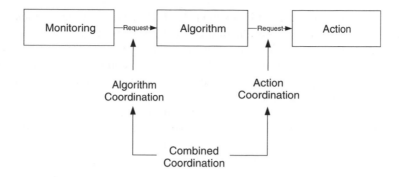

Figure 9.20   Interaction Points for different Coordination Schemes.

- **Action Execution Phase:** This is the phase where the computed configurations are enforced. This can either happen by direct interaction between the SON function and the targeted network elements or through interaction between SON function and a network configuration sub-system.

The information required to take a coordination decision that resolves a coordination conflict is not always available during each of the phases. Depending on the information required to perform to deal with a given conflict the time point for the coordination decision has to be selected carefully.

Coordination makes most sense at the transition between each of these phases since they mark important stages of the function execution. Theoretically coordination could also take place before the monitoring phase. Based on the conflict definitions (cf. Table 9.1) it is questionable if there are any benefits in coordinating at this point of the SON function lifecycle, especially, because the monitoring parts of SON functions are often continuously running.

Figure 9.20 shows the separation of a SON function into Monitoring, Algorithm and Action execution phases together with the requests at the transitions between the different phases. It shows also at which points the different coordination schemes, presented in the following sections, interact with the SON function.

The analysis of a SON function and its possible interactions and conflicts with other functions gives already an indication which information is required to evaluate the conditions within the decision tree of the SON function.

9.2.4.1   Action Coordination

The most obvious way of coordinating SON functions is to coordinate the action execution, since the actions show the largest conflict potential. Actions perform configuration changes at the network elements; therefore it is beneficial to be able to block conflicting actions.

If action coordination is used, a SON function has to request the approval to execute its action from the coordination function. It sends a request including all required information to the coordination function and acts according to the response of the coordination function. Only in case of an acknowledgement the actions are triggered and the changes are performed.

A major benefit of action coordination is the possibility for the coordination function to access all results of the algorithm and evaluate the configuration changes that are going to be performed in case the action request is acknowledged. This possibility in combination with a long relevance time allows the coordination function to base the coordination decision on detailed information not only from the current request but also on previously applied configuration changes. Sanity checks can be performed, abrupt or subsequent opposing configuration changes suppressed and oscillating behaviour detected and avoided before it actually happens in the live network.

Such possibilities are especially important for SON functions that influence important properties within the network by changing different parameters as, for example, SON functions that have an influence on the coverage area of cells. If repeatedly the transmission power of a base station is increased followed by a downtilt of the antenna this is a strong indication of oscillating reconfigurations. Such behaviour can easily be detected if action coordination is used.

Also from an efficiency viewpoint, action coordination is very appealing, as it reduces the time interval during which the SON function is considered to be active. The impact time for functions that are coordinated with action coordination does not include the algorithm execution time.

9.2.4.2 Algorithm Coordination

In case action coordination is used, the algorithm part of each SON function is always executed regardless of a following approval or rejection of the action request. For several reasons this is not the best mode of operation.

- Computational overhead should not be underestimated: depending on the number of rejected action requests there could be a large number of algorithms that have been executed without any benefit. This is especially the case whenever the algorithm results are irrelevant for the coordination decision. Analysis has shown that for many combinations of conflicting SON functions the only information required for the decision taking is whether there is another SON function with an intersecting impact area active.
- The soundness of the input parameters that are either used to decide whether an algorithm should be executed or which are used as a basis to compute the configurations that should be enforced is something that is neglected or at least hard to reach if action coordination is used. The monitoring part of each SON function monitors a set of measurement values or KPIs in order to determine when the function has to be executed. Coordination, as stated previously, is used because it is impossible to completely co-design SON functions in a way that there are no conflicts. SON functions that enforce configuration changes within the network will have an effect on measurement values monitored by other SON functions. Usually those changes will not have an immediate effect but it will take some time for that effect to manifest itself. Enforcement and visibility time are used to guarantee that all changes that have been performed are fully visible in the measurement values of conflicting SON functions before they are allowed to be executed. It may be required that algorithms of SON function instances execute for very long time intervals. The impact time of another function that has been active in the moment the algorithm has been triggered can easily time out until action coordination is performed. That means that it is possible that measurement values that caused the other algorithm to be executed or that are used as input to the other algorithm did not fully reflect the already performed changes. Since the impact time of the first SON function already timed

out, the potentially erroneous changes requested by the action of the later function will be enforced. Therefore the requirement that a monitoring function does not trigger a SON function algorithm based on measurement values that do not fully reflect the current state of the network cannot be met with action coordination.

If algorithm coordination is employed, the coordination function operates on the algorithm execution request, which allows it to take coordination decisions *before* the algorithm execution.

A well chosen enforcement and protection time will therefore prevent the algorithm execution until the monitored values have stabilised and give a correct impression of the current state of the network. In this mode of operation, a SON function is considered to be active from the moment the algorithm execution is acknowledged. It is thus important to choose appropriate visibility, enforcement and protection times.

The major disadvantage of algorithm coordination is the lacking ability to perform a coordination based on the detailed requested configuration changes. Algorithm coordination has therefore no possibility to protect the system against contradicting or even oscillating configurations. As soon as an algorithm execution has been acknowledged the action will also be executed.

9.2.4.3 Combined Algorithm and Action Coordination

If the coordination function is given the ability to coordinate both, algorithm and action execution the benefits of both coordination schemes can be utilised at the price of higher overall coordination complexity. SON functions are not executed before reliable input values to both monitoring and algorithm part of the SON function are available but also the network can be protected against contradicting reconfigurations.

The main intention with SON function coordination is to reduce the risk of processing erroneous input values. Therefore, most of the coordination is performed by the algorithm coordination. Action coordination is only performed for a small number of functions. Typically if multiple functions are used which change identical characteristics, as for example subsequent tilt and transmission power changes for coverage and capacity optimisation. For some SON functions also intra-function coordination requires a combined coordination. The same function could be executed multiple times for the same target network elements, but some extended requirements have to be fulfilled in order to allow the configuration changes requested by all function instances to be enforced. Often such extended requirements try to allow only gradual configuration changes but prevent abrupt configuration changes or oscillating reconfigurations.

The algorithm coordination part is designed to allow parallel execution of potentially conflicting SON functions. Those potential conflicts are then handled by the action coordination. It is important to note, that algorithm coordination still protects against the possibly erroneous input values.

The usage of a combined coordination scheme has also effects on the used decision trees. While for algorithm or action coordination a single decision tree that is traversed completely, if a respective request is received, is sufficient, an adaptation is required for the combined coordination. Each function still has a single decision tree but this tree is partitioned into two parts (one part for the algorithm and action coordination respectively as

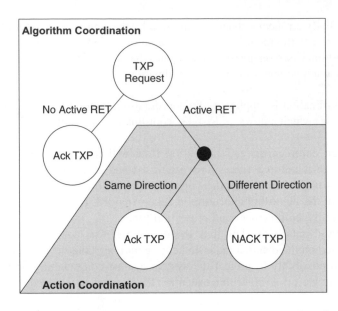

Figure 9.21 Separated decision tree for Algorithm and Action Coordination.

shown in Figure 9.21). The upper part of the decision tree is processed in the moment the algorithm coordination request is received. The second part requires information that is only available during action coordination. Therefore this part is not processed until the algorithm execution has been successfully completed and an action coordination request which contains the required information is received by the coordination function. It is important to keep in mind that such a coordination scheme evaluates the context twice (once for each request). When designing the coordination logic that should be used with a combined coordination scheme, it is important to keep in mind that start and runtime will affect the coordination decisions more compared to the situation when only a single coordination scheme is used. The SON function designer has to choose the constituents of the impact time long enough to avoid negative effects that could arise if the impact times timed out during algorithm execution.

Another difference to the single phase coordination schemes is that a pair of SON functions has more interaction points where they can affect each other. While for a single phase coordination scheme there is only one point of interaction when the function that has been triggered later requests a coordination decision, the combined coordination scheme has up to three interaction points. The first takes place when the later-triggered function requests algorithm coordination and then potentially when both functions request their respective action coordination.

This section presented different coordination schemes. Algorithm and action coordination as well as a combined coordination scheme have been introduced. The general advantages and disadvantages of each of the coordination schemes have been described. Based on this information the best coordination scheme for each SON function can be chosen at design-time. More information on the process how to select an appropriate coordination scheme is given in the next section.

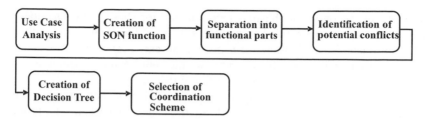

Figure 9.22 SON function design process with regards to function coordination.

9.2.4.4 Application of Coordination Schemes to SON functions

Selecting the appropriate coordination scheme for a SON function is the last step to be done at design-time in order to prepare SON function coordination.

At design-time the same basic process is performed for each function, which is visualised in Figure 9.22:

- In the beginning the SON use case is evaluated to determine the trigger for the use-case and which tasks have to be performed for which targets.
- From the use case analysis the SON function is created. The function designer separates the function into the functional parts monitoring, algorithm and action part.
- After the definition of the SON function it is analysed to detect potential conflicts with other SON functions. This step is performed separately for all of the three parts of the function.
- The result of the function conflict analysis is used to create the function specific decision tree. In addition the decision trees of other potentially conflicting SON functions need to be adapted.
- As a last step, based on the decision tree and the conflict analysis combined with information on the functions' requirements the coordination scheme can be chosen and the constituents of the impact times be defined.

After the steps of this process have been performed all the information is available that is required to realise SON function coordination within a specific system.

The selection of the appropriate coordination scheme for each SON function is quite important for the quality and the efficiency of the resulting coordination. It needs to serve the specific requirements on the coordination for the function therefore it is highly dependent on the SON function characteristics. In case there are not only A2 conflicts possible, algorithm coordination is chosen. A good indicator for algorithm coordination is the possibility to resolve conflicts based on function priorities. In case there are only dependencies between the subsequent configuration changes, the A2 conflicts, of multiple SON functions, action coordination is used. If both types of conflicts are possible the combined algorithm and action coordination is selected.

It is possible to apply different coordination schemes to different SON functions. Two functions that are coordinated against each other do not have to use the same coordination scheme. This also depends on which type of conflict causes the need for coordination. For example, two Functions A and B need to be coordinated. Analysis of the functions has lead to the following requirements:

- Execution of Function A may not be triggered while Function B is active.
- The action of Function B depends on the configuration change previously performed by Function A, but it can run in parallel with Function A.

In case these are the only coordination requirements for those SON functions different coordination schemes can be applied to the SON functions. Function A is coordinated using algorithm coordination, which allows to it prevent an execution while B is active. Action coordination is used for Function B, which allows the coordination function to evaluate the configuration changes performed by either of the functions.

In case new functions are introduced during run-time, a change of coordination schemes for already existing functions could be required. Such changes will usually be to move from algorithm or action coordination to a combined coordination scheme, as the previously used functionality must be retained. The combined algorithm scheme complements the already deployed coordination scheme.

9.2.5 Related Work

The problem of SON function conflicts is clearly visible, therefore different approaches have been proposed to target the problem through SON function coordination. SOCRATES is an EU FP7 Project (see also Section 3.3) where SON function coordination has been addressed (SOCRATES, 2008), (SOCRATES, 2011). The project has performed an extensive classification and categorisation of potential SON function conflicts. While parts of the resulting classification can be directly matched to the categorisation shown in Table 9.1, some of the conflict types presented in the project deliverables are specific to the SOCRATES system architecture.

Based on this classification and a SON system architecture, the proposed coordination approach is separated into two parts, 'heading' and 'tailing' coordination. The theoretical concept covers both SON function co-design and also SON function coordination, although both approaches are referred to as a type of coordination.

- Heading coordination tries to remove potential conflicts already at design-time. This is similar to SON function co-design.
- Tailing coordination performs the coordination between conflicting functions at run-time. There is a set of functions that performs a wide range of tasks that sum up to a complete function coordination.

Apart from acknowledging or rejecting function execution requests, tailing coordination has the ability to change the incoming requests in order to harmonise requests from multiple SON functions concerning an identical target.

Another approach to SON function coordination is presented in (Bandh *et al.*, 2011; Sanneck *et al.*, 2010). The concept is based on an event based execution environment for SON function instances and employs a policy based framework to perform coordination. It does not include SON function co-design but assumes that a proper separation of SON functions is done at design-time and therefore targets only the coordination. Section 9.2.6 shows an example how that framework is used to coordinate CCO SON function instances.

9.2.6 SON Function Coordination Example

The experimental system presented in (Bandh *et al.*, 2011) is based on the SON function coordination concept introduced in Section 9.2.3. It is capable to do algorithm and action coordination with a very powerful policy based coordination function. Policy based systems provide the benefit that their behaviour can easily be changed at run-time through introduction, activation, deactivation or adaptation of policies. These characteristics are beneficial since they allow run-time introduction of new coordination logic whenever new SON functions are deployed to the network or to adapt existing logic whenever needed.

The used system is fully event-based, which is an important design decision if run-time adaptation of the system is required. Events are used to transmit all the coordination requests and replies. Coordination request events trigger the policies which evaluate the context in order to take the coordination decision. In the coverage and capacity optimisation example a small cluster of cells with two coverage holes and two active SON functions are used to demonstrate the efficiency of the coordination function. Coverage and capacity optimisation (cf. Section 5.4) aims at a well balanced cell load and a full coverage without any coverage holes in a network. The goal is reached with two active SON function instances CCO(RET) and CCO(TXP) which have to be coordinated to reach the best results. The decision logic provided by decisions trees was mapped into a set of policies and a combined coordination scheme for algorithm and action coordination was selected for both functions.

In order to show the benefits of function coordination a single event trace was applied to a network layout, which is reset between the first and second run. For the first run the SON function coordination is disabled and subsequently enabled for the second run. Without active SON function coordination all requests are acknowledged and the respective SON functions are all directly executed. The result shows that the CCO goals are not reached; Figure 9.23 shows the result with an unbalanced distribution of cell sizes and the coverage holes still not fully covered.

The second run is performed with activated SON function coordination. The difference is that instead of all function requests directly being acknowledged one request is rejected and two previous changes are rolled back. The used event sequence and the coordination decisions taken by the coordination function are shown in Table 9.2. The first column contains the sequence of CCO(RET) and CCO(TXP) functions that are performed on the indicated cells (Column 2). The third column shows the coordination decisions for the SON function execution requests, in case there is a rollback decision the arrows indicate which previously acknowledged functions are targeted by the rollback and in case there is a reject decision the arrows indicate the reason for the rejection.

After all functions have been executed an equal distribution of cell sizes with a complete coverage is reached. All previously existing coverage holes are closed (Figure 9.24).

This example shows the benefits of SON function coordination. SON function-specific coordination logic is specified at design-time in solution agnostic decision trees, which is then mapped into a concrete coordination function and enforced at run-time. Network operators gain the possibility to influence the behaviour of the coordination function by including their operational presets into the decision trees for example to enforce context specific behaviour for example at different times or for different network deployment scenarios.

During the design-time analysis of the SON functions, the coordination requirements of a SON function are determined. After the analysis it is possible to determine which kind of coordination scheme is required for a successful coordination, whether it is sufficient to know whether another SON function is operating on the target cells or if a more detailed knowledge about previously performed actions is required. In the example with the selection of a combined coordination scheme, that performs both algorithm and action coordination, it is possible for

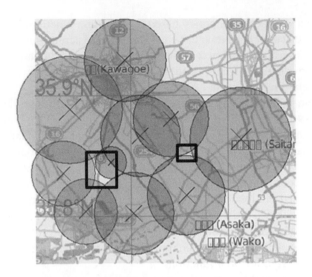

Figure 9.23 'Without coordination' case: unbalanced cell sizes with coverage holes (Bandh *et al.*, 2011) Reproduced with permission of © 2011 IEEE.

the coordination function to detect that a previously executed function did not have the expected effects. Therefore a previously executed function is rolled back twice in order to allow another function to reach the goals.

The operator gains the possibility to control the intended behaviour of the coordination function by creating SON function specific decision trees. Those operator decision guidelines are then automatically enforced at run-time, which allows more detailed coordination decisions than only acknowledging or rejecting execution request. Combined with action coordination it provides the possibility for the coordination function to recognise that a previously executed function did not have the expected effects and therefore has to be rolled back to allow a better suited function to solve the problem.

Table 9.2 Event sequence and coordination decisions (Bandh *et al.*, 2011) Adapted with permission of © 2011 IEEE

Change	Cell	Coordination
RET	18	ACK
RET	17	ACK
RET	16	ACK
TXP	20	ACK
TXP	23	ACK
TXP	25	ACK
TXP	16	NACK
TXP	17	ACK, Rollback
TXP	19	ACK
TXP	18	ACK, Rollback

Figure 9.24 'With coordination' case: well balanced cell sizes and full coverage (Bandh *et al.*, 2011) Reproduced with permission of © 2011 IEEE.

9.3 Conclusions

The focus of this chapter lies on interactions between SON functions, especially on conflicts between functions concurrently active in the network. The chapter shows both: SON function conflicts and how they can be handled.

Section 9.1 provides a detailed introduction to SON function interactions in general and conflicts in particular. It explains when an interaction should be seen as a conflict and which effects a conflicting function can have on another function. This allows also to classify a given conflict in order to find a way how to deal with it.

It is important to note that not only concurrently active functions can be in conflict but also dependency chains or function execution chains that originate in two given functions can cause conflicting behaviour.

In order to give the reader a better understanding on the origin of SON function conflicts typical conflict reasons are not only discussed but also shown on the basis of SON function examples that have been introduced by the NGMN forum.

After the introduction to SON function conflicts, Section 9.2 shows several ways how those conflicts can be handled in order to maintain a reliable and efficient network operation.

SON function co-design and coordination are discussed. While the focus lies on SON function coordination, the co-design is considered to be an important part to guarantee stable network operations as it helps to reduce the dependencies between individual function. A careful SON function design helps to reduce the number of conflicts that have to be considered in order to take a coordination decision.

Section 9.2 specifies the goals intended with function coordination and shows a general concept of function coordination. This concept is based on basic SON function character- istics and does not presuppose any special technical solution. It provides an understanding on how the analysis of the SON functions leads to a coordination logic that can be used to perform run-time coordination. Apart from low-level function coordination, the presented concept additionally allows full network operator control of the network behaviour without the need of constant operator-network interaction. From this basic analysis of the SON functions and coordination logic basic coordination schemes are derived and described with their pros and cons.

A closing example shows how the theoretic SON function coordination concept can be applied to existing SON functions. Which coordination scheme is chosen based on the coordination requirements derived from the SON function analysis. The different results of the network optimisation with and without SON function coordination are shown based on a sequence of SON function execution requests issued within the system.

Chapter 9 shows the dangers that come from independent conflicting SON functions being concurrently active within a network. It shows how those dangers can be handled through a detailed function analysis at design-time which results in enhanced functions and a well balanced coordination at run-time. A fine-tuned function coordination does not only increase the overall efficiency by preventing function conflicts or service-quality degradation caused by opposing or even oscillating reconfigurations but also by allowing a high degree of parallel function execution. A coordination function will enforce the parallel execution of potentially conflicting functions while their impact areas are spatially separated. This contributes highly to a higher overall network operation efficiency.

References

Baliosian, J., Sailhan, F., Devitt, A. and Bosneag, A.N. (2006) The omega architecture: towards adaptable, self-managed networks. Proceedings of the 1st Annual Workshop on Distributed Autonomous Network Management Systems, June, Dublin, Ireland.

Bandh, T., Romeikat, R., Sanneck, H. and Tang, H. (2011) Policy-based coordination and management of Self-Organising-Network (SON) Functions. IFIP/IEEE Symposium on Integrated Management, May, Dublin, Ireland.

Bandh, T., Sanneck, H. and Romeikat, R. (2011) An experimental system for SON function coordination. IEEE International Workshop on Self-Organising Networks, May, Budapest, Hungary.

Barth, U. and Kuehn, E. (2010) Self-Organisation in 4G mobile networks: motivation and vision. 7th International Symposium on Wireless Communication Systems, September, York, UK.

Calder, M., Kolberg, M., Magill, E.H. and Reiff-Marganiec, S. (2002) Feature interaction; a critical review and considered forecast. *Computer Networks*, **41**, 115–141.

Cleaveland, R. and Smolka, S. (1996) Strategic directions in concurrency research. *ACM Computing Surveys*, **28**(4), 607.

Doettling, M. and Viering, I. (2009) Challenges in mobile network operation: Towards self optimising networks. Proceedings of IEEE International Conference on Acoustics, Speech and Signal Processing, April.

Laiho, J., Wacker, A. and Novasad, T. (2006) *Radio Network Planning and Optimization for UMTS*, 2nd edn, John Wiley & Sons, Inc., New York, USA, pp. 505–569.

Sanneck, H., Schmelz, C., Bandh, T. *et al.* (2010) Policy-driven workflows for mobile network management automation. 6th International Wireless Communications and Mobile Computing Conference, June, Caen, France.

Schmelz, C., Amirijoo, M., Eisenblaetter, A. *et al.* (2011) A coordination framework for self-organisation in LTE networks. IFIP/IEEE Symposium on Integrated Management, May, Dublin, Ireland.

SOCRATES (2008) Use Cases for Self-Organising Networks Deliverable D2.1, INFSO-ICT-216284 SOCRATES, March. Available at http://www.fp7-socrates.eu/, [accessed 30 August 2011].

SOCRATES (2011) Final Report on Self-Organisation and its Implications in Wireless Access Networks. Deliverable D5.9, INFSO-ICT-216284 SOCRATES, January. Available at http://www.fp7-socrates.eu/, [accessed 30 August 2011].

Strassner, J., Agoulmine, N. and Lehtitet, E. (2006) FOCALE – A novel autonomic networking architecture. Latin American Autonomic Computing Symposium (LAACS), Campo Grande, MS, Brazil.

10

SON for Heterogeneous Networks (HetNet)

Cinzia Sartori, Henning Sanneck, Klaus Pedersen, Johanna Pekonen and Ingo Viering

10.1 Introduction

As introduced in Chapter 2, mobile operators acknowledge that the Heterogeneous Network ('HetNet') is an attractive solution, able to cope with the enormous data traffic increase.

In fact migration from HSPA to LTE brings additional improvements in the terms of higher spectral efficiency, but those alone are expected to be insufficient when compared to the traffic growth predictions from the industry. This essentially means that other performance boosters such as macro-site densification, improved receivers, higher order sectorisation and addition of small cells are likely to be needed. As the spectral efficiency per link for LTE is approaching the theoretical Shannon limit, it is postulated that the addition of small cells is amongst the most promising solutions for building improved spectral efficiency per area. Thus, the migration from macro-only networks to multi-layer topology networks, often referred to as Heterogeneous Networks or simply HetNet (see also Figure 10.1), is expected to further accelerate in the future. This chapter mainly focuses on LTE multi-layer networks and outlines some of the supportive mechanisms that help enable easy rollout and operation of such multi-layer deployments. In particular, we consider cases where the different LTE cell types as summarised in Table 10.1 are deployed. Notice that the main characteristics in Table 10.1 shall only be considered as an example, since vendors of course have the freedom to also develop cell types with other power settings and antenna gains as listed in this table. From Table 10.1 it is worth noticing that there are huge differences in the Equivalent Isotropic Radiated Power (EIRP),

LTE Self-Organising Networks (SON): Network Management Automation for Operational Efficiency, First Edition.
Edited by Seppo Hämäläinen, Henning Sanneck and Cinzia Sartori.

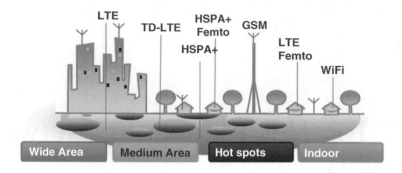

Figure 10.1   Multi-RAT and Multi-Layer network.

which essentially means that the coverage area of each cell type differs significantly, as well as the generated interference to surrounding cells.

The macro, micro and pico eNBs are characterised by the same architecture, while the Home eNBs (HeNB) have different architecture attributes as explained in more detail in Section 10.2. In addition macro eNBs are typically installed by the operator after careful radio network planning considerations, pico eNBs may be installed on an ad hoc basis without prior detailed radio network planning. On the other extreme, HeNBs are typically installed by an end-user who is not a technical expert and they can be moved to a different location and/or switched on or off at any time by the end-user. HeNBs are therefore deployed in an uncoordinated manner without any direct control through the operator. Furthermore, HeNBs may likely be deployed with a restricted access configuration that, amongst others, can results in rather challenging interference scenarios calling for autonomous interference management solutions.

A major precondition for a successful LTE multi-layer deployment is that especially the small eNBs (pico, HeNB) are self-configuring and self-optimising, as conducting configuration and performance management operations manually is simply impossible for the deployment of a large number of small base stations. In Section 10.2 first an overview about the main HeNB characteristics is provided, in order to further set the scene and define this particular small cell type. In Section 10.3 the main self-configuration mechanisms relevant for LTE multi-layer deployments are outlined, while Section 10.4 and 10.5 deal with key self optimisation Sections 10.4 and 10.5, the key self-optimisation techniques for LTE multi-layer deployments such as interference management, mobility optimisation and load balancing. All of the aforementioned techniques are considered to be key enablers for efficient and successful LTE multi-layer deployment and operation.

SON for LTE Relay nodes, SON for Multi-RAT and Energy Saving are not directly addressed in this chapter, since they have already been described in Chapters 4 and 5, respectively.

Table 10.1   Example of approximate main characteristics of different LTE cell types @ 10 MHz bandwidth

Cell type	Transmit power	Antenna gain	EIRP	Range
Macro	46–49 dBm	14 dBi	60–63 dBm	>100 m
Micro	37–40 dBm	5 dBi	42–45 dBm	100 m
Pico	24–37 dBm	4 dBi	28–41 dBm	20–50 m
HeNB	10–20 dBm	0 dBi	10–20 dBm	10–20 m

10.2 Standardisation and Network Architecture

The 3GPP standardisation for LTE has addressed the HetNet deployment scenarios already from the first LTE Release onwards (i.e. 3GPP Release 8). As an example the E-UTRAN base station class specifications in (3GPP TS36.104, 2011) defines pico and HeNB in addition to the macro BS class. Additionally the Closed Subscriber Group (CSG) concept was introduced from the first LTE Release onwards in relation to the HeNB subsystem (see also Section 2.1.8.5).

While pico eNBs can simply be described as small base stations (the typical downlink TX power for a micro eNB is 36 dBm, and for a pico eNB it is 24 dBm, both for 10 MHz bandwidth), having the same architecture as a macro eNB (e.g. the X2 interface is present between macro and pico eNBs and amongst pico eNBs), this is definitely not the case for HeNB. Specifications of the HeNB are created by 3GPP with clear directions given by the Femto Forum (http://www.femtoforum.org).

The Next Generation Mobile Networks (NGMN) Alliance also puts requirements to the multi-layer scenario with 'Interaction between Home and Macro BTS' in the SON session of the (NGMN, 2010). In particular, NGMN addressed the interference management between macro eNB and HeNB.

The 3GPP SON features introduced originally for the macro environment in 3GPP are all applicable for micro and pico eNBs, too. The HeNBs, however, are required to support self-management procedures, which do not necessarily require signalling interactions with the macro, micro or pico eNB neighbours located in the same HetNet environment. Figure 10.2 gives an overview of the HeNB related 3GPP standardisation topics from Release 8 to Release 10 which are explained in more detail in the following.

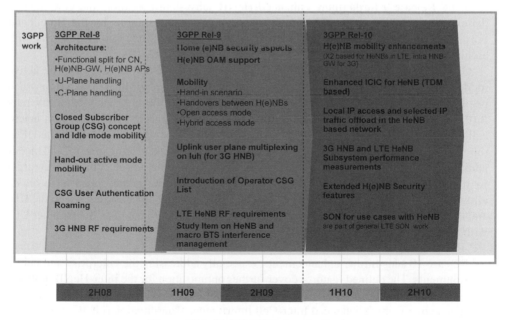

Figure 10.2 HeNB 3GPP Standardisation.

3GPP Release 8

3GPP Release 8 defines the HeNB basic architecture and *closed access mode* concept (which only allows CSG users to be served, see Section 2.1.8.5) focusing especially on the functionalities needed in residential HetNet deployments. HeNB specific functionalities like Network Listening Mode (NLM, which is responsible for acquiring information about the surrounding network environment, see also Section 10.3.2) and Autonomous HeNB power calibration were introduced to support self-configuration capabilities for HeNBs deployed in an uncoordinated manner. Defining the idle mode mobility rules for the HetNet scenarios with CSG cells from the first LTE Release onwards solved the legacy terminal issues, because based on the 3GPP Release 8 definitions any LTE terminal, which has a CSG subscription (i.e. it is allowed to camp on a CSG cell) is aware of CSG specific rules and functionalities.

The active mode mobility (handover) from a HeNB to a macro cell is based on the same principles as the macro layer internal mobility. The support of this *outbound mobility* (from an uncoordinated HeNB to a macro/micro/pico eNB) is a necessary feature for residential scenarios in order to maintain the ongoing call or data connection when the subscriber is moving out from the indoor HeNB coverage to outdoor macro cell coverage. Additionally the CSG subscriptions are supported for the roaming scenarios already from 3GPP Release 8 onwards.

3GPP Release 9

Self-optimisation use cases and supporting procedures were introduced in 3GPP Release 9 for macro eNB scenarios. Those procedures are also applicable to HetNet deployments with macro; small cell or inter-small cell scenarios if an X2 interface is present. The macro eNB–HeNB interference management was addressed with dedicated study items, for FDD mode in (3GPP TR36.922, 2011) and for TDD mode in (3GPP TR36.922, 2011), which finally led to the corresponding specification work in 3GPP Release 10.

3GPP Release 9 furthermore enhanced HeNB subsystem operation and CSG related definitions, including the OAM system (based on recommendations by the Broadband Forum, for example, BBF TR-069, 2010 and BBF TR-196, 2011) and H(e)NB security architecture and requirements.

One of the key features for HetNet environments is the handover support. The active mode *inbound mobility* (from a macro/micro/pico eNB to an uncoordinated HeNB) has also been introduced in 3GPP Release 9. In a typical scenario the high HeNB density may furthermore require Physical Cell Identity (PCI) reuse amongst the HeNBs within macro cell coverage, which may lead to PCI confusion in the serving macro cell when the UE is reporting neighbour cell measurement results; therefore the serving macro cell needs additionally the cell identity of the measured HeNB cell in order to select the correct target for the handover request. Two new access modes for HeNBs were defined: *open access mode* allowing all UEs to use the HeNBs (i.e. the cell appears as any other normal cell to the UE) and the *hybrid access mode* allowing all UEs still to use the cell, but the CSG member UEs with higher priority.

3GPP Release 10

3GPP Release 10 adds further enhancements, especially addressing the enterprise and public environments. The defined features cover further optimisation for the inter-HeNB mobility (over X2 connection), Local IP Access (LIPA) from HeNB, and methods for interference management, of which enhanced Inter-Cell Interference Coordination (eICIC) is certainly the most relevant feature (see Section 10.4).

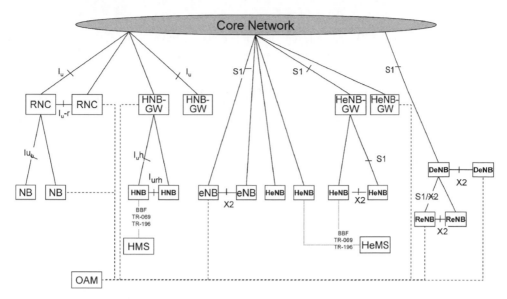

Rel-10 specs support X2 between two HeNBs if served by the same MME

Figure 10.3 Network architecture for Multi-RAT HetNet.

As for macro cells, specifications for HeNBs solutions assure that they can work in real multi-vendor environments with open interfaces between the network elements, and can be fully integrated with other 3GPP based cellular systems. Simultaneously to the LTE HeNB specification work similar standardised concepts have been defined for 3G Home NodeBs (HNB).

The functional split of the LTE HeNB subsystem and the CSG mobility principles are defined in (3GPP TS36.300, 2011). The 3G HNB subsystem Stage 2 descriptions are described in (3GPP TS25.467, 2011).

10.2.1 Network Architecture for HetNet

Figure 10.3 shows a complete multi-RAT Heterogeneous Network architecture which, for completeness, includes 3G and LTE macro (e)NBs, H(e)NBs and LTE Relays.

10.2.1.1 HeNB Deployment Scenarios

The LTE HeNB subsystem architecture, as defined in 3GPP Release 8, allows three different architecture deployment scenarios:

- HeNB with direct S1 interface connection to the Evolved Packet Core (EPC): the HeNB connects to the EPC like a macro eNB and may have S1 connections to multiple Mobility Management Entities (MME)/Serving Gateways (SGW).
- HeNB with EPC connection via a HeNB Gateway (HeNB GW): the HeNB GW serves as a concentrator for the C-Plane in case of high number of HeNBs and may terminate the User

Plane towards the HeNB and towards the Serving GW. The HeNB GW appears towards the
MME during S1 setup like an eNB with multiple cells and towards the HeNB like an MME.
• HeNB with EPC connection via a HeNB GW for the control plane (C-Plane) only.

There are some HeNB specific functionalities (e.g. paging optimisation), which are part of
HeNB GW. In HeNB GW-less scenario these functionalities are supported by the MME.

The CSG subscription concept is supported in LTE specifications from 3GPP Release 8
onwards. Therefore it has been possible to define the MME to be responsible for the CSG access
control in all scenarios. The LIPA support in the HeNB subsystem requires the integration
of Local-GW functionality into the HeNB, which is described in (3GPP TS36.300, 2011).
The HeNB is also required to support the S5-interface towards the EPC.

10.2.1.2 Residential and Enterprise Scenarios

The HeNB features required for HetNets with enterprise scenarios are different from the HeNB
features required for residential scenarios:

• The residential HeNB is typically operated in closed or hybrid access mode, in some seldom
  cases in open access mode. The handovers with the macro layer are performed via the S1
  interface. Inter-HeNB mobility is not foreseen.
• The enterprise HeNB should support handovers with the macro layer (inbound and
  outbound), and with neighbouring HeNBs. For the inter-HeNB mobility the handover
  procedures over X2-connection are recommended in order to reduce the signalling load in
  the MME caused by the enterprise internal handovers. The enterprise HeNBs are typically
  operated in hybrid or open access mode to ensure coverage also for visitors; however, the
  closed access mode is also possible.

10.2.1.3 3G HNB Subsystem

In the 3G HNB subsystem concept the presence of legacy systems and terminals requires
particular solutions for the CSG subscription control. Hence, the 3G HNB Gateway (HNB GW)
has some dedicated functionalities, which are not required for the HeNB GW. The main
differences of the 3G HNB subsystem compared to the LTE HeNB subsystem are:

• HNB GW is a mandatory network element of HNB subsystem.
• Support for legacy terminal base, CSG access control related functions in HNB GW.
• Support of Circuit Switched User Plane.
• The Iu based interface between the HNB and HNB GW contains some HNB subsystem
  specific extensions for HNB and legacy UE registration at HNB GW.

10.3 Self-Configuration

Self-Configuration is a key requirement for small cells. The newly added base station (BS) has
to 'integrate' itself into the network with minimum human intervention: the BS must be able to
boot, connect to the network with proper security credentials, and get the appropriate software

and parameters. The acquisition of radio parameters is called 'Dynamic Radio Configuration' DRC (see Section 4.2), which includes parameters such as neighbour list, handover configuration and pilot power settings. In addition to compute the parameters for a new BS, also parameters for existing BSs need to be adapted.

10.3.1 Auto-Connectivity and -Commissioning

Upon being placed into the network and powered on, any new base station must be able to connect in a secure way to its supervising OAM system to obtain any network element-specific or location-specific configuration parameters (see Section 4.1). Even more than for macro networks, heterogeneous network nodes will be 'off-the-shelf', that is, any required customisation will be done on-site just before going into operation rather than in the factory. With regard to the 'Auto-Connectivity and -Commissioning' phase those dependencies need to be taken into account in the preceding network planning phase. In the actual commissioning phase, the generated 'static' configuration data set are downloaded to the nodes. As HeNB nodes are deployed in an 'uncoordinated' manner, whereas the macro network is planned, it is obvious that the number of dependencies to be addressed here between the HeNB and the 'regular' (macro/micro/pico) eNB is rather low (e.g. aligning the frequency bands in which the eNB and the HeNB nodes should operate respectively). The auto-connectivity and –commissioning processes are rather independent with regard to the content (e.g. SW versions of the nodes) and time of execution.

The HeNB and the eNB have their respective self-configuration infrastructures (see Figure 10.4): the HeNB network has a standardised Itf-S (3GPP 'Type 1' OAM interface)

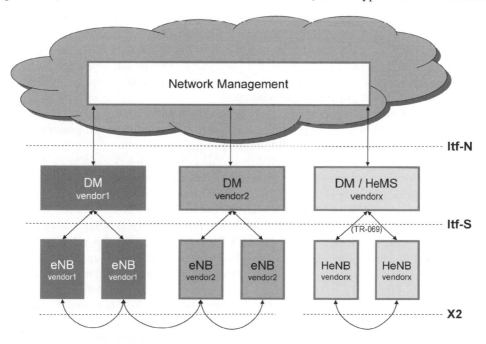

Figure 10.4 LTE network architecture in multi-vendor scenario.

based on the Broadband Forum TR-069 standards family (BBF TR-069, 2010), whereas for eNB the Itf-S is not standardised and only the security-related part of the auto-connectivity process is standardised (see Section 4.1).

The self-configuration processes can be controlled jointly for eNB and HeNB across the Itf-N (3GPP 'Type 2' OAM interface). This requires that both, HeNB and eNB domain, are attached to a common Network Management system and implement the agent functionality of the Self-Configuration Management IRP (see Section 4.1.2).

DRC comprises mainly the allocation of Physical Cell IDs (PCI), the assignment of Tracking Area (TA) Code and setup of neighbourships. For the case of HeNB/eNB interworking, fixed ranges of parameters can be pre-assigned to each domain in the planning phase (for the case of a co-channel deployment), while the assignment of PCI and TA is, however, dynamic within a domain. Inter-domain neighbourships can be established by means of Automatic Neighbour Relationship (ANR) actions (see Section 4.2.3 and 10.3.3), provided that an X2 interface is available.

10.3.2   Automatic Site Identification and Hardware-to-Site Mapping

Section 4.1.2 explains how to map a base station to the site designated in the network plan and describes several mechanisms to do so, ranging from manual interaction to fully automatic site identification. In HetNet scenarios, a manual mechanism cannot be applied anymore. Hence only the following methods are relevant:

- Satellite-based positioning (e.g. using GPS): this method is known to work very reliably in macro scenarios with high inter-site distances. In HetNet scenarios, however, the positioning inaccuracy starts to become problematic which may lead (in connection with an overall high site density and many sites having co-located equipment of different RATs/radio layers) to potential ambiguous site identification. For indoor deployments satellite-based positioning is rather not applicable due to missing/bad reception of the satellite signals.
- Radio network-based positioning/identification:
  - A location-based service available via an already existing radio network (e.g. the available macro layer) can be used to position the (H)eNB. The location-based positioning is of course less accurate than the satellite-based positioning method.
  - HeNB Network Listening Mode (NLM): in this case the HeNB initially switches to a listener mode and hence is able to obtain coarse information about cells in its proximity.
- Radio Frequency Identification (RFID) tags: RFID transfers data between a reader device and an electronic tag which is attached to an object using radio waves. If the site preparation process (providing backhaul connectivity and power) includes the possibility to define a site by marking it with an RFID tag, the Network Element can acquire this information when deployed with a built-in or temporarily attached RFID reader. This is a method particularly suited to indoor deployments, because only reception of close range signals is required and the sites are potentially defined 'on the fly' by suitable locations with regard to LAN connectivity and power. A newly deployed network element will then send the described site-related information and its hardware ID to the OAM system (cf. Figure 10.5). This server will respond with the required commissioning steps needed for the network element, such as an inventory update and the, software download

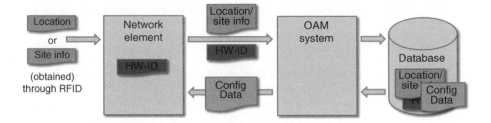

Figure 10.5 Automatic site identification and hardware-to-site-mapping.

One of the major challenges for site identification in case of HetNets is how to avoid identification ambiguities in very dense deployments. In Bandh and Sanneck (2011) it is proposed to extend the above methods by exploiting all related available information in the OAM system which can be acquired on-site by the NE. These are the site functions (RATs, cell types, frequency/sector configuration) which can be evaluated against the actual NE hardware inventory. The site status (already installed versus not installed) can be used to remove sites from the established set of potential sites. Furthermore (to the extent supported by the actual hardware), info acquired by the Network Element with regard to the antenna and auxiliary equipment connected to it, can be taken into account.

10.3.3 Automatic Neighbour Relations (ANR)

ANR is a feature which does not require much interaction amongst the nodes in a network neighbourhood. Part of the functionality is located in the terminal and thus is not affected by the network structure. As explained in Section 4.2.3, a UE connected to a given cell (Cell A), measures an unknown neighbour cell (Cell B). Within the same LTE frequency layer, the terminals would do that continuously. Outside the frequency layer (inter-frequency or inter-RAT), the serving cell needs to configure the UE to conduct inter-frequency measurements. Upon reception of an unknown PCI (of Cell B), Cell A asks the terminal to measure and report the E-UTRAN Global Cell ID (EGCI) of Cell B. Alternatively, similar information can be supplied by the Domain Management; however, this method is not applicable if the involved cells belong to different domains (e.g. Cell A belongs to a macro eNB and Cell B belongs to a HeNB). Once the neighbour cell has been identified, the automatic X2 setup procedure takes place with the support of the MME.

Hence, the UE-based ANR function is applicable to the HetNet scenario. This is due to the fact that detection and reporting of the neighbour cell is delegated to the UE. The only exception is for the transport part, that is, the automatic X2 interface set-up procedure, which of course does not work in scenarios where the X2 interface is not foreseen (between eNB and HeNB).

10.4 Self-Optimisation: Interference Management

10.4.1 Interference Characteristics in HetNet Scenarios

When migrating from networks with planned macro cells towards HetNet with a variety of different base station types, the overall interference foot print also changes significantly. As an example, in macro-only networks, the first step in managing the interference is often done by network planning before installing new sites. Network planning is an efficient instrument for ensuring that adjacent sites do not create too much interference for each other, by selecting appropriate site locations, using efficient sectorisation, optimised antenna downtilt, and so on. However, when migrating towards HetNet scenarios, it is not possible to rely on the same degree of interference shaping via initial radio network planning and selection of site locations. Although small cells such as pico eNBs are assumed to be operator installed, they are typically installed without using same detailed degree of radio network planning for site selection as it is the case for macro eNBs. Furthermore, the installation of privately owned HeNBs is by definition an uncoordinated deployment, where the operator has no control about the location where those small HeNBs are installed. Thus, user deployed HeNBs could in principle be placed in the close vicinity of operator installed pico eNBs, or multiple HeNBs may experience very short inter-site distances if installed by different users, and so on. Uncoordinated deployment of HeNBs is therefore often said to result in chaotic interference scenarios, which requires active use of autonomous self-adapting interference management techniques to ensure satisfactory overall system performance for all users. Depending on the exact HetNet scenario, the experienced downlink and uplink interference levels are therefore likely to be significantly different compared to macro-only scenarios, and also the Dominant Interference Ratio (DIR) is often more significant. The DIR basically expresses the ratio of the strongest experienced interferer as compared to the sum of other interferers. Thus, the higher DIR for HetNet scenarios (as compared to macro-only) is an indicator of potentially higher relative gain potential from applying interference mitigation techniques where the dominant interferer(s) is avoided, or reduced by using Inter-Cell Interference Coordination (ICIC) and/or advanced receiver architectures with capabilities of suppressing or cancelling dominant interferers.

In this chapter the focus is on ICIC techniques for HetNet scenarios. Basic interference management techniques in the power- and frequency-domain are first outlined in Section 10.4.2 followed by the description of the more elaborate 3GPP Release 10 time-domain enhanced ICIC (eICIC) in Section 10.4.3. The discussion of ICIC techniques for HetNet scenarios is closed by giving pointers to further innovative solutions that are under development for 3GPP Release 11.

10.4.2 Basic Interference Management Techniques

This section outlines the basic interference management techniques that are applicable for current LTE Releases, by effectively utilising the available degrees of freedom for selecting reasonable carrier deployments for the different base station types, as well as power control techniques. Thus, although many of the methods outlined in the following are explained for LTE, they are in many cases equally applicable to 3G/HSPA. The baseline techniques are outlined in the following for two different scenario examples; namely for macro environments where either HeNBs or pico eNBs are added for additional capacity and coverage improvements.

10.4.2.1 Scenarios with Macro eNBs and HeNBs

The downlink interference scenario for co-channel deployment of macro eNBs and HeNBs is illustrated in Figure 10.6. In this context, co-channel deployment refers to the case where both base station types are deployed on the same carrier frequency and bandwidth. The most challenging interference scenario occurs for the case where the HeNBs are configured as CSG,

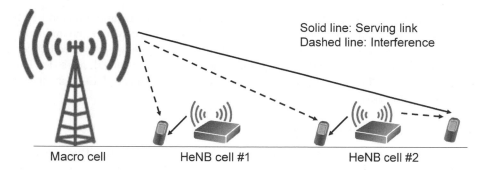

Figure 10.6 Downlink interference scenario for co-channel deployment of macro eNBs and HeNBs.

so only a small set of dedicated terminals with matching CSG IDs are allowed to be served. Thus, assuming that HeNB Cell #2 in Figure 10.6 is CSG, the UE at the macro cell-edge will continue to be served by the macro layer, although being very close to HeNB Cell #2. Only if the UE had matching CSG ID it would be allowed to connect to HeNB Cell #2. This basically implies that macro UEs in the close vicinity of non-allowed HeNBs will experience high interference from the HeNB(s), and therefore also high DIR. The latter is often said to result in macro-layer coverage holes, since macro UEs will experience too high interference in the dominance area near non-allowed HeNBs. The second downlink interference situation is the case where HeNB Cell #1 is placed near the macro cell. For this case, the macro cell will act as the aggressor node, causing potentially high interference for the UE served on HeNB Cell #1. However, this case is considered less critical as compared to the aforementioned problem of macro layer coverage holes as a result of interference dominance areas from CSG HeNBs.

The corresponding uplink interference paths are illustrated in Figure 10.7. Here it is observed that the situation is reversed since the cell-edge macro UE close to HeNB Cell #2 has now become the aggressor for the uplink, while it was the victim for the downlink. Due to active use of fractional path-loss compensation uplink UE power control, the macro UE far from its serving cell is likely to transmit with relative high power, resulting in potentially strong interference for the nearby HeNBs. However, in most scenarios, the macro UEs are typically only scheduled on a subset of the available bandwidth, meaning that HeNB Cell #2 is still having bandwidth portions without excessive interference from nearby macro UEs. The second uplink interference case that requires attention is the one where the UE served in HeNB Cell #1

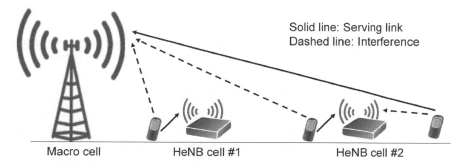

Figure 10.7 Uplink interference scenario for co-channel deployment of macro eNBs and HeNBs.

is causing uplink interference to the macro cell. Uplink interference at the macro cell from HeNB users shall be carefully controlled to avoid unwanted effects, where the uplink macro cell performance is harmed, or in worst case results in reduced uplink macro cell coverage. However, also here the 3GPP standardised uplink power control is beneficial, since HeNB users typically transmit with much lower power levels as compared to macro UEs, since they have lower path-loss towards their serving cell.

10.4.2.2   Power Domain Solutions

One of the simplest methods for reducing the effects of interference from HeNB(s) to macro UEs is to apply autonomous HeNB power calibration solutions. This is a technique where the HeNB transmit power is reduced to minimise the interference to macro UEs while still providing sufficient performance to the HeNB UEs. Autonomous HeNB power calibration is often based on local sensing, where the HeNB use NLM based on its built-in UE receiver capability to conduct measurements on signals received from other co-channel deployed eNBs and/or HeNBs. The most popular autonomous HeNB power calibration method is based on measuring the signal strength from the strongest co-channel deployed macro eNB, followed by adjusting the HeNB transmit power as a linear function of this measurement. Hence, if the HeNB is at the macro cell edge, the HeNB will transmit with lower power as compared to HeNBs positioned closer to a macro eNB. By using such techniques the HeNB transmit power is adjusted to avoid dominance over the local experienced macro cell signal level. The autonomous HeNB power calibration solution could be assisted by further fine-tuning of the HeNB transmit power level based on, for example, UE measurements, and other collected measurements and network performance statistics. The autonomous HeNB power calibration method may therefore also consist of additional configuration parameters that may be subject to SON optimisation, based on for example, collection of overall network KPIs from both the macro layer and HeNB-layer.

It should be noted that adjustment of the downlink HeNB transmit power level also affects the coverage area of the HeNB, assuming that serving cell selection is based on downlink measurements such as Received Signal Reference Power (RSRP) or Reference Signal Received Quality (RSRQ). Thus, if reducing the HeNB transmit power, the service area of the HeNB is also reduced. This implies that connected HeNB UEs will in general experience smaller path loss towards the HeNB, as compared to cases where the HeNB is transmitting with full power. Due to the use of uplink UE fractional path loss compensation power control, the latter effect translates to having lower HeNB UE transmit power as a result of using autonomous HeNB power calibration. Therefore, the use of autonomous HeNB power calibration also has positive effects on the uplink interference from HeNB cells to macro cells. In addition it is generally recommended to use different uplink UE power control parameterisation for users served by macro eNBs and HeNBs, respectively. Thus, calling for an additional dimension of parameter optimisation when introducing co-channel deployed HeNBs in macro cell networks.

10.4.2.3   Frequency Domain Resource Partitioning

Although autonomous HeNB power calibration solutions are attractive for reducing the probability of macro layer coverage holes from non-allowed CSG HeNBs, such techniques cannot completely eliminate the risk of coverage holes. Thus, in order to guarantee no macro layer coverage holes as caused by non-allowed CSG HeNBs, some resource partitioning between macro and HeNBs is needed, such that some macro layer resource are free of

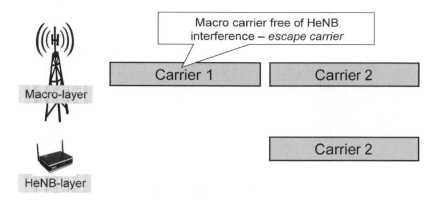

Figure 10.8 Simple illustration of so-called escape carrier configuration for deployment of macro eNBs and HeNBs, assuming two available carriers.

interference from HeNBs. It is therefore recommended to exploit the carrier deployment dimension for implementing such resource partitioning by assigning only a sub-set of the available carriers for HeNBs, such that at least one of the carriers used by macro is free of HeNB interference. The macro carrier free of HeNB interference is often called the escape carrier, as macro UEs close to non-allowed HeNB(s) can always 'escape' to this carrier to avoid the potentially damaging HeNB interference. Similarly, macro UEs that are further distant from non-allowed HeNBs can use any of the available carriers. An escape carrier configuration is depicted in Figure 10.8.

The so-called escape carrier configuration and use of autonomous HeNB power control is therefore recommended for deployment of CSG HeNBs. Further optimisation of such configurations can naturally be achieved via the use of inter-frequency carrier load balancing at the macro layer, taking into account that some of the macro UEs may experience significantly different interference levels on the available carriers, depending on whether they are in the close vicinity of non-allowed CSG HeNBs or not. Furthermore, users configured with a CSG ID should preferably be served on the same carrier as their HeNBs with matching CSG ID to ensure that they will be served on HeNBs whenever possible, that is, maximising the offload from macro to HeNBs whenever possible.

10.4.3 Scenarios with Macro eNBs and Micro/Pico eNBs

Operator deployments of pico (or micro) eNBs together with macro cells are less problematic from an interference management point of view, as compared to deployment of CSG HeNBs. If a macro UE is approaching a co-channel deployed pico eNB it will simply experience handover to that eNB when, for example, the RSRP or RSRQ of the pico eNB becomes dominant over the same measurement at the macro eNB, corresponding to handover event A3 (3GPP TS36.331, 2011). Thus, co-channel deployment of pico or micro eNBs does in general not result in the risk of creating coverage holes. It is therefore a generally accepted conclusion that co-channel deployment of macro and pico eNBs can work without any use of active inter-cell interference management. However, as for the case with macro eNB and HeNB, it is recommended to use

Figure 10.9 Range extension.

different uplink UE power control parameterisation depending on whether the UE is served by macro or small eNBs with reduced transmit power.

Although co-channel deployment of macro and pico eNBs is possible without any interference management, the system performance can be further optimised by using various techniques as explained in the following. For optimising the overall system performance, it is often beneficial to push more users to the pico eNBs. This is possible by using so-called range extension (RE) for the pico eNBs, where a cell individual offset is applied to the RSRP or RSRQ for the pico eNB allowing more users to be served (see Figure 10.9). However, care should be taken on how the cell individual offset is set for pico eNBs, as using too high cell individual offset can result in downlink interference problems for the pico UEs, as those will then start to experience too high interference from the co-channel deployed macro eNBs. It has therefore been established that there can be additional benefits from reducing the experienced interference from the macro eNBs to the pico UEs, as this basically results in higher offload potential and therefore a gain in overall systems performance. The simplest method for achieving the latter is to deploy the macro and pico eNBs on different carriers. By doing this, the picos will not suffer from macro interference and therefore the coverage area of each pico eNB is increased. However, in order to experience an overall gain from deploying the macro and pico eNBs on different carriers, there needs to be sufficient amount of pico eNBs, otherwise it would result in inefficient frequency resource utilisation if there are only a few picos sharing one frequency carrier. Hence, for initial deployment of pico eNBs, co-channel deployment with macro eNBs is recommended.

10.4.4 Enhanced Time-Domain Interference Management: eICIC

10.4.4.1 Basic TDM eICIC Concept Description

3GPP Release 10 introduces a new downlink time domain (TDM) enhanced inter-cell interference coordination (eICIC) concept for HetNet cases (3GPP TS36.300, 2011). The TDM eICIC concept is mainly designed for operators having only one carrier available for LTE deployment, and therefore don't have the option of for applying the simpler frequency domain interference management techniques described in the previous section, with using different carrier configurations for macro and small eNBs. As the name indicates, the TDM eICIC concept relies on interference coordination in the time domain on a subframe resolution (1 ms) as illustrated in Figure 10.10. The basic principle is that some subframes are partially muted by using Almost Blank Subframes (ABS). An ABS is a subframe where only Common Reference Symbols (CRS) and potentially other mandatory common channels are transmitted, while no data transmission occurs. Thus, during time instants where a cell transmits ABS, it generates much less interference to its surrounding cells. Referring to Figure 10.9, it basically means that during subframes where CSG HeNBs are transmitting ABS, nearby macro UEs can be served as they experience reduced HeNB interference. Similarly, when the macro eNBs are using ABS, pico UEs experience reduced interference and thus the picos can serve users from a larger geographical area as this would otherwise be possible in the presence of full interference from macro eNBs.

10.4.4.2 Cell Associations and ABS Coordination

In order to exploit the full benefit of TDM eICIC it is required that base stations are fully time synchronised, and sub-frames are also aligned illustrated in Figure 10.10. As there are no strict 3GPP requirements for inter-eNB time synchronisation (for FDD cases), it is up to the network vendor to select the methods and accuracy of the time synchronisation. Secondly, it is important to have the ABS muting patterns well-coordinated and configured between different cell types to achieve the performance benefits of TDM eICIC. The latter requires establishment of cell associations, so for example, it is known:

- Which macro eNB(s) a certain pico eNB is experiencing dominant interference from.
- Which macro-cell different HeNBs are located at.

For the first case, it would be beneficial to know if a pico is, for example, located at the cell border between two macro eNBs. If this is the case, it would basically mean that those two macro eNBs should in principle use ABS at the same time to effectively reduce the experienced

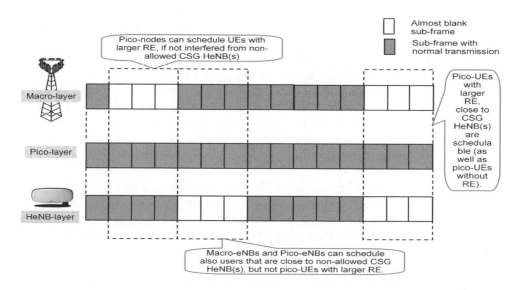

Figure 10.10 TDM eICIC concept, with ABS alignment on different HetNet layers.

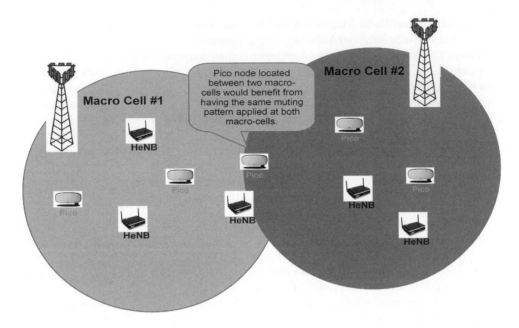

Figure 10.11 Sketch of cell associations for a typical heterogeneous network scenario with various cell types.

macro cell interference at the pico eNB in a coordinated manner as indicated in Figure 10.11. In 3GPP Release 10, the cell associations are assumed to be configured from OAM (note here that cell associations are closely related to neighbour cell lists). This allows using centralised SON functionalities for semi-statically building cell associations based on various collected network statistics such as, for example, UE handover statistics and other measurements. For HeNBs, reported NLM measurements can furthermore be used to estimate in which macro cells they are deployed, and so on.

As it can generally not be assumed that HeNBs have an established X2 interface towards other HeNBs and macro/pico eNBs, the ABS muting pattern for HeNBs is assumed to be semi-statically configured by OAM in a centralised way. Thus, the ABS pattern for HeNBs can be slowly adjusted from OAM as a function of collected network statistics for overall network performance optimisation. Notice that for optimal performance, the macro eNBs would need to also know the ABS muting pattern applied at the HeNBs. Such knowledge will be used by the macro eNBs to only schedule macro UEs in the dominance area of non-allowed CSG HeNB(s) when the HeNB(s) are using ABS.

For macro and pico eNBs with established X2 interface, the configuration of the ABS muting pattern can be more dynamically negotiated and optimised. For this purpose, new information elements (IEs) have been added to the X2 application protocol (AP) for 3GPP Release 10 to facilitate collaborative configuration of ABS muting patterns between eNBs. (3GPP TS 36.423, 2011) describes those new IEs as follows:

- **ABS information IE:** IE in the Load Information message, which can be used by an eNB to signal which ABS muting pattern it is currently using to another eNB.

- **Invoke IE:** IE in the Load information message, which provides an indication that the sending eNB would like to receive ABS information.
- **Downlink ABS status IE:** IE in the Resource Status Response message, which expresses the percentage of resource blocks in ABS allocated for UEs protected by ABS from inter-cell interference. This includes resource blocks of ABS unusable due to other reasons. The denominator of the percentage calculation is indicated in the Usable ABS Information. Such an update can be requested by the Resource Status Request message where the Report Characteristic 'ABS Status' was added.

In order to further illustrate the usage of those new X2 IEs consider the following example with macro and pico eNBs. As described earlier, the TDM eICIC concept is designed such that some subframes are configured as ABS at the macro eNBs, allowing the pico eNBs to serve users from a larger geographical area during those subframes. The macro eNBs therefore decide in how many subframes they want to use ABS. Once decided, the macro eNB sends ABS information to its neighbouring macro eNBs and pico eNBs such that they all know the applied ABS muting pattern at the macro eNB. Such information is beneficial for the other macro eNBs so they can try to use similar ABS muting patterns, as well as for the pico eNBs so they know in which subframes they can expect reduced interference from macro. A pico eNB can send the Invoke IE to a macro eNB to indicate that it would like to receive ABS information from the macro. If a pico eNB sends the Invoke IE, it can be used to indicate to the macro eNBs that it would be desirable for the pico eNB if the macro eNB starts configuring more subframes as ABS. The pico eNB can also send a Resource Status Update message with ABS status, providing additional knowledge for the macro to better determine the best ABS configuration. An example for the exchange of X2 signalling messages between a macro eNB and a pico eNB for dynamic coordination of ABS muting pattern is depicted in Figure 10.12.

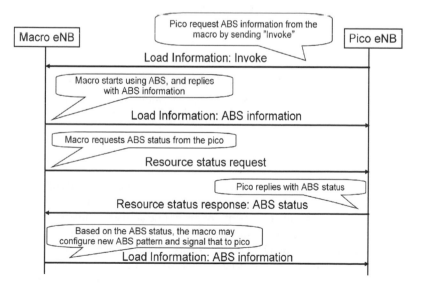

Figure 10.12 Example of inter-eNB X2 signalling for distributed dynamic coordination of ABS muting pattern.

Although the new X2 signalling for inter-eNB coordination of ABS muting patterns is standardised for 3GPP Release 10, there is still some implementation freedom on how to use it in practice. This basically means that the exact eNB algorithms for, for example, deciding on the ABS muting pattern to be used at the macro eNB are largely vendor specific.

10.4.4.3 Terminal Aspects of TDM eICIC

The use of TDM eICIC with ABS muting patterns results in potentially large interference fluctuations at the terminals, depending on whether the interfering (H)eNBs are using normal transmission or ABS. Therefore 3GPP Release 10 terminals support configuration of time-domain restrictions for the following measurements:

- **Channel state information (CSI):** Measurements from the UE used by the network for packet scheduling and link adaption decisions. CSI include Channel Quality Indicator (CQI), Precoding Matrix Indicator (PMI), and Rank Indication (RI).
- **Radio resource management (RRM):** Measurements from the UEs used by the network for handover decisions. Examples of RRM measurements include Reference Signal Received Power (RSRP) and Reference Signal Received Quality (RSRQ).
- **Radio link monitoring (RLM):** Measurements to determine Radio Link Failure (RLF).

The network can configure these measurement restrictions for 3GPP Release 10 UEs with dedicated Radio Resource Control (RRC) signalling. The measurement restrictions therefore only apply for RRC Connected mode UEs. Configuration of restricted RRM measurements is especially beneficial for having optimised mobility triggers from UEs in networks with TDM eICIC enabled.

As the aforementioned UE measurement restrictions cannot be configured for 3GPP Releases 8 and 9 legacy UEs, such terminal types may to experience lower performance than 3GPP Release 10 UEs in networks with TDM eICIC enabled.

10.4.4.4 Summary of the TDM eICIC Concept

The main characteristics of 3GPP Release 10 TDM eICIC are summarised in Table 10.2. As mentioned, this interference management concept is mainly developed for operators having only one LTE carrier for deployment of multiple base station types.

10.4.5 Outlook on Further Interference Management Innovations

Further innovations for interference management and mitigation are currently under study in 3GPP for 3GPP Release 10. Amongst others, autonomous carrier-based interference management techniques are under development, for providing additional instruments for reaching the optimal usage of available spectrum assets for different eNB types depending on various factors. Further enhancements of TDM eICIC are also under discussion for 3GPP Release 11, including the introduction of UE advanced receiver types that can further help improve the overall network performance. As mentioned in Section 10.4.1, HetNet scenarios are often characterised by high DIR, so having receiver capable of cancelling the most dominant interferers could yield attractive benefits.

Table 10.2 Summary of LTE Release 10 TDM eICIC benefits, requirements, and characteristics

TDM eICIC benefits	Protect UEs from interference originating from non-allowed CSG cells.
	Increased offload to pico nodes via reduced co-channel macro interference.
ABS characteristics	ABS are subframes with reduced transmit power (including no transmission) on some physical channels and/or reduced activity. The eNB ensures backwards compatibility towards UEs by transmitting necessary control channels and physical signals (e.g. CRS) as well as System Information
Time synchronisation	Requires subframe strict time synchronisation between base station nodes.
	No mandatory 3GPP time-synchronisation requirements for FDD. Several implementation alternatives; e.g. GPS and backhaul based solutions such as the IEEE Precision Time Protocol standard IEEE1588.
ABS Muting pattern configuration	Static configuration of ABS muting pattern from OAM for HeNBs.
	Distributed dynamic coordination of ABS muting pattern between macro and pico layers via standardised X2 signalling. The muting pattern shall be carefully selected to also protect downlink common control channels (e.g. paging and PBCH) that always appear in certain subframe numbers as well as uplink HARQ.
Cell associations	Configured from OAM. Amongst others, cell associations are needed to configure which macro and pico eNBs shall exchange X2 signalling when dynamically coordinating ABS muting patterns.
Network RRM	eNB RRM functions such as the MAC packet scheduler shall be updated so UEs requiring protection from aggressor nodes are only scheduled when those use ABS.
UE requirements	Best performance achieved with 3GPP Release 10 UEs supporting configuration of restricted RLM, RRM, and CSI measurements. 3GPP Release 8/9 legacy UEs may potentially have reduced performance as compared to 3GPP Release 10 UEs with eICIC supporting mechanisms.

10.5 Self-Optimisation: Mobility Aspects; MRO and Traffic Steering

10.5.1 Mobility Robustness Optimisation

The general goal of MRO is to minimise RLFs and call drops due to bad HO decisions, minimise unnecessary HOs (e.g. ping-pongs) and minimise idle mode problems (UE must always be 'paged').

The main description of MRO is given in Chapter 5. In this paragraph only the differences with regard to the standard case are discussed. H(e)NBs will be treated first, since they involve the main specialities. Pico and enterprise scenarios are shortly discussed at the end.

A very important question is to what extent proper mobility (with seamless handovers) is required from macro (e)NB to H(e)NB, and vice versa. As a matter of fact the interruption through an RLF can be almost negligible if the re-establishment procedure is successful. Furthermore, the mobility between macro (e)NB and H(e)NB is much less challenging since the velocity on those cell boundaries is typically very small.

Irrespective of this question here the applicability of existing MRO, as introduced in Section 5.1, solutions to the H(e)NB-macro (e)NB cell are analysed.

Recall that 3GPP provides the following tools for MRO:

- RLF indication via X2 (3GPP Release 9): for root cause evaluation of occurred RLFs.
- HO Report via X2 (3GPP Release 9): for root cause evaluation of occurred RLFs.
- Mobility Change procedure via X2 (3GPP Release 9): for negotiation of HO parameters.
- Too early/late and so on. KPIs via Itf-N (3GPP Release 9): sending statistics to NM.
- Exchange/negotiation of interRAT parameters via RIM: may be in 3GPP Release 11.
- RLF reports from the terminal (3GPP Release 9, updated in 3GPP Release 10): the terminal provides measurements preceding the RLF.

The definition of standard messages does not mandate where the actual MRO corrections should be located. In fact different solutions can be undertaken:

- *Centralised*: in this case the root cause evaluation is via X2, but the actual MRO correction runs on DM level.
- *Distributed*: the root cause evaluation is via X2 and the actual MRO correction runs locally in the base stations (possibly using the mobility change procedure via X2.

Both approaches will obviously have problems to run between macro eNB and H(e)NB (no X2 interface, and different Domain Managers). As a consequence the mobility parameters between macro (e)NB and H(e)NB:

- *Can stay with their default values.* It is still possible to configure dedicated and 'good' defaults, tailored for macro (e)NB -H(e)NB. Furthermore, we have already mentioned that some mobility problems between macro (e)NB and H(e)NB might even be accepted.
- *Can be optimised manually* (at least for H(e)NBs which are critical and/or show extreme mobility problems.
- *Can be optimised by local algorithms* (similar to the distributed approach). This would require a H(e)NB-specific extension of the current MRO solution in the macro (e)NBs (for inbound mobility). For outbound mobility, the H(e)NBs would also need such a local algorithm. Note that this solution would need to live without the root cause evaluation specified in 3GPP Release 9 via X2. But still, some adaptation should be possible, for instance depending on the position of the H(e)NB within the cell.

This latter remedy certainly has much more commonalities with the distributed approach.

The previous description holds for the intra-LTE case, both intra- and inter-frequency. As far as *inter-RAT* is concerned, specification is ongoing. Information exchange will probably be done via the RIM procedure which, in contrast to the X2 interface, would also be available for H (e)NBs. As a consequence, distributed interRAT MRO solutions are applicable to the H(e)NB case as well, although tailored for the macro case. On the other hand, centralised MRO solutions are again a matter of interaction between the H(e)NB and macro (e)NB DM, equivalent to the intra-LTE case explained in more detail previously.

3GPP Release 10 supports X2 in the *enterprise* scenario, so distributed MRO solutions will work. Centralised solutions are highly dependent on the (BBF TR-069, 2010) data model extension. At least inside the enterprise, that is, between cells of the same enterprise, there can be a large reuse of existing centralised MRO functionality as well.

Pico cells follow the same architecture as macro cells, that is, the X2 interface is available such that distributed MRO solutions are applicable. Centralised solutions depend on the interaction of the macro and pico DM.

10.5.2 Multi-Layer Traffic steering and Load Balancing

Traffic Steering mechanisms described in Section 5.2 are also applicable to LTE multi-layer networks. A multi-layer is a network with overlaying cells, usually with one underlay macro cell for coverage, and one or more overlay small cells for additional capacity (see Figure 10.13). The overlay small layer may than share the same carrier (co-channel scenario has gained very much attention by operators which have scarce network resources) or be located on different frequencies.

Task of Mobility Load Balancing (MLB) is to optimally distribute traffic over the different layers. Section 5.2.4.4 describes the tools and methods to do so for both, LTE intra-frequency and inter-frequency (and inter-RAT). In addition to those mechanisms, the co-channel case needs special attention since both layers use the same frequency. In this case the overall system performance (i.e. UE/cell throughput) is improved by hosting as large part of the traffic demand, which otherwise would be served by macro cells, over small cells as possible.

The *range extension* (or *expansion*, RE) is a technique used to push more users to small cells by simply adding a cell individual offset to RSRP or RSRQ (recommendation is to use a quite small RE, in the order of 3dB). As addressed previously in this chapter (see Section 10.4.3), RE brings limited potential for basic co-channel case if not used in combination with eICIC.

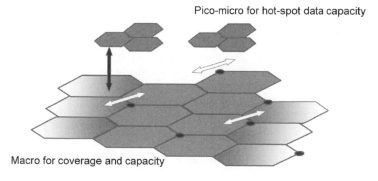

Figure 10.13 LTE Multi-layer network.

10.5.3 IEEE 802.11 (WiFi) Integration

In addition to 3GPP cellular standards, the deployment of wireless networks such as IEEE 802.11 family networks (often referred to as WiFi) in unlicensed or license-exempt frequencies is definitely an interesting alternative to cope with enormous traffic increase.

Offloading data traffic from cellular air interfaces to WiFi is highly attractive for cellular operators from a cost point of view as it provides higher bandwidth to the end-user and deployment based on local initiative.

The terminal- or consumer-driven WiFi offload is already taking place today, that is, where consumers manually connect their smartphone or laptop to their WiFi hotspot; cellular operators might often prefer a setup where they also own the WiFi infrastructure and where the system decides when offload is to be triggered.

As described in Section 2.2.5, ANDSF (Access Network Discovery and Selection Function) is a solution which leverages on 3GPP (3GPP Release 9) for user authentication and authorisation, and which gives the possibility to operators to prioritise WiFi hot-spots in desired areas.

ANDSF contains data management and control functionality necessary for providing network discovery and selection assistance data to the User Equipment as per operators' policy. ANDSF is located in the subscriber's home operator network and the information to access it should be either configured on the UE or discovered by other means.

With such a solution WiFi can now be integrated into an MNO network, giving operators the possibility to set up policies for the access network selection and basic inter-system mobility policies. Although it is a good step to integrate 3GPP and non-3GPP technologies, this solution is not providing evolved inter-system features as the ones amongst different 3GPP RATs, for example, seamless handover and QoS. As a consequence flexible and self-optimising mechanisms as specified and possible for 3GPP technologies are missing. Traffic steering and load balancing, as described in Chapter 5, are not possible in scenarios involving WiFi.

References

3GPP TS25.467 (2011) *UTRAN Architecture for 3G Home NodeB (HNB)*; Stage 2, ver.10.1.0.
3GPP TS36.300 (2011) *E-UTRA and E-UTRAN; Overall Description*, Stage 2, TS36.300, ver.10.3.0.
3GPP TS36.104 (2011) *Evolved Universal Terrestrial Radio Access (E-UTRA), Base Station (BS) Radio Transmission and Reception*, ver.10.3.0.
3GPP TS36.331 (2011) *Evolved Universal Terrestrial Radio Access (E-UTRA) Radio Resource Control (RRC)*, Protocol specification.
3GPP TS36.423 (2011) *E-UTRAN X2 Application Protocol* (X2AP) ver.10.1.0.
3GPP TR36.921 (2011) *FDD Home eNode B (HeNB) Radio Frequency (RF) Requirements Analysis*, ver.10.0.0.
3GPP TR36.922 (2011) *TDD Home eNode B (HeNB) Radio Frequency (RF) Requirements Analysis*, ver.10.0.0.
Bandh, T. and Sanneck, H. (2011) *Automatic Site Identification and Hardware-to-Site Mapping for Base Station Self-Configuration*, International Workshop on Self-Organising Networks (IWSON), May, Budapest.
BBF TR-069 (2010) *TR-069 Amendment 3, CPE WAN Management Protocol, Broadband Forum*, November.
BBF TR-196 (2011) *TR-196 Amendment 1, Femto Access Point Service Data Model, Broadband Forum*, May.
NGMN (2010) *Top OPE Recommendations, NGMN P-OPE PROJECT*, September (ed. F. Lehser) NGMN Ltd., Reading.

11

Future Research Topics

Christoph Frenzel, Henning Sanneck and Seppo Hämäläinen

SON provides effective and efficient means to overcome the operational challenges that the introduction of LTE imposes on the operation of mobile networks (cf. Section 1.1). Thereby, the focus lies on relieving human operators from time-consuming, recurring standard tasks through automation. As the research in mobile networks continues, new radio network architectures and corresponding technologies evolve enabling enhanced mobile services which improve the user experience. However, these advantages come at the cost of an increased technological complexity which challenges SON-based network management.

This chapter introduces an evolved network management paradigm, called Cognitive Radio Networks (CRN), whose goal is to increase the level of automation and flexibility of network operations beyond that of SON. In this way, CRN enables the effective and efficient operation of future mobile networks.

11.1 Future Mobile Network Scenarios

This section presents the two most prominent future radio network architecture blueprints and technologies, and outlines their network operations specific requirements.

11.1.1 Heterogeneous Networks

Multi-layer and multi-RAT networks, introduced as Heterogeneous Networks (HetNet) in Chapter 10, are an architectural paradigm with great potential to improve the capacity and coverage of mobile networks, especially in urban environments. However, HetNets also increase the complexity of network management. First, the number and variety of managed cells is increasing considerably as a HetNet architecture introduces lots of additional cells of diverse sizes. Second, Home eNBs (HeNB), a new type of network element, are located at the premises of the end users. So, operators can only partially control them because they have no

LTE Self-Organising Networks (SON): Network Management Automation for Operational Efficiency, First Edition.
Edited by Seppo Hämäläinen, Henning Sanneck and Cinzia Sartori.
© 2012 John Wiley & Sons, Ltd. Published 2012 by John Wiley & Sons, Ltd.

influence on their location or their operation times. This uncoordinated deployment scenario requires a flexible and adaptive network management. Third, the operation of network domains consisting of HeNBs and eNBs has to be coordinated.

11.1.2 Cloud RAN

Current research in mobile network architectures is, in general, directed to a distributed deployment of functionality. The aim is to move features from centralised entities in the RAN and the core network to the eNBs in order to improve performance and reduce costs.

Recently, a contrary trend, called Cloud Baseband or Cloud RAN (C-RAN), emerged. Its goal is to move functionality, especially baseband processing, from the antenna site to a central location as shown in Figure 11.1. At this central location, which can be several kilometres away from the antenna site, the baseband system modules of several antenna sites are pooled. Consequently, the antenna site consists only of a relatively simple RF unit which is connected to the central location via optical fibre.

C-RAN has some potential advantages over a conventional RAN architecture:

- Reduction of the Total Cost of Ownership (TCO) of the RAN because the simple antenna sites are less expensive and require fewer maintenance visits.
- The central location enables centralised software updates in the network which simplifies operations.
- Pooling of baseband hardware at the central location enables the distribution of traffic load, for example, between residential areas and business areas, which allows reducing the initial investment for and increasing the utilisation of costly hardware.

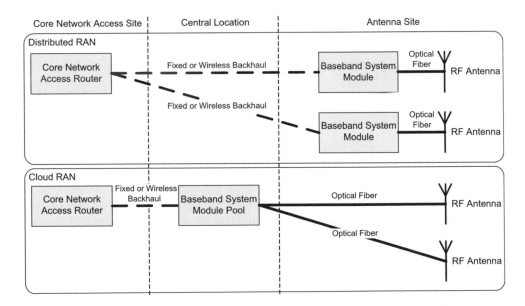

Figure 11.1 C-RAN moves the baseband system modules away from the antenna site and pools it at a central location.

- Enhanced scalability of the network through improved load balancing and mobility because handovers between the cells of one central location can be handled like intra-site handovers and, thus, require less core network and inter-element signalling.
- Increased data rates can be achieved because it enables sophisticated Coordinated Multipoint transmission and reception (cf. Section 2.1.8.4).

However, C-RAN is also facing severe technical and economic challenges due to the required optical fibre connection between the central location and the antenna sites. Besides the fact that this connection is very expensive, it also imposes challenging requirements concerning data rates, latency jitter, and latency asymmetry. Even worse, for redundancy reasons, one antenna will likely be linked to the central location through several connections in order to avoid a single point of failure.

The emergence of C-RAN will pose new requirements on network management. In general, it is foreseen that if an operator decides to introduce C-RAN, this will proceed in a gradual way. Therefore, future Operation, Administration and Management (OAM) systems have to support mixed RAN setups which add to the overall network complexity.

11.1.3 Requirements for Future OAM Systems

Several mobile network technologies are currently researched. All of them are aiming to increase the performance and decrease the TCO of future mobile networks. However, they also add to the complexity of mobile network management either by increasing the number of network elements, for example, HetNet, or by introducing new technologies which have to coexist with previous systems, for example, C-RAN. The increased complexity is challenging for technical personnel which inevitably leads to more human-induced errors. However, at the same time, high availability of the network has to be assured since customers are not willing to suffer from poor quality of service due to these network-internal issues. Furthermore, it is foreseen that the revenue per bit in mobile networks will continue to decline and, hence, the Operation Expenditure (OPEX) of future mobile networks has to be even more dramatically reduced than with SON. In summary, a future OAM system has to, on the one hand, aid humans to operate the network by reducing the complexity, and, on the other hand, reduce the OPEX and improve the availability of the mobile network through more automation.

11.2 Cognitive Radio Networks (CRN)

This section introduces CRN, the successor of SON, which is capable of satisfying the requirements of future mobile networks. Therefore, it also presents a framework for CRN and outlines important technologies from Artificial Intelligence (AI) which enable this evolution.

11.2.1 From SON to CRN

Chapter 3 describes that SON defines a set of automated operations and management processes for mobile networks, called SON functions. Each one of these is concentrating on a specific aspect and achieves its goal by employing complex algorithms. The behaviour of a SON function is configured through high-level parameters and policies which are determined by human operators according to the operational context, that is, the operator goals, the network environment, and the technical properties of the network. However, this turns out to be cumbersome if the context changes regularly since the adaptation of the configuration is challenging for the human operator with regards to skill and workload level. However, future network scenarios, like the ones mentioned in Section 11.1, are highly dynamic and induce frequent changes of the operational context. Therefore, the SON concept is regarded as not fully meeting the challenges imposed by these new scenarios. In consequence, SON has to evolve in two directions.

First, SON must be extended to allow the management of mobile networks from an *end-to-end* perspective through high level goals. In this way, instead of breaking down the operational goals into values for SON function configuration parameters, the human operator just defines the high level goals of the mobile network as a whole. This requires the mobile network to understand the goals and adapt itself in order to achieve them. As a result, the complexity of the management of ever more complex mobile networks is significantly reduced.

Second, the mobile network must adapt itself to changes in the operational context, thus, increasing the level of automation and leading to a reduction of human effort. This can be achieved through sophisticated AI technologies, especially *machine learning*.

These two extensions lead to the concept of CRN.

11.2.2 Definitions

The term *cognitive* refers to an entity which performs a conscious intellectual activity which involves thinking, reasoning, and remembering (Merriam-Webster, 2003). In the network domain, it is most prominently known from the term Cognitive Radio (CR) which was first coined by Mitola (2000) and is now commonly used as defined by the Federal Communications Commission (2003): a CR is a reconfigurable radio which allows dynamic spectrum access and usage.

Another related term which recently gained a lot of attention, for example, through the book by Mahmoud (2007), is Cognitive Networks (CN). Originally, the term was coined by Thomas *et al.* (2005):

> A cognitive network has a cognitive process that can perceive current network conditions, and then plan, decide and act on those conditions. The network can learn from these adaptations and use them to make future decisions, all while taking into account end-to-end goals.

It should be noticed that CN is not mobile network-specific but instead refers to all kinds of networks.

In order to limit the scope of this chapter, we define the term Cognitive Radio Network (CRN) in a similar manner as CN, that is, a *radio network* which employs a cognitive process and learning capabilities in order to achieve end-to-end goals.

The three terms, CR, CN, and CRN, mainly differ in their scope, that is, the layers of the network stack and the elements of a network they concentrate on (Fortuna and Mohorcic, 2009). Figure 11.2 illustrates this fact: CR is concerned with the Physical Layer and Medium Access Control Layer of wireless links, whereas CRN spans the complete protocol stack of wireless networks, that is, radio access and backhaul, and CN targets the complete protocol stack of networks which consist of wireless access, core, and transport parts.

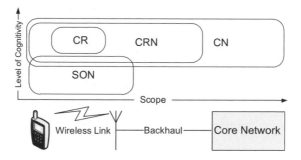

Figure 11.2 The relation between SON, CR, CRN, and CN concerning their scope and level of cognition.

The remainder of this chapter concentrates on CRN. Regarding the scope, it is apparent that CRN is the most suitable technology for succeeding SON since both are concerned mainly with RANs. In this sense, CRN can be seen as an evolved SON which provides self-configuration, self-optimisation, and self-healing extended by an end-to-end view and the ability to learn from previous experiences (Mahmoud, 2007). This relation is depicted in Figure 11.2 as well.

11.2.3 Framework

Section 3.1 outlined the vision of SON and introduced a framework for network management with closed-loop automation. Its idea is to automate the management by defining a set of SON functions which are performing their tasks without human intervention. However, the functions are still configured by human operators through detailed, fine-grained parameters and policies. Nevertheless, the human operator is relieved from routine and recurring tasks, and can concentrate on higher layers of network management.

CRN extends this vision by substituting the SON functions and workflows with cognitive processes which are controlled by high-level goals. Therefore, the SON network management loop is replaced by a cognitive network management loop. Figure 11.3 depicts a simplified cognitive network management loop as presented by Thomas *et al.* (2005). This layered cognitive framework links the end-to-end goals, the cognitive process, and the underlying network elements together. So, a CRN has to know and understand the operational end-to-end goals, and accordingly has to adjust the configuration of the network in a cognitive process. The topmost layer represents the end-to-end goals of the network, for example, cell capacity requirements of the operator. These have to be made machine understandable for the cognitive process by expressing them in a Cognitive Specification Language (CSL). The cognitive process contains a cognitive loop which senses the network conditions through Network Status Sensors and changes the configuration of the network through a generic Network API in order to achieve the given end-to-end goals. The API is a generic interface, provided by the Software Adaptable Network (SAN), which allows the adaptation of the behaviour of the network elements without human intervention. Hence, SAN requires flexible network elements which go beyond Software Defined Radio (SDR) in the sense that all parameters have to be software adjustable, for example, the antenna tilt.

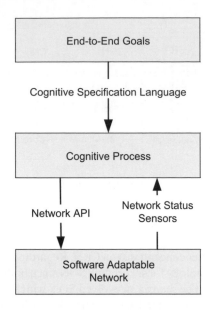

Figure 11.3 The cognitive network management loop (Thomas *et al.*, 2005). Adapted with permission from © 2005 IEEE.

In general, the cognitive loop of the cognitive process is a control loop which senses the environment, decides on necessary adjustments of the network to fulfil given goals, and enacts these changes. However, in order to make this general control loop cognitive, the ability to learn from former actions and adapt the decision accordingly is added. In this way, the cognitive process gains the ability to continuously improve its effectiveness and efficiency.

Several reference loops have been proposed as a starting point for the development of a specific cognitive loop, for example, the Observer-Orient-Decide-Act loop (Thomas *et al.*, 2005), and the Cognitive Cycle for CR containing the states observe, orient, plan, decide, act and learn (Mitola, 2000). However, the proposal by Fortuna and Mohorcic (2009) as, depicted in Figure 11.4, seems to capture the important steps of a cognitive loop particularly well. The Cognitive Process continuously monitors the environment (Sense) though Network Status Sensors. On the one hand, this information is used to create several potential strategies,

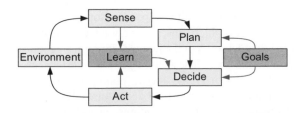

Figure 11.4 The cognitive loop (Fortuna and Mohorcic, 2009). Adapted with permission from Elsevier.

that is, courses of action, how the network configuration should be changed (Plan) based on the given end-to-end goals (Goals). On the other hand, the sensor information is also used for learning in order to build up knowledge of the effects of the actions (Learn). That is, the system continuously validates and benchmarks previous actions, and learns from this. After planning, the system has to decide which plan should be put into action (Decide) based on the end-to-end goals and its experience. This strategy is then enacted using the Network API (Act). The system can learn at several steps in this loop and adapt the knowledge base accordingly. Since this cognitive loop is very complex, some steps can be bypassed if the system is required to react quickly. For instance, it is possible to declare a sensor output as critical and assign an action which causes the system to skip the plan and decide phase and act immediately.

CRN do not assume any specific architecture of the network. Therefore, functionality of the cognitive processes can be distributed in the same ways as SON functions, that is, centralised, decentralised, or hybrid. As described in Section 3.4, each architectural option has its advantages and disadvantages which apply here in the same way.

11.2.4 Artificial Intelligence

CRN relies on sophisticated AI technologies to fulfil the challenging vision. AI is a research area in computer sciences which is concerned with intelligent behaviour of computer systems (Russell and Norvig, 2003). Thereby, knowledge representations, planning and decision making algorithms, and learning technologies are of special interest for the development of cognitive loops.

11.2.4.1 Knowledge Representations

A knowledge representation is a language which can be used to express information. However, more importantly, it also allows expressing the semantics of concepts in order to facilitate reasoning, that is, the inference of implicit, new information from explicitly, given knowledge. The ability to express semantics distinguishes knowledge representations from simple data formats.

Knowledge representations can be separated into the ones able to represent and reason with certain knowledge and the ones able to represent and reason with uncertain, probabilistic knowledge (Russell and Norvig, 2003). In the former case, appropriate conclusions can be drawn from some knowledge with certainty. For instance, if a light is not on then it is certainly off. Representatives of this group are the Web Ontology Language (OWL) and derivation rules. In the latter case, a conclusion from some knowledge cannot be drawn with certainty but solely with a specific probability. For instance, if an animal is a bird then it is very likely that it can fly, however, not certain (e.g. penguins). A famous representative of this category is the Bayesian network (cf. Section 6.3).

11.2.4.2 Machine Planning and Decision Making

Machine planning is concerned with the creation of a plan, that is, a sequence of actions, which achieves a given final goal, that is, a desired state of the world. For certain environments, there exists a variety of so called classical planning approaches (Russell and Norvig, 2003). For instance, planning with forward state-space search is creating a plan by, beginning from an initial state, consecutively trying all actions until a goal state is reached. In order to limit the complexity of this algorithm in real life applications, it is usually necessary to employ a heuristic to guide the search. Another approach is to translate the planning problem into a set of propositional axioms and use a satisfiability solver to find a valid plan. In uncertain environments, planning gets more complex, but there exists a number of algorithms as well. A prominent approach is continuous planning. Thereby, the system initially creates a plan assuming a deterministic environment. After the execution of each action, it performs a replanning in order to adapt the plan to the current situation (Russell and Norvig, 2003).

Usually, various plans differ in their quality. For instance, it could be the goal of the user to perform as few actions as possible. In this case, a plan is more desirable, that is, it has higher utility, the fewer actions it contains. In the same way, several plans can also differ in the likelihood to be effective. The problem of choosing the best plan under multiple goals and uncertainty is addressed by decision making based on decision theory, which combines probability theory and utility theory. It can come up with a rational plan that takes into account the importance and likelihood of the goals (Russell and Norvig, 2003). In certain environments, the problem can be solved using some algebraic optimisation algorithm. In uncertain environment, however, the problem becomes more complex and can be solved using influence diagrams or Markov decision processes. It should be noticed that planning and decision making can also be combined, that is, planning continuously uses decision making in order to guide the planning process. A special case exists, if there are several interacting entities in an environment. Then, game theory can be utilised to create a strategy, that is, a plan, how every entity should behave to maximise the overall utility (Russell and Norvig, 2003).

11.2.4.3 Machine Learning

Machine learning investigates mechanisms which aim to improve the performance of an algorithm from previous experiences. As Dietterich and Langley (2007) show, machine learning can be used for various tasks and is of special interest in the development of CRN. In the framework presented in Section 11.2.3, machine learning is used to improve the effectiveness and efficiency of the decision stage by taking into account new experiences gained during operation. Typically, there are three types of machine learning algorithms which are classified according to the type of feedback available (Russell and Norvig, 2003):

- Supervised learning methods assume a set of cases which contain direct feedback, that is, the set contains premises and according conclusions. For instance, the system is presented a set of cases, each with a set of measurements (premise) and a failure state of the network (conclusion), and learns a mapping function between measurements and failure states.
- Unsupervised learning assumes that the system is not presented any feedback at all. Hence, the system cannot infer any conclusions from the given premises. However, it can still learn statistical properties, for example, distributions or clusters, from the cases and, thus, create a statistical model of the environment.
- Reinforcement learning is based on trial and error, that is, the system learns from received rewards for the performed actions. Hence, the system is presented a case, decides on an

action, and then receives a reward for the action depending on its effectiveness and efficiency. This feedback can be used to learn a strategy for choosing the adequate action for some case.

11.3 Applications

CRN is the most promising technological approach to ease the management of future mobile networks beyond SON. As outlined in Section 3.1, SON attempts to bring self-organising behaviour, especially self-configuration, self-optimisation, and self-healing, in form of SON functions into mobile networks. This relieves human operators form frequent routine work, however, the control of the SON system and its adaptation to changes is still very complex. CRN is supposed to solve this problem and ease the work of operators even more. This section shows where SON concepts can be improved for each of the areas of self-organising behaviour separately, and outlines the envisioned CRN approach.

11.3.1 Self-Configuration

SON introduces self-configuration capabilities into mobile networks (cf. Chapter 4 and Section 10.3). These are mainly concerned with loading a specific configuration to an initially not configured base station. Nevertheless, this does not obviate the need to determine the configuration parameters in a complex, manual network planning procedure. Since future mobile networks will consist of lots of base stations which will be just partially under the control of the operator, more flexible, cognitive planning methods are desirable.

The idea of cognitive self-configuration is that the OAM system creates initial configurations for new base stations automatically. Thereby, the creation of the configuration is performed through a planning and decision making phase which is guided by measurements of the environment from the new base station, the location of the base station, and operator policies. After that, the initial configuration is loaded to the node, for example, using SON self-configuration algorithms. In order to continuously improve the initial configurations created through cognitive self-configuration, the efficiency of the initial configuration is evaluated by the new base station and its neighbours. Thus, the performance of the configurations in different scenarios can be learned and taken into account in the next deployment. As a result of this approach, the human operator does not have to create an initial configuration for each new base station manually, but instead defines the operational goals for new nodes.

11.3.2 Self-Optimisation

Chapter 5 shows that significant effort has been put in the development of complex, algorithmic SON functions for self-optimisation which increase the efficiency of mobile networks enormously. However, they still can be improved because they are mainly static and do not adapt to the operational context. For instance, SON self-optimisation usually corrects a sub-optimal configuration by slightly changing the parameter values and monitoring the reaction of the network. In case that a specific configuration problem occurs twice, the optimisation could be sped up by directly applying the same parameter values as before. Furthermore, current algorithms cannot handle uncertainty properly leading to inferior results. As a consequence, well-trained experts are necessary to configure the optimisation functions in order to ensure optimal performance.

Self-optimisation in CRN improves SON self-optimisation by employing sophisticated AI technologies. An uncertain knowledge representation allows inferring the most likely current system state and probable reasons for performance issues, for example, inferior capacity or sub-optimal handover parameters, from measurements of the network performance. Based on this analysis, the system can create several optimisation plans and use decision making under uncertainty in order to come up with an optimal strategy comprising various configuration parameters which achieves the end-to-end goals. During the enactment of the plan, the cognitive process monitors the network continuously to evaluate the performance of the actions. This information is learned in order to improve the effectiveness and efficiency of cognitive self-optimisation.

11.3.3 Self-Healing

SON self-healing currently focuses on the detection and diagnosis of problems using some given knowledge (cf. Sections 6.2 and 6.3) and the compensation of a cell outage by neighbouring cells (cf. Section 6.4). The former can be seen as the first step towards a cognitive self-healing cycle because the detection and diagnosis already makes considerable use of learning techniques, for example, to learn the mapping between measurement and root causes of failures in a supervised manner. However, the second part of a cognitive cycle, the determination of the most suitable recovery procedure, is still manual work and requires a lot of human effort. Since future mobile networks will consist of more base stations than today, the number of incidents is likely to increase and renders this approach insufficient. Therefore, besides the automated detection and diagnosis of network failures, future mobile networks also require an automated advisory process determining which recovery actions to execute. Thereby, the SON cell outage compensation function can be seen as a possible recovery action.

The main goal of an automated recovery advisory process is to plan and decide on the most effective and efficient countermeasures, that is, actions, for a diagnosed fault given some end-to-end goals. End-to-end goals can be expressed as objective functions like minimising the costs for the recovery or constraints like a cell cannot be restarted at daytime. Consequently, the recovery advisor has to take into account the context in which the mobile network operates. This context includes but is not limited to configuration data, performance data and operational data like date and time. Since troubleshooting of mobile networks is highly probabilistic, continuous planning and decision theory seem to be very promising techniques for this complex planning and decision problem under uncertainty. Learning plays an important role in the outlined approach: it is used to continuously monitor the effects of the performed actions and learn the effectiveness and efficiency of them in different fault situations. This information can be used to adapt decision making. As a result, the performance of the compensation will improve over time.

11.3.4 Operation

Chapter 9 describes that uncoordinated execution of several SON functions in a mobile network can lead to conflicts and inconsistencies. Therefore, SON proposes the use of a coordination function in order to control the operation of the SON functions towards the operator goals. The coordination logic is expressed in fine-grained, human-defined policies and parameters for each pair of SON functions and each set of goals. The management of these policies is costly, error-prone, and time-consuming. If the operational goals of the mobile network change, then the policies and parameters have to be adapted manually in order to reflect the new goals and ensure their fulfilment. In the same way, new or adapted SON functions force human operators to change the policies and parameters in order to ensure conflict-free network operations.

CRN overcomes this problem by employing a sophisticated knowledge representation (Räisänen and Tang, 2011) which allows representing the semantics of operational goals, network properties, and historical and current network statuses. Using this information, the system can perform an automated reasoning in order to come up with a coordination result for a specific set of CRN functions at runtime which achieves the given end-to-end goals. In this way, a change of the goals can easily be reflected in the knowledge of the system and is put into action immediately. Furthermore, the sophisticated knowledge representation allows detecting conflicts between functions which, currently, can only be determined manually (cf. Section 9.1, Table 9.1, Category C).

11.4 Conclusion

Future mobile network technologies will confront OAM systems with new and challenging requirements: ever more and ever diverse network elements have to be managed while ever less OPEX should be spent. SON provides an initial step to achieve these competing goals. However, the requirements go beyond the capabilities of the established SON concept. Therefore, SON has to be extended to increase the automation in network operations and to allow managing a mobile network through end-to-end goals. This leads to a new paradigm: CRN.

CRN is envisioned to put a cognitive loop into network elements and the OAM system. It perceives the current network conditions, plans and decides on reactive actions in order to achieve the given end-to-end goals, and enacts them. During this process, the system learns from the reactions of the network in order to improve later decisions. Several technologies from artificial intelligence research have been identified to be useful to accomplish this vision.

Although CRN is still in its infancy, it is possible to outline how CRN can improve the self-organising behaviour of SON. First, self-configuration can be improved by learning initial configurations over time, thus, making manual planning almost needless. Second, cognitive self-optimisation can use learning and decision making in order to reduce the time to find an optimal configuration. Third, CRN can provide self-healing capabilities by extending the detection and diagnosis features of SON with recovery planning, thus, reducing manual effort. Finally, the operation in CRN is eased compared to SON since semantic knowledge representations allow to resolve conflicts automatically.

References

Dietterich, T.G. and Langley, P. (2007) Machine learning for cognitive networks: technology assessment and research challenges, in *Cognitive Networks: Towards Self-Aware Networks* (ed. Q.H. Mahmoud) John Wiley & Sons, Ltd., Chichester.

Federal Communications Commission (2003) Notice of Proposed Rule Making and Order (03-322), Technical report, Federal Communications Commission.

Fortuna, C. and Mohorcic, M. (2009) Trends in the development of communication networks: Cognitive networks. *Computer Networks*, **53**(9), 1354–1376.

Mahmoud, Q.H. (ed.) (2007) *Cognitive networks: Towards Self-Aware Networks*, John Wiley & Sons, Ltd., Chichester.

Merriam-Webster (2003) *Merriam-Webster's Collegiate Dictionary*, 11th edn, Springfield, MA.

Mitola, J. III (2000) Cognitive Radio – An Integrated Agent Architecture for Software Defined Radio, PhD thesis, Royal Institute of Technology (KTH) - Teleinformatics.

Russell, S.J. and Norvig, P. (2003) *Artificial Intelligence: A Modern Approach*, Prentice Hall, Upper Saddle River, NJ.

Thomas, R., DaSilva, L. and MacKenzie, A. (2005) Cognitive networks, *in* '2005 1st IEEE International Symposium on New Frontiers in Dynamic Spectrum Access Networks', Baltimore, MD pp. 352–360.

Räisänen, V. and Tang, H. (2011) Knowledge modeling for conflict detection in self-organized networks, 3rd International ICST Conference on Mobile Networks & Management, Aveiro.

Index

LTE Self-Organising Networks (SON): Network Management Automation for Operational Efficiency, First Edition.
Edited by Seppo Hämäläinen, Henning Sanneck and Cinzia Sartori.
© 2012 John Wiley & Sons, Ltd. Published 2012 by John Wiley & Sons, Ltd.